Conscious Choice:

The origins of slavery in America and why it matters today and for our future in outer space

by

Robert Zimmerman

©2021 copyright Robert Zimmerman

All rights reserved. Except for the inclusion of brief excerpts for use in a review, no part of this book may be reproduced or transmitted in any form or by any means, electronic or mechanical, including photocopying, recording, or by any other information storage or retrieval system, without permission in writing from the author.

Published by eBookIt.com

http://www.eBookIt.com

ISBN-13: 978-1-4566-3919-8 (hardcover)

ISBN-13: 978-1-4566-3918-1 (paperback)

ISBN-13: 978-1-4566-3738-5 (ebook)

Cover background: Taken by the rover Curiosity in Gale Crater on Mars, looking towards Mount Sharp.

The ways of man are passing strange,
He buys his freedom and he counts his change,
Then he lets the wind his days arrange,
And he calls the tide his master.

—Fisherman's shanty, as sung by
Gordon Bok, Ann Muir, Ed Trickett

About the Author

Robert Zimmerman is probably the world's foremost historian on history of space exploration since the launch of Sputnik in 1957. His books not only document all of the most important events in that history's first half century, they also put that history into the wider context of world affairs.

Zimmerman is also an award-winning independent science journalist whose articles have been published in *Science*, *Air & Space*, *Sky & Telescope*, *Astronomy*, *The Wall Street Journal*, *USA Today*, and a host of other publications. He also reports on space, science, and culture daily on his website, Behind the Black (http://behindtheblack.com).

Also by Robert Zimmerman

Genesis: The Story of Apollo 8

The Chronological Encyclopedia of Discoveries in Space

Leaving Earth: Space Stations, Rival Superpowers, and the quest for interplanetary travel

The Universe in a Mirror: The saga of the Hubble Space Telescope and the visionaries who built it

Capitalism in Space: Private enterprise and competition reshape the global aerospace launch industry

Table of Contents

Dedication 7

Acknowledgements 8

Introduction 9

1. The First Thirty-five Years 15
2. The World of William Berkeley 29
3. England's First Premise 39
   - Fig. 3.1: Virginia, 1620-1682, Black slaves as a % of population 45
   - Fig. 3.2: Maryland, 1620-1680, Black slaves as a % of population 46
   - Fig. 3.3: Massachusetts, 1620-1680, Black slaves as a % of population 46
   - Fig. 3.4: Rhode Island, 1620-1680, Black slaves as a % of population 47
   - Fig. 3.5: Connecticut, 1620-1680, Black slaves as a % of population 47
4. Berkeley's First Term 53
   - Table 1: The Grand Assembly, April 1642 57
   - Table 2: The Grand Assembly, March 1643 58
   - Table 3: The Grand Assembly, March/June 1644 59
   - Table 4: The Grand Assembly, October 1644 60
   - Table 5: The Grand Assembly, February 1645 61
   - Table 6: The Grand Assembly, November 1645 62
   - Table 7: The Grand Assembly, October 1646 63
   - Table 8: The Grand Assembly, November 1647 64
   - Fig. 4.1: Slave-Owners in the Assembly, 1642-1652 70
   - Fig. 4.2: Present and Future Slave-Owners in the Assembly, 1642-1652 71
   - Fig. 4.3: Percentage of Assembly members raised in Va., 1642-1652 74
5. A Royalist Safe Haven 77
   - Table 9: The Grand Assembly, October 1649 80
   - Table 10: The Grand Assembly, March 1651 to March 1652 81
6. Virginia Under the Commonwealth 87
   - Table 11: The Grand Assembly, April 1652 89
   - Table 12: The Grand Assembly, November 1652 94
   - Table 13: The Grand Assembly, July 1653 95
   - Fig. 6.1: Slave-Owners in the Assembly, 1642-1659 96
   - Fig. 6.2: Present and Future Slave-Owners in the Assembly, 1642-1659 97
7. A Royalist Resurgence 103
   - Table 14: The Grand Assembly, November 1654 104
   - Table 15: The Grand Assembly, March 1653 105
   - Table 16: The Grand Assembly, December 1656 106
   - Table 17: The Grand Assembly, March 1658 113
   - Table 18: The Grand Assembly, March/April 1658 115
   - Table 19: The Grand Assembly, March 1659 117
8. The Consequence of Family 119
   - Fig. 8.1: Percentage of Assembly members raised in Virginia, 1642-1677 127
   - Fig. 8.2: Slave-Owners in the Assembly, 1642-1677 128
   - Fig. 8.3: Present and Future Slave-Owners in the Assembly, 1642-1677 128
   - Table 20: The Restoration Assembly, 1660 130

9. The Restoration Assembly — 133
Table 21: The Restoriation Assembly, 1661 — 136
Table 22: The Restoriation Assembly, October 1661 to March 1662 — 139

10. The Choice of the Electorate — 143
Fig. 10.1: Incumbancy in the House of Burgesses, 1642-1662 — 143
Table 23: The Long Assembly, October to December 1662 — 144
Table 24: The Long Assembly, 1663 — 150

11. The Long Assembly — 153
Fig. 11.1: Incumbancy in the House of Burgesses, 1642-1676 — 154
Table 28: The Long Assembly, October/November 1666 — 163
Table 29: The Long Assembly, 1667 — 164

12. Bacon's Rebellion — 167
Table 41: The Long Assembly, 1675 — 173
Table 42: The Long Assembly, February to March 1676 — 174
Table 43: Bacon's Assembly, June 1676 — 177
Table 44: The Court Martial Councils, 1677 — 182

13. A Matter of Choice — 185
Fig. 13.1: Massachusetts, 1630-1790, Black slaves as a % of population — 188
Fig. 13.2: Connecticut, 1640-1790, Black slaves as a % of population — 189
Fig. 13.3: Rhode Island, 1650-1790, Black slaves as a % of population — 189
Fig. 13.4: New York, 1630-1800, Black slaves as a % of population — 191
Fig. 13.5: Pennsylvania, 1680-1800, Black slaves as a % of population — 191
Fig. 13.6: New Jersey, 1670-1800, Black slaves as a % of population — 193
Fig. 13.7: Delaware, 1650-1790, Black slaves as a % of population — 193
Fig. 13.8: Virginia, 1620-1790, Black slaves as a % of population — 199
Fig. 13.9: South Carolina, 1660-1790, Black slaves as a % of population — 201
Fig. 13.10: North Carolina, 1660-1790, Black slaves as a % of population — 201
Fig. 13.11: Maryland, 1640-1790, Black slaves as a % of population — 203

14. A Proclamation for Civilization in the Far Future — 205

Appendix A: Additional Tables — 223

Appendix B: Source Information for Tables — 239

Bibliography — 255

Endnotes — 271

Index — 307

Dedication

This book is dedicated to Elon Musk,
who has done more than any other person
in the last half century to make possible
the eventual exploration and settlement
of the solar system.

I offer this book to him as a guide for
when he builds those first colonies on Mars.

Acknowledgements

    The publication of this book would not have happened without the help and assistance of David Vidonic, Rand Simberg, and Dave Truesdale. Their support and advice helped me improve the book immeasurably, while forcing me to finally get it finished and in print.
    I must also thank John Batchelor, David Livingston, Sarah Hoyt, and Robert Pratt for their years of support. Their willingness to share my work with their audiences and readerships has been one of the main reasons my work has reached a wider audience, something for which I am endlessly grateful.

Introduction

> There is one calamity which penetrated furtively into the world, and which was at first scarcely distinguishable amid the ordinary abuses of power: it originated with an individual whose name history has not preserved; it was wafted like some accursed germ upon a portion of the soil; but afterwards nurtured itself, grew without effort, and spread naturally with the society to which it belonged. This calamity is slavery.
> —Alexis de Tocqueville, *Democracy in America*.[1]

This is the story of a failed effort to make a new society where none had existed before, a story that has direct bearing on the future effort of humans to establish colonies in space.

The establishment of the North American English colonies in the 1600s was a tale that involved individual sacrifice, courage, a great deal of foolhardiness, and, most importantly, some inspired social experimentation. In the north, first the Pilgrims and then the Puritans attempted to establish small, democratic, self-governing religious communities in Massachusetts. In Pennsylvania the Quakers attempted their own religious variation, while the Catholics did the same in Maryland. In New York the English found themselves learning to live with many other cultures. And in Virginia, the first British colony on the North American continent, the leadership formed a secular and centralized social order, centered on individual personal gain and governed firmly by the governor and legislature in Jamestown.

From these various social experiments developed many complicated and sophisticated ideas about creating new societies in a wilderness, including the government's role within that new social order. Issues ranging from religious freedom to taxation were passionately debated for more than a century, and from those debates eventually came the birth of the United States, a nation formed on what at the time was the startling premise "that all men are created equal, that they are endowed by their Creator with certain unalienable rights, and among these are life, liberty and the pursuit of happiness."

Yet, in Virginia, home to the man who wrote these words, the social experiment to create a new society ultimately failed, and it failed very badly. From 1607 to 1690, the British colony of Virginia degenerated from a culture that opposed slavery and honored respect for the individual and individual freedom to a culture that eagerly and cruelly imposed slavery on a certain portion of its population for the benefit of everyone else. By the middle of the eighteenth century almost half the population of Virginia was permanently enslaved, more than 140,000 people working in perpetual and unrelenting servitude. Even now, more than a century and a half after the abolishment of slavery, American society is still torn asunder by the terrible aftermath of what was decided by a select group of men in a small brick statehouse in swampy Jamestown between the years 1642 and 1677.

Future space colonists trying to create their own new societies on other worlds would be wise to pay attention to this history.

This book has literally taken decades to write. I first took up this subject in the 1990s as a filmmaker who was interested in writing a screenplay. It had occurred to me that I had never been taught anything about the very beginnings of slavery in America, despite the fact that for slavery to flourish in a British society seemed at first glance to be a very strange paradox. I knew that the colony of Virginia had had a democratic legislature, and that the colony had formed from the same legislative ideals that had established both the English parliamentary system as well as the other northern colonies. Furthermore, slavery had not existed in British culture for centuries. That a group of British citizens in a British colony were able to buy slaves and legalize the custom seemed to me to be a strange mystery, and hence fertile ground for some crackling good drama.

So I went to the library, assuming that there would be numerous books written on the men and women who first brought slaves to Virginia. Much to my surprise, I could find almost nothing. In fact, considering the significance of slavery to American history, it was astonishing how little had been written about the individuals who made these early, seventeenth-century events happen.

I found that prior to the twentieth century, few historians were interested in the subject. Though a handful made an attempt after the Civil War to describe the origins of slavery, most ignored the subject, influenced as they were by regional roots and the lingering emotions of the Civil War. The only ones really interested in studying the south, southern historians, focused their efforts proudly describing how the early southern pioneers settled Virginia, pretending that the growing dominance of slavery was only a minor and inconsequential part of that history.

After World War II a new wave of historians tackled the problem, wielding with great skill the modern research tools available since the invention of the computer. Yet, even their work I found unsatisfying, as it only partially revealed the story of slavery's beginning in America. These twentieth century historians, under the new concept of social history, focused on issues of racism, economics, and culture as the causes for slavery, and therefore had little to say about the individuals who actually made slavery a legal and popular custom in Virginia. History was defined by the general sweep of mass movements and large populations, the specific actions of individuals being generally ignored.

To my dismay and frustration, I discovered that in the more than three hundred years since the birth of American slavery, no historian has ever been willing to name names. Studies of slavery would not discuss who established it. Studies of who established Virginia would not discuss slavery. My fundamental question, who first sanctioned American slavery and how it was possible in a democracy for them to do it, had never been addressed.

Therefore I felt compelled to try to answer the question myself. In the process I learned that this story about distant past events in colonial Virginia has a significant and direct relationship to both our society today as well as the future establishment of colonies in space. All the issues of right and wrong, of individual freedom, of government corruption, of high taxation, of racial friction and harmony, of economic trouble, and of religious belief and toleration that were being argued in colonial America have the same direct bearing on the present *and* the future. The creation of the British colonies in North America, beginning with Virginia, were test cases on how humans establish new societies in an untamed wilderness. When humans begin to establish colonies on Mars, on the Moon, on the asteroids, they will be faced with exactly the same issues that colonial Virginians and New Englanders faced. What the British learned in trying to create their colonies in North America were fundamental lessons. They will apply to future space colonists, and by drawing on those lessons, those future space-farers will increase the chances they will build healthy and vibrant societies, while avoiding evils such as slavery.

These lessons are also applicable to present day America, where we are now tragically seeing their abandonment. A rereading of this history by today's Americans might help them stave off that abandonment, and encourage this country to once again embrace these lessons, for the good of ourselves and future generations.

In writing this work, I have taken an approach to history somewhat different than many recent historians. First of all, I bring to my writing a deep, heartfelt desire to re-assert the power of the individual in making history. While I recognize the importance of mass movements, social forces, and economics in determining human action, I consider the personal influence of each unique person to be as important. People do matter, and their individual actions can clearly make a difference.

My approach to history differs as well from many modern historians in that I have given my work a strong narrative line. History does not happen instantaneously in large chunks, but sequentially over time, moment by moment, hour by hour, day by day. Each event influences people, changing their approach to life and its problems. And as time passes, men and women die, and are replaced by new individuals motivated by different life experiences. Just as we today recognize the vast differences between the 1940s and the 1970s, Virginia in the 1640s was a vastly different place from Virginia in the 1670s.

Moreover, as Virginia was changing, so was England. The events of the English Civil War had direct bearing on the events in Virginia, especially because most of the colonists were raised in England and had some personal stake in what happened there. I have therefore tried to connect Virginia's leaders with their English roots and the revolution taking place there during these very years.

Finally, in writing this history I made it a point to describe the ethical and religious beliefs of the men and women who lived at the time. In this context, I probably differ most from the majority of historians who have studied this issue since World War II. Rather than consider ethics or morality as a factor, most recent historians have instead focused almost entirely on various economic and sociological factors as their explanation for the appearance of slavery. Considerations such as the drop in the price of tobacco, the shortage of labor, the existence of racism, the difficulties of colonization, and the competition between different economic classes, have all been cited by historians in the past half century for explaining the growth of slavery.

While I grant that all of the above were important factors in the rise of slavery, I also believe that none were more than contributory, and that a healthy society with strong moral values could have successfully resisted the temptation of slavery, despite any and all of these problems.

The evil of slavery sprung from the social order of Virginia in the middle decades of the seventeenth century because underlying all these problems was a devaluation by Virginians of the fundamental importance of ethics and morality. The colony's most influential members considered morality and ethics important, but were easily willing to put these issues aside if such ideas caused inconvenient difficulties for them.

George Washington, raised in a slave society that forbade him from freeing his slaves, knew more than anyone the terrible consequences that could result from the denial and rejection of ethics and morality. As he said in his farewell speech to the nation in 1796,

> Of all the dispositions and habits which lead to political prosperity, religion and morality are indispensble supports. . . . [W]here is the security for property, for reputation, for life, if the sense of religious obligation *desert* the oaths, which are the instruments of investigation in the courts of justice?[2]

I could probably spend the next two hundred pages trying to define exactly what I think Washington meant by "religion and morality" and "religious obligation." I actually however consider such a discussion beside the point. To me, a healthy and just democracy is best guaranteed when it freely allows its citizens the ability to debate such moral issues as widely as possible, even when such debate challenges the dominant beliefs of the time. Such freedom and intellectual openness must then be combined with a willingness of all citizens to participate, considering questions of right and wrong in both their daily lives as well as the more public institutions of government and church. When society allows such free discourse, and the citizens *use* that freedom to passionately and peaceably advocate their convictions, I believe society will surely prosper.

This is not to say that morality is merely in the eyes of the beholder, or that the avocation and practice of any human behavior is acceptable. On the contrary, I believe that some cultures are more humane, that some moral concepts are more just, and that some human actions are more ethical than others. It is in the free discourse of ideas that a healthy democracy learns to make the distinctions.

However, because democracy is a form of government based on choice and freedom, it is inherently without morality. Citizens are free to choose both wisely and badly, and in fact, they actually might find it easier to choose badly. The idea of democratic rule only achieved a dominant place in most world cultures during the twentieth century, and thus it was then that we first began to see the range of success and failure that this form of government is able to achieve. The successes have been exhilarating. The failures have been frightening and profound.

More than any other social order, I believe democracy requires the full participation of its citizens in the decision-making process for it to function properly. Since the system does not depend on any form of coercion to impose moral rules, a voluntary, positive involvement of everyone is required, combined with a conscious effort by all to learn and practice the moral rules that humans have developed over the centuries for promoting healthy societies. Each law and social custom must be reviewed by the entire society and agreed to. Laws cannot be passed nonchalantly, for petty personal reasons. Ideas and moral issues *must* be weighed in the balance each time by all citizens, honestly and without legal restraint. The willingness or unwillingness of the individuals of a democratic society to do this carefully will decide whether or not that society will live or die.

As Edmund Burke said, "For evil to triumph good men need only do nothing." It is my belief that democracies will always fail when too many citizens simply stop paying attention to issues of right and wrong and let the worst tendencies of human nature dominate. We are all individually responsible.

This is I think what happened in Virginia. The leaders of the colony, both in America and in England, did not consider moral issues fundamental in the creation of a new democratic society. They not only strongly opposed giving such ideas a dominant place in the establishment of the colony, they actually restrained dissent and the avocation of alternative religious views. Left without an ethical framework for their society, Virginians instead placed on their altar of worship the obtaining of economic wealth and power. Just as the Marxists, the Soviets, and many modern scholars in our own day have focused exclusively on economic causes, considering this to be the sole and primary motive for human action, so did the citizens of Virginia.

While economics is an essential factor in any human decision, it is hardly the only factor, and in fact will often take second place to a person's moral ideals and beliefs. Just as people today are willing to tolerate inconvenience and added expenses to recycle their paper, to lose money divesting from fossil fuel companies, to buy more expensive tuna because it is "dolphin-free," to contribute large quantities of money to any number of charitable causes, past generations could also make the same kind of difficult moral choices. The Puritans did not come to New England for added profit. Many were settled middle class citizens in England who experienced significant financial loss in emigrating. Instead, "they believed that, by emigrating, they followed the will of God and that their obedience would not escape divine notice."[3] They came for moral reasons, even if those moral reasons were eccentrically their own.

The willingness of Virginians to discount the importance of morality in their economic decisions eventually rippled down through the decades until their children and their children's children could no longer tell the difference between right and wrong when it came to the enslavement of others. Every event in Virginia from 1607 until 1690 presented each generation with a choice concerning issues of universal freedom and black slavery, and with each event the members of each new generation chose the direction leading towards oppression and institutionalized privilege.

Americans today, as well as future generations building colonies on new worlds beyond Earth, would be wise to study this history. Otherwise, they may find themselves reliving it.

1. The First Thirty-five Years

> The six and twentieth day of April. About four o'clock in the morning we descried the land of Virginia . . . There we landed and discovered a little way, but we could find nothing worth speaking of but fair meadows and goodly tall trees, with such fresh-waters running through the woods, as I was almost ravished at the first sight thereof.[4]
> —George Percy, describing his first vision of Virginia

The year was 1607, and the English had once again arrived in North America. Like the Spanish and Portuguese before them, as well as the earlier English failure on Roanoke Island more than twenty years before, a small band of British settlers had sailed to the continent of North America, hoping, as written by one of those early arrivals, "to plant a nation where none before hath stood."[5]

These men, or "adventurers" as they called themselves, included thirty-six gentlemen, one surgeon, four carpenters, two bricklayers, one barber, one sailor, one blacksmith, one mason, one tailor, one drummer, fourteen laborers, four boys, and about thirty-eight unnamed others (most of which were probably footmen/butlers for the thirty-six gentlemen).[6] Of the gentlemen, George Percy, the writer of the above narrative, was probably quite typical. A member of the aristocracy by birth, he was the eighth son of Henry, Earl of Northumberland and Catherine, daughter of Lord Latimer. As the eighth son, Percy was cut off from his father's wealth. In order to prevent the break-up of a noble's property, English law required that only the oldest living son could inherit it. In the past the father would have purchased a second estate for his younger sons. Now, English land was becoming scarce, and a younger son either had to join the military, clergy, or legal profession to make his way in the world.

Becoming a lawyer or minister did not appeal to Percy, so he joined the military, fighting for the Low Countries against Spain at the turn of the century. When war ended, Percy was left dependent on his older brother, Henry, the ninth Earl of Northumberland. He was twenty-six years old, owned no property, had no source of income, and few prospects. When the opportunity to join the expedition to Virginia in

1606 came up, Percy grabbed at it, becoming a shareholder in the Virginia Company and its future profits.[7]

Like the other gentlemen adventurers, Percy journeyed to Virginia in search of glory, unexplored wilderness, and the dream of building something new for the world. Much like today's astronauts, Percy and his fellow gentlemen surely saw their expedition as going "where no man has gone before." As Elizabethan poet Samuel Daniel said in 1599 in a poem celebrating the English language,

> And who in time knows whither we may vent
> The treasure of our tongue, to what strange shores
> This gain of our best glory shall be sent,
> To enrich unknowing nations with our stores?
> What worlds in the yet unformed Occident
> May come refined with the accents that are ours?
>
> Or who can tell for what great work in hand
> The greatness of our style is now ordained?
> What powers it shall bring in, what spirits command,
> What thoughts let out, what humors keep restrained,
> What mischief it may powerfully withstand,
> And what fair ends may thereby be attained.[8]

Or possibly we can sense the passion for exploration in seventeenth-century Britain by the tone of the anonymous poem, "Tom a Bedlam:"

> With a host of furious fancies
> Whereof I am commander,
> With a burning spear and a horse of air,
> To the wilderness I wander.
> By a knight of ghosts and shadows
> I summoned am to a tourney
> Ten leagues beyond the wide world's end:
> Methinks it is no journey.[9]

That they called themselves "adventurers" was no accident: their sincere goal was to create a new English society, to spread the ideas of English culture to a new land, bringing what Percy and others obviously believed to be a good way of life to others.

When asked, these adventurers would most often proclaim religion as their primary purpose for colonization: they would go to the New World to convert the natives to God and Christianity. "The principal and main ends . . . were first to preach and baptize unto Christian religion, and by propagation of the gospel, to recover out of the arms of the devil a number of poor and miserable souls, wrapt up unto death, in almost invincible ignorance . . . to add to our might to the treasury of heaven."[10]

The adventurers also saw colonization of the New World as a way to strengthen the English nation, first by siphoning off "the rankness and multitude of increase in our people,"[11] second by creating a buffer colony against their country's enemies, and third by bringing the resources and wealth of that colony to England. These men were passionately loyal to their British culture, and wished to restrict the growth of Spain, Portugal, and the Catholic Church, what they would call "Popish priests." A strong colony in North America would give the British a foothold that could prevent the Spanish and Portuguese from occupying more New World territory. The colony would also provide a place for the poor and indigent of their own country to find a home and place to live, expanding British culture beyond the British Isles.

Colonization could also bring wealth to the Crown, the English nation, and the adventurers themselves. These men hoped to find gold, silver, and jewels, as the Spaniards had done in Mexico. Lacking this, the wood, iron, and steel of the new land would be of use to England. Or as noted by historian Warren Billings, by "recovering and possessing to themselves a fruitful land," they would enrich both England and themselves.[12]

The key word however is "themselves." For though George Percy and the other gentlemen adventurers might have sincerely wanted to bring their religion and beloved English culture to the rest of the world, the need and desire of all was to bring to themselves a return on the pound. The expedition was funded and managed by a joint stock company for the purpose of earning a profit for its investors and the king. Though moral and philosophical ideas of community and expansionism as well as goals of religious conversion were cited as reasons for building the colony, few if any of these gentlemen adventurers came to Virginia with these ideas as their primary motives.[13] Instead, profit was what they came for, and this colored their every decision.

Hence, they brought only one clergyman on this first expedition. Since missionary work would not directly improve profits, there was no incentive to stock the expedition's company with preachers. Minister Robert Hunt would attend to the settlers, but no more.

Nor were women or families included in this first expedition. British society, as with most pre-Industrial societies, restricted women to the home and child care, work that could not have contributed to any immediate profits.[14] Men would fight the battles, mine the gold and silver, harvest the timber, or negotiate trade with the natives. Just as Cortez and the Spaniards had become rich in Mexico by stealing the wealth of the local cultures, so too would the adventurers of the Virginia Company. "There was no talk, no hope, nor work, but dig gold, wash gold, refine gold, load gold," said Anas Todkill in Jamestown in the winter of 1607.[15] To have included women or families in this first expedition would have interfered with this work, requiring a longer investment than the Virginia Company desired.

Nor had the Company carefully thought out the kinds of men needed to establish a new community in a New World. Noble gentlemen like George Percy brought with them their concept of their place in society, and how that society functioned. Gentlemen did not use their hands in labor. They had others do it. Gentlemen instead issued orders for the management of the community. Hence, the thirty-six gentlemen adventurers honestly expected their thirty-eight footmen and the other thirty laborers to do the work for them. As Todkill reminded all when he described how the President of the colony had chopped down a tree,

> Let no man think that the President or these gentlemen spent their times as common wood-hackers at felling of trees, or such like labors, or that they were pressed to anything as hirelings or common slaves; for what they did . .. it seemed, as they conceited it, only as a pleasure and a recreation.[16]

Or as Todkill noted wearily at another time when several of these gentlemen departed the colony,

> Captain Newport . .. set sail for England, and we, not having any use of Parliaments, plays, petitions, admirals, recorders, interpreters, chronologers, courts of plea, nor Justices of peace, sent M. Wingfield and Captain Archer with him, for England, to seek some place of better employment.[17]

John Smith put it more bluntly. "They would rather starve and rot with idleness, then be persuaded to do anything for their own relief."[18]

Born of wealth and privilege, these gentlemen expected to pick from the ground whatever wealth the New World offered in order to continue their status of wealth and privilege. Though I will discuss at greater length the British aristocracy's concept of rule and order in the next chapter, suffice it to say that these men came to Virginia intending to re-establish for themselves a privileged position similar to that held by their father and their father's father: a community in which they stood at the apex of leadership and power, responsible for providing that community with a just and wise rule while retaining rights of privilege and power over all others.

And so, they invested in the Virginia Company and adventured to the New World, expecting the laborers among them to do all the work so that they could then reap the benefits. And in reaping these benefits they dreamed of planting "a nation where none before hath stood."[19]

The result was an unmitigated failure. Those in charge bickered and fought. By 1609 and 1610, a time period later dubbed "The Starving Time," the colonists were dying in horrible numbers, either sickening from the poor water and swampy conditions surrounding Jamestown, or simply starving. They had not planted crops, and could not hunt for food, and when the Indians refused to feed them, they could only starve. As Percy himself said as early as the summer of 1607, "Our men were destroyed with cruel diseases, as swellings, flixes [sic], burning fevers, and by wars, and some departed suddenly, but for the most part they died of mere famine."[20]

Privileged rank and obsessed with profit describes this first expedition's leaders. High mortality, forced labor, and a skewed sex ratio describes the colony that leadership created. As a scouting mission or an exploratory base camp such circumstances might have been expected, tolerated, or even required. As a foundation for a new nation, however, such conditions could only horrify.

In 1611 to 1614 the leaders of the Virginia Company imposed martial law, brutal and strict. Though it kept the approximately 300 settlers from starving, it also condemned them to a miserable form of permanent boot camp.[21] Deputy governor Sir Thomas Gates, chosen by Lord De la Warr, set the rules, under the premise that "no good service can be performed, or war well-managed, where military discipline is not observed."[22]

Virginia became the worst sort of company town, organized like a military operation. Everyone there worked for the Virginia Company. Everything there was owned by the Virginia Company. And all the profits there were supposed to go to the Virginia Company, to be re-distributed to investors. Though now no one starved, the Virginia Company made no money. In fact the colony barely produced the food and staples needed to survive. Additionally, too many men continued to die from disease, and few English families were willing to come and settle in such a colony. None of the institutions of British culture which these adventurers were so passionately proud were being transferred to Virginia.

So the Company re-organized again, though Gates did so under the continuing assumption that the colony should be ruled from above for the sake of profit. In 1614 he decided to allot "every man three acres of clear ground" and the freedom to work one month out of twelve on this land for themselves.[23] He did this partly because, during the previous three years, settler and plantation owner John Rolfe had discovered that tobacco was not only easy to grow in Virginia, it was extremely profitable. The result was that, by 1617, John Smith described Jamestown as consisting of

> but five or six houses, the church down, the palizado's broken, the bridge in pieces, the well of fresh water spoiled; the store-house they used for the church; the market-place, and the streets, and all other spare places planted with tobacco . .. the colony dispersed all about, planting tobacco.[24]

Mortality was still high and the population, only about 350 settlers, remained mostly men under harsh leadership.[25] Given a little freedom to plant for themselves, the colonists, having discovered the ease in which they could make money from tobacco, had single-mindedly grabbed for its quick profit, letting the colony crumble around them.

In 1619 the Virginia Company reorganized once more in an effort to encourage British men and women to immigrate. Using the previous experiments as his guide, Sir Edwin Sandys, the newly elected treasurer of the company in London, further liberalized the land ownership policy, allowing individual settlers the right to own as much land as they could earn or buy. The company would reward those who paid for the transportation of new British settlers with a "headright" of fifty acres per settler. Furthermore, under a document dubbed "The Great Charter," the company authorized the creation of Virginia's very own legislative body, the House of Burgesses.[26]

Even these changes, which did provide some freedom and individual rights to Virginia's general population and was followed by an increase in immigration and a doubling of the population to more than 800 souls,[27] had little to do with creating a viable and stable community of men, women, and families, where people could live their lives and prosper. Instead, Sandys' reorganization repeated the early pattern of rule-from-above with the focus on profit.

For example, the new headright system, rather than building a community, was actually designed to financially benefit the wealthiest patrons of Virginia. The Company did not award each *immigrant* fifty acres for coming to Virginia, which seems the most direct way to encourage settlement. All Englishmen coveted the ownership of land, and such an offer would have fired the imagination of many.

Instead, the fifty acre headright was specifically awarded to those who *paid* the transportation costs. "And for every person which they shall transport thither . . . it shall be to the transporters of fifty acres."[28] For the younger sons of the aristocracy, unable to inherit land in England but able to pay their own way to Virginia, the headright became a direct incentive to emigrate. For the poorer citizens of England, however, the fifty acre reward really didn't exist, since few could afford the cost of immigrating. Instead, they had to contract for someone else to pay the cost (usually those same younger sons), who would thereby get the fifty acres. This arrangement, though insuring the survival of the Virginia colony, specifically concentrated land and power into the hands of those with power and wealth.[29]

Sandys' headright system also made it easy and convenient for Virginia's wealthy and powerful to turn this contract into a very distorted and oppressive form of indentured servitude for these poor British immigrants. In any negotiation, knowledge is always power. Since the gentlemen in Virginia were usually the only educated and literate party to the contract, they could always negotiate excellent terms for themselves. Soon, the man who paid the transportation costs not only got the fifty acre headright, he got several years (from four to seven) of forced labor. The servant would be given transport, work for his master for several years for free, and then be given, *at most*, his freedom dues of some barrels of corn, clothing, and a small sum of money.[30]

In England indentured servitude was usually a system for the education and apprenticeship of the young as well as an attempt by different guilds to control labor supply. Such servitude fell under the concept of a contractual agreement, and the law and society put strict controls on how a master could treat a servant within such an arrangement.[31] In Virginia, however, no such controls existed, and so the leadership expanded indentured servitude beyond mere apprenticeship, turning it into a cheap form of forced labor. To the men in power the financial advantages of using forced, unpaid labor extensively far outweighed its social destructiveness.[32]

The single-mindedness in which the Virginian settlers pursued tobacco also demonstrates their focus on profit over building a viable community. Developing and exporting a crop for profit in a newly settled colony is extremely difficult. Numerous historical examples, not just in North America, indicate that new settlers are either required to base their economic system on that single exportable crop and use forced labor to grow it, or be content, like the Puritans in New England, with merely making do as pioneer settlers in small, self-sufficient communities. Virginian planters chose the one crop system and forced labor, since this would bring them large and quick profits.[33] And they did this despite King James' hatred of tobacco, calling it

> a custom lothsome to the eye, hateful to the nose,
> harmful to the brains, dangerous to the lungs, and in
> the black stinking fume thereof, nearest resembling

the horrible Stygian smoke of the pit that is
bottomless.34

Virginians grew tobacco everywhere, in the fields, on the streets, and in their homes, wherever they could find the space.35 Nor is this surprising, considering that the average income of an ordinary farmer was usually between 500 and 750 shillings a year, while a farmer planting tobacco alone could instead make at least 2000. And with the help of unpaid servants he could bring in much, much more.36

The focus on profit can also be seen by an examination of the Great Charter, designed as part of Edwin Sandys' reorganization and sent over with Governor George Yeardley in 1619. This document, which established Virginia's legislative system, was less interested in establishing a government and more concerned with better managing the employees of a company. The new legislature, called the Grand Assembly and made up of a Council of fourteen men and a House of Burgesses made up of representatives from every plantation in Virginia, was intended to be under the control of the Company's leaders in England and their appointees in Virginia. The Company Board chose the Governor and the Governor appointed the councilors. Furthermore, both the Company Board in England and the Governor could nullify any law passed by the House.37

Though this Charter gave many Virginians the right to vote, the government was surely not of, for, and by the people. Everyone understood and expected decisions to come from the company's managers at the top, and acquiesced to those decisions, either because they believed that this was the best and most practical system or because they did not have the power to change it.

This 1619 re-organization by Edwin Sandys surely revitalized Virginia. Offered the false promise of owning land, thousands of settlers flocked to the colony, and in the coming decades Virginia would consistently draw the most immigrants from England. By 1624 the population of the colony had increased to over 2,200.38

Among the immigrants to take advantage of Sandys' re-organization was merchant Edward Bennett, a stockholder in the Virginia Company and ally of Sandys. In 1621 he obtained a large plantation in exchange for the transport of a hundred settlers, and sent his son Robert and nephew Richard to Virginia to manage "Bennett's Welcome" for him.39 What happened to Robert we don't know, but Richard Bennett would spend the rest of his life in Virginia, playing a crucial dissenting role in the establishment of slavery. A Puritan with close ties to many religious leaders in England, Maryland, and Massachusetts, Bennett would be elected burgess in 1629; in 1639 he would be appointed to the Council; by 1652 he would begin a three year term as Governor.[40]

Also arriving just after Sandys' re-organization were the colony's first black slaves. Numbering only two dozen before 1627[41] and usually appearing when Spanish and Portuguese traders would dock in Jamestown for supplies, the status of these black slaves was not clearly addressed by British society. Many believed that British law followed the Bible, which defined all Christian souls as equal under God. Hence, as early as 1569 it was declared in a British court "that England was too pure an air for slaves to breath in."[42] Others, however, felt that the law was unclear, and that the enslavement of such alien people was perfectly reasonable.

Of these first two dozen slaves, two in particular were very fortunate, and their auspicious start as members of the British community in Virginia does demonstrate that the institutionalization of slavery in Virginia did not have to happen. "Antonio a Negro" and "Mary a Negro woman" arrived in 1621 and 1622 respectively and were purchased to work on the Bennett plantation. From the sparse records still available to us, it appears that Anthony (as he was later called) and Mary were eventually freed by Richard Bennett, married, and moved across Chesapeake Bay to the Eastern Shore to settle on their own land, taking the surname of Johnson for themselves and their children. By the late 1640's the Johnsons belonged to a small but thriving population of about fifty free blacks who were living in Virginia.[43]

Most of these early black slaves, however, were not as fortunate, remaining slaves for the rest of their lives. Despite the success of Anthony and Mary Johnson, the issue of slavery was not yet worked out in British culture. Human beings, as seemingly alien as blacks, pushed seventeenth-century British law to its limits, and since the law did not clearly address the issue, each slave's freedom depended mostly on the generosity of his or her master.[44]

Unfortunately for Virginia and the Virginia Company, the surge of immigration between 1619 and 1624 as a result of Edwin Sandys' re-organization did little to improve the colony's quality of life. The majority of immigrants continued to be single young men, at a ratio of approximately three men to one woman, most of whom were poor and uneducated indentured servants. Middle class English families had no interest in moving to Virginia.[45] Nor did Sandys' re-organization do anything to address the horribly high death rate for these new immigrants. In fact, it probably contributed to their deaths by pouring thousands of people into a colony unprepared to feed them. From 1619 to 1622, 3,570 immigrants came to the colony, joining the approximately 800 already there. Yet, the 1624 census found only 2,200 settlers still alive. Almost half the population had died in less than two years.[46]

Despite its success in saving the colony, Edwin Sandys' effort totally failed to bring any profits to the Virginia Company. While individual Virginia settlers, especially the wealthiest, were making fortunes exporting tobacco, the company was getting nothing. By 1624 it was bankrupt and King Charles I was forced to dissolve it, placing Virginia directly under his control.

The king did not, however, change the basic structure of Virginian society as already established. The headright, the labor system of indentured servants, the rush to grow tobacco, the poorly structured legal system based on arbitrary rule by those in power: these ingredients the King left untouched. Charles also left the running of the colony almost entirely in the hands of those who already controlled it, and from 1624 until 1642 Virginia became the territorial possession of England's aristocracy.

George Yeardley, whom Edwin Sandys had appointed governor in November 1618, was re-appointed governor. Lacking clear instructions from the king, Yeardley continued the same headright policies created by Sandys. When Yeardley died in 1627 the next two governors, Francis West and John Pott, maintained these same policies. Both were already established in Virginia with their own large tracts of land, and saw no reasons to rock the boat.

In 1629 the King appointed Sir John Harvey as Governor. For the next six years it appears he tried to rein in the headright system, and would not issue new patents for land. His reasons were hardly all altruistic. He, like the Virginian planters, believed in rule-from-above, but, because his wealth and power was not based in Virginia, he took his lead from his superiors in England. When the Privy Council in England would not confirm the patents issued by the previous governors, Harvey decided to stop issuing patents. Furthermore, he antagonized the elites in Virginia by refusing to represent their interests in England and by supporting the establishment of the Catholic colony of Maryland, carved from a portion of Virginia.[47]

By 1635 the powerful planters in Virginia had had enough.[48] At a stormy Council meeting, Harvey threatened to arrest several planters, vowing for example,

> that he would not leave [Samuel] Mathews worth a cow tail before he had done with him, and that if [Harvey] stood the other should fall, and if he swam the other should sink.[49]

Faced with such a threat, the Virginia planters literally picked Harvey up by the scruff of the neck, marched him to a ship, and sent him packing to England. Having their own powerful allies in the King's court, they knew they could convince the King their action was justified, thereby avoiding punishment for this rebellion.[50]

Francis Wyatt, a former resident of the colony and a friend to these planters, was next appointed Governor. The King also confirmed the previously issued land patents as well as the headright and labor system. In 1635 Wyatt began issuing large numbers of patents, clearing the backlog from the last six years.[51]

As the 1630s drew to a close, the colony's tragic future can already be dimly seen. During these early decades, as historian Edmund Morgan has noted, "we can see not only the fleeting ugliness of private enterprise operating temporarily without check, not only greed magnified by opportunity, producing fortunes for a few and misery for many. We may also see Virginians beginning to move toward a system of labor that treated men as things."[52] Here was a place that an aristocrat could send his younger sons to create, for themselves, an estate with servants ready-made to rule over.

The heart of the problem was the legal framework that had been created by Edwin Sandys in London, and warped by the powerful planters living in Virgina for their advantage.

First, because the Great Charter had not been written to create a true social order and was therefore very incomplete, it set a precedent throughout these early years where laws were generally written in a haphazard and inconsistent manner, often allowing the law to be enforced according to the mere whim of a judge or powerful planter. And since the colony's leaders assumed that all rule should be imposed and dictated from above, such a system made easy the worst types of abuse of power.[53]

Second, by giving the fifty acre headright not to the settler but to the wealthy person who paid that settler's transportation, the system served to concentrate power to those wealthy individuals, and to encourage them to create large dispersed plantations focused solely on growing tobacco. The immigrants scattered across the Chesapeake, forming neither communities nor towns, only large plantations, each of which was run not unlike a company town, with the plantation owner ruler of all. As an anonymous report described in 1661, Virginia

> contains above half as much land as England . ... the families whereof are dispersedly and scatteringly seated upon the sides of rivers; some of which running very far into the country . ... above a hundred miles.[54]

As historian Phillip Bruce noted, "The only place in Virginia previous to 1700 to which the name of a town could, with any degree of appropriateness, be applied, was Jamestown, and even this settlement never rose to a dignity superior to that of a village."[55] Throughout the seventeenth century, Virginians lived in isolated lonely plantations without the neighborly society found in villages, towns and cities.[56]

The headright system also served to unnaturally skew the sex ratio of immigrants by encouraging Virginian planters to bring to the colony male laborers for harvesting tobacco, rather than families who would settle on their own farms and thus form a more natural society. As a result, throughout the seventeenth century the colony averaged an abnormal immigration rate of approximately three men to one woman. This shortage of women also amplified the lack of a stable family life, for both children and the swarms of men that lived there.[57]

Finally, and most importantly, the headright system as designed ended up encouraging the growth of a violent and oppressive form of labor. As already discussed, the majority of Virginia's citizens in these early years lived in servitude, at the whim of those in charge. As servant Thomas Best said in a letter in 1623, "My Master Atkins hath sold me for a 150 pounds sterling like a damnd slave."[58] Against this background of forced unpaid labor, waged workers could not gain a foothold in Virginia. Furthermore, the extensive use of forced, unpaid labor during the colony's first three decades served to teach Virginians that this kind of labor was culturally acceptable, and that, just like George Percy and the gentlemen in that first expedition, it was actually demeaning for free Virginians to work with their hands.

Other problems existed that were not directly caused by Sandys' legal reorganization, but they were related nonetheless. For example, because the focus by Sandys and the Virginia planters had been to make Virginia profitable, neither Sandys nor those early planters made any effort to establish religious institutions or schools within the colony, despite a strong desire by many of the colonists for such institutions. Furthermore, the colony's early leaders were never very interested in bringing clergymen to Virginia, nor giving sufficient freedom or power to any who came, since that would have allowed such clergymen to question the social order and power structure of the colony. The result was a colony with no schools, no churches, and few ways for the poorer immigrants to educate their children.

All of these issues were then amplified by something that Virginia's leadership could not fix, the colony's distressingly high mortality rate. Throughout the first six decades of settlement, it was consistently expected that at least 20 percent of all new immigrants would die from some form of disease in their first season in Virginia. This "seasoning," as Virginians called it, meant that despite the immigration of approximately 15,000 people between the years 1625 and 1640, the population of the colony grew by only 7,000.[59] Because of this, children could reasonably expect the loss of at least one parent before they reached adulthood; almost 20 percent would be orphans before their thirteenth birthday, and over 30 percent before their eighteenth.[60]

Still, there was a chance to fix the colony's social problems, had the colony's leadership and ruling class had the will to do it, since most of the problems stemmed from the headright system as established by Edwin Sandys. Change the headright system to give the land to the settler, and then families might have started immigrating to Virginia, creating small farms and towns with their accompanying schools and churches.

The problem was that the headright system was already established, and once such systems exist they create vested interests (in this case the planters who owned large plantations) who resist any change. The revolt against Governor Harvey, who might have actually been trying to change the headright system, illustrates this difficulty.

That it was difficult however does not mean it was impossible. From 1639 to 1641 Frances Wyatt presided over the colony. Then, in 1641, a new governor was appointed. This man would lead the colony for most of the next four decades, during which he and the wealthy planters who ruled Virginia would try again and again to address the initial problems outlined above.

Why they failed, with that failure helping to make Virginia the largest slave-owning community in the English-speaking world, is the question that still concerns us today.

The First Thirty-five Years

2. The World of William Berkeley

The winter of 1637 was fast approaching, and for almost a year and a half the London theaters had been closed for fear of the plague. With the weekly death toll rising into the many hundreds, it had become a routine to post bills of mortality each Thursday morning at the Parish Clerks' Hall.[61] Even the King's Men, the theatrical company of Shakespeare and a favorite of both King Charles and Queen Henrietta, had been forbidden to perform because of the plague, with both their outdoor Globe Theater and their indoor theater in the Blackfriars neighborhood of London shuttered for almost seventeen months.

Now it was September, and for most of the summer the weekly deaths had been dropping, finally dipping below 50 per week in August. With this decline, the many different theatrical companies of London petitioned the king and his Privy Council to allow them to perform, "having no other employment nor means to maintain themselves and their families."[62] Finally, on October 2nd the Council felt it safe enough to open the theaters, and by November the King's Men were staging plays in Blackfriars.

The plague's easing also allowed the resumption of private Royal performances, and throughout that winter the King's Men presented a series of court performances at Whitehall for the king and queen, totaling fourteen plays. Of these, only two are known definitively, John Suckling's *Algaura* and William Berkeley's *A Lost Lady*. While John Suckling is still sometimes remembered for his light and witty poetry, William Berkeley's legacy as a poet and playwright has been largely forgotten. His historical significance, however, must not be, for it was he who presided over the legalization and establishment of slavery in America.

A Lost Lady opens with a "Prologue to the King," in which Berkeley expressed his heartfelt loyalty to King Charles:

> Most royal Sir could I yet lower bend
> I should, by his commission that did send
> Me to this duty, who stands trembling now,
> Expecting what you like or disallow.[63]

The prologue further expresses Berkeley's wish that "his fortune shall so far comply/With his desire to serve." Though such a dedication might appear no more than the routine flattery for a patron as powerful as the king, Berkeley was probably quite sincere. Not only had he been surrounded from birth by privilege, wealth, and the elite interests of royalty, whenever he expressed his mind on the righteousness of royal rule, he indicated a willingness to fight and die for it.

Born in 1605 as the sixth child of Sir Maurice Berkeley and Elizabeth Killigrew, William Berkeley had grown up in the substantial estates of his parents in Somerset, Gloucester, Middlesex, and London.[64] Educated, well-connected, loyal to the king, Berkeley was in many ways the typical "Royalist." He had earned a baccalaureate degree at Oxford, was made a fellow at Merton College, and studied law at the Middle Temple. In 1630, he did what most young British gentlemen did during the Stuart era when they finished college, crossing the channel to explore the continent, visiting France, Italy, and Rome, as well as joining his brother in Holland for a short stint of soldiering.[65]

In 1632 Berkeley returned to England. Though born of nobility, Berkeley was a younger son, and like George Percy was cut off from his father's estate under England's law of primogeniture. Twenty-seven years old, Berkeley had to find a way in the world. Though he had earned a law degree, Berkeley preferred to put himself in the center of court politics. He obtained an appointment as a gentleman of the king's Privy Chamber, a ceremonial post in the king's household that provided Berkeley with room and board at Whitehall for one month in four, and placed him within the high circles of the royal entourage.[66] Encouraged by Queen Henrietta-Maria, he started writing plays.

When *A Lost Lady* was performed in 1637, William Berkeley had spent almost half a decade living and working within this royal court, participating fully in an elite social life that was becoming increasingly isolated and distant from the interests, desires, and cultural experience of most British citizens.

Charles had become king in 1625 upon the death of his father James. Three times in the first four years of his reign he had convened Parliament in order to obtain revenues. Three times Parliament had resisted, either by refusing his requests or by demanding that the king's powers be restricted significantly. Three times Charles had quickly dissolved these Parliaments. After the third dissolution, the king decided to do without Parliament entirely. As he announced in a royal proclamation in 1629,

> Howsoever we have showed by our frequent meeting with our people our love to the use of parliaments, the late abuse having for the present driven us unwillingly out of that course, we shall account it presumption for any to prescribe any time unto us for parliaments.[67]

From 1629 until 1640 Charles ruled Great Britain without the approval of any parliamentary body. No elections were held and no new laws were passed. The king raised money by selling knighthoods and other positions, by attempting to widen the imposition of old forms of tribute to the king, such as forced loans and wardships, and by enforcing the disputed tax called ship money, imposed unilaterally by the king on coastal cities.[68]

While ruling without Parliament, Charles also attempted to emulate the elite nature of palace life as illustrated by the kings on the European continent.[69] He surrounded himself with a cadre of loyal supporters, members of the English aristocracy, and barricaded himself within the halls of Whitehall, devoting much of his time and effort to buying art and staging elaborate masques and plays.[70] William Berkeley was a very minor but willing member of Charles' court, closely linked to the various factions within that court.

The elitist nature of that court can be gleaned by the actions of its members. Of the hardline Royalists with whom Berkeley had contact, the already mentioned John Suckling was probably quite typical. A gentleman of the king's Privy Chamber like Berkeley, Suckling was known for his sharp wit, smooth poetry, ability at cards, and spendthrift nature.

More than a witty writer of verse, however, Suckling was so fiercely loyal to the king and what that royalty represented that he was willing to use force to defy Parliament. Together with other members of the Royal court, men such as William Davenant, Henry Jermyn and others, he helped gather Army troops in 1641 for a planned march on Parliament, a rebellion later earmarked "the Army Plot."

Jermyn, who was Berkeley's first cousin,[71] was an important member of Charles's court. Closely aligned with Queen Henrietta-Maria, he served for many years as one of the king's most loyal henchman, often doing his dirty work. In 1641, he helped plan and instigate the Army plot, pursuing his scheme even after it was publicly rejected by the king.[72]

William Davenant, meanwhile, was a writer of plays and court masques (a kind of variety show of dance, music, poetry, and drama) that glorified the king and the concept of royal rule. His father, who at his death in 1621 was mayor of Oxford, owned an inn and tavern in which Shakespeare often stayed, and of whom Davenant almost certainly met as a child. Educated at Oxford, Davenant joined the king's court upon reaching adulthood, where he wrote several well known masques, in which Queen Henrietta-Maria and other court nobles performed. When Ben Johnson died in 1637, the queen got him the title of poet laureate of England.[73]

The Army plot that these men engineered was quickly discovered and squelched, with the king refusing to publicly support it. When the

House of Commons started asking questions, Suckling, Jermyn and Davenant were forced to flee England to France. Suckling, stranded without money or support, is thought to have quickly killed himself.[74] Jermyn and Davenant, however, had little problem surviving, and when war finally broke out between the king and Parliament in 1642 they worked closely with the queen, also exiled in France, to provide soldiers and supplies for the king's army. As the Civil War intensified, both men returned to England to fight, and after the king's defeat they once again fled to France to the queen's protection.

Suckling, Davenant, and Jermyn were typical examples of a faction at court so loyal to the idea of royalty that they considered violence against Parliament an acceptable option.

Berkeley knew them all: throughout the decade of the thirties we continually find him and them wandering about within the same social circles. As already mentioned, both Suckling and Berkeley were gentlemen of the Privy Chamber, assigned to attend to the king's daily needs. Both had their first plays performed in the winter of 1637. When the king gathered troops to fight the Scots in the first Bishops' War in 1639 all were there: Berkeley as one of the king's guards, Suckling in charge of a troop of 100 cavaliers, outfitted and dressed extravagantly out of his own pocket, with Davenant as one of those 100.[75]

Suckling himself referred to both Berkeley and Davenant, as well as other court poets, in his poem, "The Wits, or The Sessions of the Poets."

> To Will Berkeley sure all the wits meant well,
> But first they would see how his snow would sell:
> Will smil'd and swore in their judgments they went less,
> That concluded of merit upon success.[76]

This poem was written in the summer of 1637 just as Berkeley was probably finishing *A Lost Lady*, and Suckling's commentary carries with it the attitude of a skeptic, combined with a friendly dig at what R.C. Bald described as "the frigidity, or purity" of Berkeley's play.[77]

Yet, while Berkeley surely knew Suckling, Davenant, and Jermyn, he probably only saw them as Royalist allies and acquaintances, not close friends. They represented the most extreme wing of the Royalist court, while Berkeley's strongest influences appeared to come from more moderate Royalists like Sir Edward Hyde, the first Earl of Clarendon, whom Berkeley had known when they were both students at the Middle Temple and whom was known to oppose Jermyn.[78] While John Suckling was willing to overthrow Parliament so the king could rule, men like Hyde believed that though the king was chosen by birth to rule, he must do so according to law and with the help and aid of his people. As Hyde stated in his history of these times, the king must "shelter himself wholly under the law."[79]

Another friend of Berkeley was John Tradescant, gardener to the queen.[80] He had been in Virginia in 1637 "gathering all varieties of flowers," and it was probably his gentle influence that made Berkeley so dedicated to agricultural experiments when he moved to the New World.[81]

And yet, Berkeley neither avoided contact with nor did he oppose men willing to use force in maintaining the royalty. His younger brother John was involved in a second Army plot to overthrow Parliament, and was arrested by the House of Commons in 1641.[82] Later John fought heroically in the civil war on the side of the king, winning promotion and prestige for his actions. A favorite of the queen, he also spent time in Paris during the war, having frequent contact with cousin Henry Jermyn.[83]

The Army Plot was indicative of the growing conflict brewing in England during King Charles' personal rule, a conflict partially rooted in how all these Royalists conceived the nature of society, politics, and the human condition, and most specifically how they perceived their social and economic inferiors.

Much of the Royalist philosophy of life centered on issues of degree and place, what in 1733 Alexander Pope called the "vast Chain of Being." As outlined in *An Exhortation to Obedience*, (written in 1562 and required reading by royal decree in every English church):

> Almighty God hath created and appointed all things in heaven, earth and waters, in a most excellent and perfect order. In heaven he hath appointed distinct and several orders and states of archangels and angels. In earth he hath assigned and appointed kings, princes and other governors under them, all in good
>
> and necessary order. . . .Every degree of people in their vocations, calling and office, hath appointed in them their duty and order. Some are in high degree, some in low, some kings and princes, some inferiors and subjects, priests and laymen, masters and servants, fathers and children, husbands and wives, rich and poor, and every one have need of the other.[84]

According to King Charles I and all Royalists like Jermyn, Suckling, Davenant, Hyde, and Berkeley, whether extreme or moderate, the Chain of Being was interpreted as meaning that each person had his or her place in society, and was expected to remain within that position throughout his or her life. Pigeonholed by God on the great chain, a cobbler was a cobbler, a butler was a butler, a footman was a footman, and a noble was a noble, with their children and children's children accepting their place as well.[85]

At the very top and ruling all of society was the king, born to the position of rule. For society to be stable it was essential that he be able to maintain social policy, without opposition. And because society provided him with everything he could ever possibly want, a king was assumed to be incorruptible and above any bribe. He would have no motive to favor one faction over another, and would therefore rule fairly and with justice. Disagreements between various political factions would always end with the king's final and just decision. Disloyalty and resistance to his rule could only bring about chaos and violence as various factions fought with each other for power and control.[86]

Even as he stepped up to the scaffold to be executed in 1649, Charles repeated this doctrine.

> It is not [the people] having a share in the government; that is nothing appertaining unto them. A subject and a sovereign are clean different things; and, therefore, until you [separate them] they will never enjoy themselves."[87]

Here was a philosophy that believed in the divine right of kings, and that king's right to rule based solely upon his birthright. As was said in 1629 by an anonymous defender of the rights of British kings,

> Let no man think himself a good patriot that, under a presence of liberty of subjects or the commonwealth's welfare, stands in opposition to the king's pleasure.[88]

Nor did the Royalist doctrine of rule-from-above end with the king. Below him stood the aristocracy and the landed gentry, who supported the king and helped administer his policies. These men ruled the local communities, responsible for enforcing the law and setting community standards. Consider Berkeley's own family. His father Maurice served four times in Parliament, representing variously Sommerset, Truro, and Minehead.[89] Charles, Berkeley's oldest brother,[90] did the same. Both ruled over substantial estates in the Somerset area, serving the community as its titular head.[91] Lord of the largest property in a small local community, an estate like the Berkeley's would typically become the center of communications, entertainment, administration, employment, and justice, providing money and facilities for local schools, for picnics, for public fox hunts. Charles would sit on the local courts, administering justice at both Quarter and petty sessions. In the national government he as well as other family members would serve in Parliament, a costly responsibility because it was unpaid and required the member of Parliament to maintain a residence in London as well.[92]

Below this ruling caste stood everyone else, the yeomen, husbandmen, laborers, and footmen. Within these groups there was enormous mobility, and in fact it was even possible in sixteenth and seventeenth-century England for such lower class individuals to move upwards into the landed gentry.[93] The Royalists, however, considered such movement dangerous, to be approached with caution and reluctance. As also stated in the same *Exhortation to Obedience* that outlined the Royalist vision of an ordered nature of life,

> Take away kings, princes, rulers, magistrates, judges, and such states of God's order, no man shall ride or go by the highway unrobbed, no man shall sleep in his own house or bed unkilled, no man shall keep his wife, children and possessions in quietness: all things shall be common, and there must needs follow all mischief and utter destruction.[94]

Even a moderate Royalist such as Hyde believed that "God hath placed [kings] over his people"[95] and that the actions of those who wished "to introduce an equality into all conditions, and a parity among all men . . . would presently have produced all imaginable confusion in the parliament, army, and kingdom."[96]

We will see how these beliefs would have important ramifications when black slaves became available to Virginian Royalists in the next three decades.

Because the king and his Royalist followers believed that society could not function peaceably unless all moral and political power emanated from above, they also believed most strongly that church and government had to be combined, a concept they called Church Government. The moral teachings of Christianity should be controlled and dictated by the king. The king would pick the bishops, the bishops would pick the ministers, and the ministers would manage their local parishes, reading royally-imposed homilies like the *Exhortation to Obedience* each Sunday before their parishioners. Control of society for the king and government would be administered through church policy. As Charles I said,

> Believe it, religion is the only firm foundation of all power; that cast loose or depraved, no government can be stable; for when was there ever obedience where religion did not teach it? But, which is most of all, how

> can we expect God's blessing if we relinquish His Church?[97]

For King Charles, his control of the nation's religious institutions was essential to his rule, not merely for power but for the good of the nation. And as all would soon learn, he was willing to die for this idea.

This willingness of the Royalists to divide society into these various permanent layers, with those on top having the power of rule over everyone else, did not mean that the Royalists were totally evil or tyrannical in their approach to government and society. On the contrary, Royalist doctrine firmly believed in the concept of noblesse oblige. Those of noble rank were expected to approach the rule of law with utmost respect, understanding that their position of power carried with it enormous responsibilities, that these responsibilities required sacrifice and commitment to rule with justice and wisdom. Furthermore, the Royalists believed that each layer of society had its own rights and responsibilities, and that each had to respect the rights and responsibilities of the others.

Still, by freezing the layers of society into place, Royalist doctrine interpreted the Chain of Being into a caste system of privilege, of each person being born to a certain place and having no freedom to change this place, regardless of that person's talents or accomplishments in life.

My use of the word caste in this context is quite deliberate. The more familiar terms of class and class struggle are too vague and open to misinterpretation. Defined by the dictionary as "a number or body of persons with common characteristics,"[98] class does not clearly specify the type or purpose of these social groupings, how they define themselves, how they interact with each other, and how they should change with time. Upper, middle, and lower classes in America today have little to do with distinctions of noble and common birth in British society in 1649.

Instead I use the term caste because its meaning is more appropriate to the Royalists' attitude toward individual freedom and the political conflicts of British society in the seventeenth century. The dictionary defines caste as "one of the hereditary classes into which a society was divided."[99] The key word here is hereditary, a personal fact no human being can ever have any control over. No one picks their parents. No one picks their birthright. By dividing society into groupings defined by heredity (that is, the royal, noble, and common groups), the Royalists preferred a society of castes, not of free individuals.

Having said all this about the general beliefs of most Royalists, both in and out of the king's court, what can we say about what William Berkeley himself believed? While his associations in court give a strong hint, both by his willingness to fight with the king in the First Bishop's

War as well as his contacts with Royalists like Hyde and Suckling, these actions are still only hints.

We can obtain a better clue, however, from his own words in *A Lost Lady*. While the play has the general air of light-hearted humor similar to most of Shakespeare's comedies, it rings with violent passion whenever issues of loyalty and revenge are mentioned.

The play was probably first performed before the king and queen in either the Cockpit-in-Court Theater at Whitehall or in Somerset House, then called Denmark House.[100] Regardless of location, the king's artistic designer Inigo Jones probably provided the play with rich and opulent costumes and scenery, an innovation increasingly encouraged in the court of King Charles.[101]

Lysicle, the play's main character, is a man overwhelmed by unbearable grief, mourning the death of his true love, Milesias, who had killed herself when her uncle, her guardian, had rejected Lysicle. Now Lysicle vows on Milesias' tomb that "a full revenge of thy death and my life's misery shall make him [the uncle] pay."[102]

Amid numerous other plot twists, Lysicle is introduced to a mysterious black woman named Acanthe, a moor from Egypt who is known for her ability to read the past and the future. She tells him that if he comes to Milesias' tomb that night, he will see her ghost. At the appointed hour the ghost arises, telling him that Acanthe herself had been the one to betray their secret love to her uncle, causing him to reject them and for Milesias to commit suicide. In rage Lysicle swears

> She will walk in Hell
> With her I will begin, then seek revenge
> Under the ruins of thy uncle's house.
> All men that dare name him, and not curse
> His memory, shall feel the power
> Of my despised hate.[103]

Lysicle then obtains and gives Acanthe a poison that will kill her slowly and in great pain. Even as the black woman dies, Lysicle stands over her and says

> . . . nor should you pity her.
> Those that do trace forbidden paths of knowledge
> The Gods reserve unto themselves . . .
> And she no doubt is conscious to herself
> Of infinite more mischiefs then is yet revealed.
> I am confident she is fled her Country for the ills
> She has done there, and now the punishment
> Has overturned her here. And for her shows

> Of virtue, they are Masques
> To hide the rottenness that lies within.[104]

As she dies, however, the black of her skin begins to rub off, and we discover that she is in fact Milesias in disguise! She had been in hiding, and using this circumstance to discover whether Lysicle truly loved her or not.

Fortunately, the doctor who provided Lysicle the poison had faked it, and it does not kill her. Lysicle and Milesias are reunited, the threat of war ends, Lysicle and his friend are appointed high positions in government, and everyone lives happily ever after.

William Berkeley's strongest beliefs, as expressed in *A Lost Lady*, can be summed up as follows: Betrayal is the worst sin, and can justify the most violent punishments. And caste and skin color can indicate a person's moral and social worth.

A Lost Lady was performed several times in 1637 and 1638 for the royal court. Then, in 1639, King Charles raised his standard in Scotland in the First Bishop's War, and Berkeley followed his king north to serve as one of the privy chambermen who guarded the king. His actions during the war must have impressed the king sufficiently to earn Berkeley a knighthood.

Unfortunately, that same First Bishops' War bankrupted the king's treasury, and after almost eleven years, Charles was finally forced to call for new Parliamentary elections. He was to discover, however, that the Royalist beliefs that moved him and his closest followers were far different from those of most Englishmen, including above all the men who now controlled Parliament.

3. England's First Premise

On a warm spring day in 1638, John Liliburne was escorted from the Tower of London for his punishment. Found guilty for printing and circulating "unlicensed books," one of which had been a rousing attack on Bishop Laud and the Anglican church, he was to be whipped through the streets of the city fastened to the back of a cart, followed by two hours in the stocks.[105]

Liliburne, however, showed no remorse, nor was he frightened by what he was about to endure. In fact, he was almost cheerful about the prospect. As he was tied to the cart and the executioner pulled out his whip, Liliburne told him, "Well, my friend, do thy office."

The executioner replied, "I have whipped many a rogue but now I shall whip an honest man. Be not discouraged, it will soon be over."

Liliburne answered, "I know my God has not only enabled me to believe in his name, but also suffer for his sake."[106]

As they pulled him through the streets he loudly proclaimed his innocence and the evil of the charges against him. When next he was fastened by his ankles to the pillory, he preached to the crowd before him that the bishops of the Church of England stood "by that same power and authority that they have received from the Pope. So that their calling is not from God but from the Devil."[107] Dramatically, he pulled from his pocket three copies of the forbidden book and tossed them to the crowd.

When the prison warden demanded he be silent, Liliburne loudly proclaimed, "I would speak and declare my cause and mind, though I were to be hanged at the gate for my speaking."[108] The warden then had Liliburne gagged for the last hour and a half of his punishment. As he was carried away afterward, well-wishers and supporters lined the streets, blessing him for his courage and sincerity.

Liliburne's act of defiance and hostility to the church of King Charles was, by 1638, no longer an unusual event. For the last two decades, British citizens in increasing numbers had been voicing disagreement and almost downright hatred to the reign of Charles and the philosophies that moved him. This hostility had expressed itself in many ways, from the flight of the first Puritans to America in 1630 to the loud martyrdom of Liliburne. As the decade of the thirties drew to a close, England was a nation primed to explode in factional violence.

The simple reality was that, while the king and the Royalists might have wished to institute a caste system with all rule (both religious and political) administered from the top of the chain, British society simply did not function this way. England was not a land of castes, but of finely defined social layers through which much contact and travel took place. No one could be prevented from moving up the ranks of society if that person demonstrated sufficient wealth and achievement to merit that rise. And while the majority of British citizens might have believed that the world was comprised of a Chain of Being, they did not by any means believe as the Royalists did that each person as well as their descendants were forever locked into his or her position on that chain. Instead, Englanders were motivated by the understanding that they *could* move up the ranks of society.[109]

This, above all, was the first premise from which all other English customs followed, despite the philosophical beliefs of the king and his Royalist followers, the undeniable stratification of British society, and the loyalty and dedication that almost all Englishmen felt to the institution of the king.[110] A laborer knew that if he could gather sufficient funds, he had the right to buy land and become a yeoman. Nothing in the culture forbade that rise. A yeoman in turn knew that, by saving his money and increasing both his landholdings and property, he could eventually move upward into the gentry. If his son continued that rise, he could enter the aristocracy, actually becoming a knight or lord. As Sir Thomas Smith said in the 1560s, "No man is a Knight by succession, not the king or prince. . . . Knights therefore be not borne but made."[111]

Furthermore, because of the rules of primogeniture, there was as much movement down the social scale. William Berkeley's own search for status and security within royal circles illustrates this. In cases where the family was not as well placed as Berkeley's, younger sons were apprenticed out to the trades. The law of inheritance guaranteed that many a lord's son would spend his life in a lower place on the social ladder than his father.[112]

Such movement up and down the ladder of society was far from easy or frequent. British citizens all viewed with suspicion the advancement of anyone up the social ladder, and generally waited a full generation at least before accepting such advancement as a *fait accompli*. As English historian J.V. Beckett has noted, "Entrance into the aristocracy was far from easy, and penetrating the uppermost reaches took careful planning, considerable good fortune, and above all, patience."[113]

Yet, this idea that all men could move up and down the ranks of society, that birthright alone did not establish a person's social ranking, was fundamental to the culture of Great Britain. Neither King Charles and the Royalists could change this, no matter how they tried, and their attempt to do so in the 1630s helped cause the Civil War that followed. For, if every citizen has the freedom to climb about on the vast Chain of Being, by what right does a king claim a privileged place on that chain?

And by what right does the son of a lord claim his nobility? Doubts about the ancient claim of divine right formed the heart of England's problem during the Civil War: the nation no longer wanted its king, but the nation was not allowed to remove him. As historian Samuel Gardiner observed, "Constitutional kingship was unattainable if [Charles I] was to continue to be king, because constitutional kingship rests on the idea that, in case of deliberate and prolonged difference of opinion, it is the nation which is to have the last word, and not the king."[114]

This first premise, that all men had the equal right to obtain property, wealth, and honor, that no man's place in society was forever guaranteed, percolated almost unconsciously through every aspect of British society. For its time, seventeenth-century England was a land of almost revolutionary freedom. There were no slaves, and unpaid servitude was restricted to the education of the young. Laborers were paid daily salaries for their work, and periodically could choose new employers.[115] All citizens had the right to buy land, if they could earn the funds.[116] The literacy rate was remarkably high for its time. At the start of the century more than 30 percent of the population could read, and this premise that advancement was possible encouraged a steady increase throughout the seventeenth century until, by the end of the century, there are indications that literacy had reached more than 70 percent.[117]

Most importantly, British citizens firmly believed that the law of England applied equally to all citizens, regardless of status or wealth. As historians John Brewer and John Styles have noted,

> It was a shibboleth of English politics that English law was the birthright of every citizen who, unlike many of his European counterparts, was subject not to the whim of a capricious individual but to a set of prescriptions that bound *all* members of the polity [author's italics].[118]

Though these laws of property and status legally applied only to men, because English men were expected to live in families the laws really applied to his wife and children as well. As William Blackstone stated in his *Commentaries on the Laws of England*,

> By marriage, the husband and wife are one person in law; that is, the very being or legal existence of the woman is suspended during the marriage, or at least is incorporated and consolidated into that of the husband.[119]

Defined as a man and woman raising their children, this family unit would rise or fall together.[120] The man was expected to be the family's breadwinner, the woman was expected to manage and run the home, and together they would reap the benefits of advancement for their children.

The importance of family to British culture cannot be understated. Hundreds of popular books and pamphlets were written and sold during this time period, extolling the virtues of family life and marital unity.[121] The divorce rate was minuscule, with less than a tenth of a percent of all marriages ending in a court-ordered separation.[122] The illegitimacy rate was as small, ranging from 1.5 to 3 percent of all births throughout most of the seventeenth century. And when it rose to 3 percent, the social pressure to punish and condemn such actions rose correspondingly.[123]

And all of this was structured around the premise that by carefully raising their children, parents could push them up the social ladder. A father would work hard, earn more money, buy land, and hence, try to establish his family among a higher social strata. A mother could provide a stylish home, efficiently managed so that more money was available to educate and train the children.

Because most British families were not wealthy and did not own land, the possibility of advancement did not generally mean entering the aristocracy. Instead, these families lived in tightknit, small, local communities, and focused their hunt for betterment within these hamlets.

The small-town nature of the British society made this competition for status very personal and domestic. Both lord and laborer would belong to the same local parish, not as master and serf, but as members of a community. Each had specific responsibilities to that community, and how they performed those tasks would help determine their reputation and status. Hence, the competition to do social work was intense, and one feels almost daunted when reading of the amount of voluntary work required of every citizen. The lord had to provide protection and sponsorship, as well as pay greater taxes and the cost of representing his parish in parliament. Yeomen had to serve as sheriffs, collect taxes, run elections, help manage the local parish, and vote in parliamentary elections. All were required to own their own personal weapons and participate in regular militia training for the defense of the nation.[124]

This small-town community of traditional families competing for status was centered around the local parish. All births were recorded there. All marriages were sanctioned there. And all deaths were mourned there. Every Sunday the community would gather at church for prayer, moral judgment, and some social entertainment.[125] And while it is true that many British citizens did not involve themselves passionately in religious issues, the importance of the parish in community life flavored this British chase for prestige with a certain moral scent. Considerations of right and wrong could not easily be dismissed, and as the seventeenth-century unfolded, such issues became even more significant to British life.

And here we find another fundamental disagreement between Royalist philosophy and the rest of British society. The Royalist belief in Church Government was a direct attack on the local, down-home nature of the English neighborhood parish. Worse, it reminded too many of the customs of the Catholic church, arousing a great deal of suspicion and distrust. The English of the seventeenth-century were passionately hostile to Catholicism and its need to dictate policy from Rome while enforcing its religious practice through the use of power and the trappings of ritual, artifacts, and icons. In such a church, the local community was expected to obey these policies and practices, regardless of local concerns. For many Englishmen, from the most ordinary to the most powerful, this form of church rule-from-above threatened their autonomy, independence, and freedom. And when Charles (with his Catholic queen) appeared to adopt religious policies that imitated this church, he helped cause a schism between himself and his subjects.[126]

In fact, the friction and anger that this issue provoked can be illustrated no better than by the willingness of tens of thousands of middle-class religious British citizens to flee England throughout King Charles' reign, risking death and poverty to settle in the untamed wilderness of America. These "Puritans," as their detractors often called them, began their exodus from Charles not long after his ascension in 1625. For the next two decades more than 30,000 fled across the Atlantic, settling throughout New England so they could practice their religion as they wished.

Strongly committed to the Old Testament concepts of family, community, and religion, as illustrated by their own British society, the Puritans passionately wished to bring to Earth this concept of God's heaven.[127] As the first half of the seventeenth century unfolded the Puritans were quite willing to use the power of government to impose their social and religious precepts on all of British society, eventually becoming a significant force in the English Civil War and dominating the governments of both the Commonwealth and the Protectorate. In the 1620s and 1630s, however, this kind of power and control did not seem possible, and large numbers of Puritans sailed to New England to establish their version of British society in America while also becoming a symbol for England to emulate.[128] As John Winthrop preached on the deck of the *Arabella* as it approached Massachusetts Bay in 1630,

> [W]e shall be as a City upon a Hill, the eyes of all people are upon us . . .we shall be made a story and a byword throughout the world.[129]

Well known for their intolerance of religious dissent in England, the New England Puritans were not afraid to use extreme measures in these first decades to maintain the uniformity of their community. They banished Quaker Anne Hutchinson in 1637, whipped several other Quaker women in 1658, and hung Mary Dyer in 1660 and William Leddra in 1661.[130]

Despite their intolerance, however, the Puritans of New England were remarkably successful in quickly recreating English culture in the New World. Unlike the settlers of Virginia, the Puritans immigrated as families and lived as families in small town communities that were centered on the construction and maintenance of a local church.[131] Their society was educated, prosperous, and well-governed. Their children were carefully socialized into this cultural system. And, unlike the situation in Virginia, few New Englanders immigrated as bonded servants.[132] Like England, apprenticeship remained a form of education for the young, and was managed with strict social controls to avoid the kind of abuses typical of the Virginia tobacco fields.[133]

This lack of forced labor, in both England and New England, must be emphasized. In recent decades it has become customary for historians to focus on how, prior to the eighteenth-century, no European country had abolished slavery, and how all saw little difficulty in participating in the African slave trade.[134] Such focus, however, distorts our picture of English society. While slavery itself was generally not considered immoral,[135] and some European nations, such as Spain and Portugal, were heavily involved in the African slave trade at this time and even had brought some slaves to the continent of Europe, English citizens simply did not participate. Though John Hawkins attempted to make money as a slave trader in the previous century, not only did he never intend to sell slaves in England, he was unable to make a go of it, and no other Englishmen were willing to join him. English voyagers to Africa throughout the sixteenth century and the first half of the seventeenth-century were just not interested in this trade for human flesh, looking instead for gold, ivory, pepper.[136] This disinterest in buying human slaves was expressed best by an English captain in 1623 along the Gold Coast. Offered a shipment of black slaves, he refused, noting that the English "were a people who did not deal in any such commodities, neither did we buy or sell one another, or any that had our own shapes."[137]

This is not to say that some Englishmen were not willing to buy and sell black Africans for profit. Hawkins clearly proves that some were willing, as does the existence of slave-traders among later New Englanders. More important, however, was the general English cultural imperative against such customs: slavery denied a person the ability to improve his life, status, and worth. To be owned by another was abhorrent and horrible, and to own another was considered equally villainous and despicable. As Thomas Smith said in 1583,

> The nature of our nation is free, stout, haughty, prodigal of life and blood: but contumely, beatings, servitude, and servile torment and punishment it will not abide.[138]

While men were legally allowed to buy and sell slaves, if they did so they invoked that worst part of man's crass and evil nature, going against that British belief that each individual should be free to pursue their dreams.

The consequences of this cultural imperative was that while it was possible for individual Englishmen to deal in slaves on the periphery of British society, it was impossible for the custom to grow popular in England, the center of society. Furthermore, none of the first three North American English colonies (Virginia, New England, and Maryland) made slavery a dominant labor option for most of their early history, when the settlers were British-born immigrants. Even as late as the 1680s, fifty to seventy years after their establishment, black slaves did not exceed 9 percent of the population in any of these colonies (see figures 3.1, 3.2, 3.3, 3.4, 3.5). Even in Virginia black slavery was not favored in these early years. When faced with the easy availability of slaves in the New World, the English settlers of America generally resisted the temptation to buy slaves.

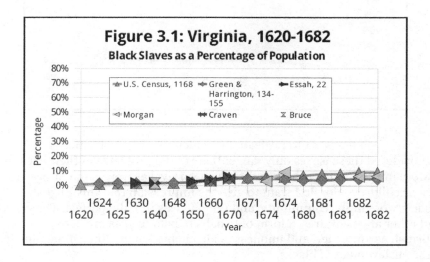

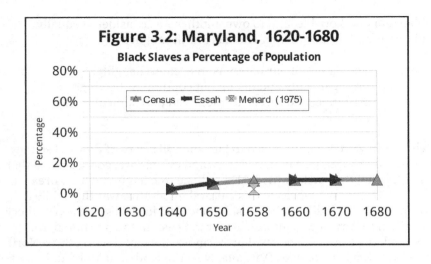

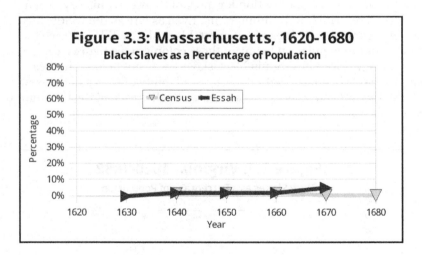

I do not want to mislead the reader. Most British citizens considered these issues only tangentially to the problems of their lives. More important to each were their own survival as well as the future of their children. Nor do I want to give the impression that this English society of family, community, and religion was morally perfect. The fear of other alien cultures, a form of tribalism, belonged as much to the British as any other, and unquestionably helped many Englishmen justify the enslavement of blacks.

This first premise did much more as well. It encouraged a kind of religious, political, and moral free-for-all that was largely unprecedented in human history, especially because this free-for-all often centered its

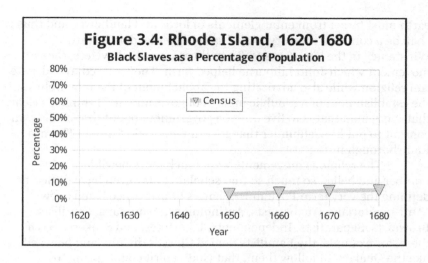

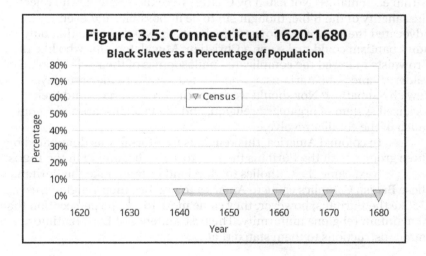

debate on transferring rule from the upper privileged classes and instead seating it in the general population of the nation.[139] This free-for-all included a wide range of religious sects, all of which were variations on Christianity (such as the already-mentioned Puritans), as well as an extensive number of radical political movements.

The political factions can sometimes sound remarkably like modern political movements. While the Royalists could be likened to today's preservers of the status quo, afraid of the chaos and anarchy that a more open system would bring, radical groups like the Diggers resembled the communists, believing that property rights were the cause of all evil, and that the land should be owned in common by all. "The earth must be set from entanglements of lords and landlords, and thus it shall be a common treasury to all," said Digger leader Gerrard Winstanley in the late 1640s.[140] Then there were the Levelers, the movement which John Liliburne helped form. They wanted to end state-run religion while also advocating the abolishment of the monarchy and the establishment of a republican form of government. "I think it's clear, that every man that is to live under a government ought first by his own consent to put himself under that government,' said Colonel Thomas Rainsborough in 1647.[141]

The religious movements were much more bewildering in number and belief, so much so that scholars today even have difficulty defining the very term Puritan.[142] Charles himself once listed a few: "Anti-trinitarians, Anabaptists, Antinomians, Arminians, Familists, Brownists, Separatists, Independents, Libertines, and Seekers."[143] Of these, for example, the Familists denied Christ's divinity and believed, like the Quakers to follow them, that God's spirit could spring from within each man.[144] Not listed by Charles were the Ranters, who rejected the sanctity of the Bible, thought sin to be impossible, and even advocated the idea of free love;[145] the Baptists, who believed that only adult baptism could make one a Christian; Muggletonians, who like the Brownists followed the complicated philosophies of their self-appointed leader;[146] and Sabbatarianists, who wanted to restore the Saturday Jewish Sabbath.[147] Nor should we forget the Presbyterians and their localized system of organized religion, and the Puritans, around whom much of the conflict revolved.

In colonial America, this religious free-for-all was dominated by three groups, with the Puritans the first to arrive beginning in the 1620s.

Next came the Catholics to Maryland in 1633. Like the Puritans, these British Catholics came to America to practice their religion freely. Unlike the Puritans, however, they came more to escape persecution than to maintain religious uniformity. The first sentence of Lord Baltimore's initial instructions to them stated that

> they be very careful to preserve unity and peace amongst all the passengers on shipboard, and that they suffer no scandal nor offense to be given to any of the Protestants . . .[and] they cause all acts of Roman Catholic religion to be done as privately as may be, and that they instruct all Roman Catholics to be silent upon all occasions of discourse concerning matters of religion; and that [they] treat the Protestants with as much mildness and favor as justice will permit.[148]

Because of the immense distrust and fear that British society felt for "popery" and the Roman Catholic religion,[149] Lord Baltimore wanted the settlers to tread very carefully, to demonstrate no example of intolerance on their part so as to avoid giving ammunition to their many enemies. They would create in Maryland their version of an ideal religious community, but they would do so without imposing church government on the colony.

In Pennsylvania the Quakers came next, arriving in large numbers in the 1680s for much the same reasons as the Catholics, having faced severe persecution during the previous half century. Unlike the Puritans or Catholics, the Quakers rejected the idea that religion should be controlled from above by a king, his ministers, or any other religious hierarchy. Instead, they believed that the divine light of God radiated from within each soul, and that it was hence possible for each person to speak directly to God. For this reason no person had the right to impose his or her interpretation upon God's word, and from this developed the very egalitarian Quaker meetings in which anyone was permitted to speak. And by the very nature of their beliefs, the persecution of another for their religious practices was considered unacceptable.[150]

Most of these political and religious factions disagreed with the Royalists' belief that power and authority automatically descended from the top of the Chain of Being. Most wished to decentralize religious and political power, though sometimes they merely objected to what the king wanted to do, not to his use of government power to do it. As an example, when Cromwell and his Puritan supporters finally took power in 1648, they very quickly tried to impose their religious values on the nation, shutting the theaters and outlawing the use of the Book of Common Prayer in Sunday service.[151]

As these political and religious factions battled for control of both English society and government, their intellectual ideas became more sophisticated, and from these debates can be drawn the two basic principles that were quickly reflected in most of the North American British colonies: religious toleration and democratic government.[152]

The idea of religious toleration, that it was not the place of government to impose any specific religious belief on society, was probably the most fundamental and original concept growing out of this political and religious debate. As stated in an anonymous paper published in England in 1644, "all former Acts which countenance persecution for matters of religion may be repealed, and liberty of conscience, which is the greatest liberty the Gospel brings, [be] restored."[153] This concept of liberty of conscience, what we today would call religious freedom, would eventually lead to the Toleration Act of 1688 during the Glorious Revolution, which permitted religious dissenters the freedom to practice their religion as they saw fit.

In America, this same evolution toward liberty of conscience also took place, though it took longer. The Quakers, Puritans, Catholics, as well as a number of other sects, were all forced to respect the wishes of others in the New World. By the end of the seventeenth century, any intolerance to religious dissent within any of those colonies was becoming increasingly unpopular. By the Revolutionary War, it was entirely unacceptable.

The second political principle that grew out of the English Civil War was the eventual rejection by all the dissenting religious sects and political factions of the premise of caste. Even the most fanatical and oppressive advocacy group eventually discarded the idea that birth alone was sufficient qualification for rule. Other considerations, such as the quality of leadership and the principles behind that leadership, were becoming as (if not more) important. As Puritan writer Henry Bullinger said in 1587, "We ought not to obey the wicked commandments of godless magistrates, because it is not permitted to magistrates to ordain or appoint anything contrary to God's law, or the law of nature."[154] Later, in New England in 1644, John Winthrop noted that

> the people, giving us power to make laws to bind them, they do implicitly give their consent to them. . ..
> [While] they put themselves into our power to bind them to laws and penalties, they can intend no other but such as are just and righteous: and although their implicit consent may bind them to outward obedience, yet it neither ties them to satisfaction, nor frees such lawmakers from unrighteousness, nor the law itself from injustice.[155]

By accepting such limitations on their rulers, Englishmen were accepting the premise that rulers of society had to bend to the will of what seventeenth-century leaders would have called "the giddy multitude." The status and birthright of those leaders would not be relevant.

Writing in England just before the execution of Charles I in 1649, Major Francis White said

> the king hath no other right to the military, regal, and legislative power than the sword did constitute and invest him with by Divine permission, the people submitting thereto for fear, and to avoid greater mischief; but now, the king and his party [the Royalists] being conquered by the sword, I believe the sword may justly remove the power from him, and

settle it in its original fountain next under God–the people.[156]

Slowly, this concept of government by, for, and of the people developed, that rule of government should spring from the bottom, not from the top. As John Liliburne asserted in 1646, "The poorest that lives hath as true a right to give a vote as well as the richest and greatest."[157] While British society was many, many decades away from establishing the democratic, parliamentary government familiar to us of the twentieth and twenty-first centuries, that parliamentarian idea had its beginnings here.

Throughout the seventeenth century each political battle in England saw the Royalists lose ground. By the Glorious Revolution in 1688, English society was quite willing to replace one king with another, based on his religion, policies and qualifications, and to establish the Toleration Act concerning the practice of religion. The king might still be different from a cobbler, but now much less so, and should he rule unjustly, with tyranny and cruelty, the citizenry had the right to remove him. It would seem that these ideas foreshadow Jefferson's Declaration of Independence, and what historian Samuel Rawson Gardiner called "the expansion of reasoning intelligence."[158]

In 1641, however, this was still all in the future. Seeing the growing unrest around him, Berkeley decided it was time to move on, to find a better position than that of a mere gentleman in the king's Privy Chamber. The incumbent governor of Virginia, Sir Francis Wyatt, agreed to sell the office to Berkeley, and the king agreed to appoint Berkeley to the post.

So, sometime in January of 1642, William Berkeley sailed to Virginia. He had left behind him in England a political situation that was about to explode into almost twenty years of revolt, war, and regicide. That same month, King Charles walked into the House of Commons and demanded the arrest of five members of Parliament, accusing them of high treason for criticizing his actions and administration. To do such a thing was to defy all the privileges and rights of Parliament customary to British government and law, and instead of surrendering these members to the Crown, the Parliament and the general population became inflamed by Charles' action. Cries of "Privilege of Parliament" rose in the streets, and Parliament was soon calling for the punishment of those who had advised the king to take such action.[159] England quickly divided into armed camps, torn apart by violent civil war.

Berkeley, however, was not in England when these events took place. He had sailed to Virginia, and there attempted to apply his Royalist concepts of loyalty, noblesse oblige, and caste privilege to the administration of the small, British colony of Virginia.[160] For the next thirty-five years, twenty-seven of which he served as its governor, he and the Royalists who followed him there struggled with the colony's myriad problems. And while New England, Pennsylvania, and the other northern British colonies increasingly turned away from the use of slaves, Virginia under Berkeley instead laid the foundation for that custom to prosper and grow.

4. Berkeley's First Term

> I, on my view of Virginia, disliked Virginia, most of it being seated scatteringly in wooden clove board houses, where many by fire were undone, and by two massacres in an instant fires, without any forts there, or retreats of safety in time of danger, and seated amongst salt marshes and creeks, where thrice worse than Essex, and Tenet, and Kent for agues and diseases, brackish water to drink and use; and a flat country, and standing waters in woods bred a double corrupt air, so the elements corrupted no wonder as the Virginians affirm, the sickness there the first thirty years to have killed 100,000 men. And then generally five or six imported died, and now in June, July, and August chiefly one in nine die imported.
>
> —Beauchamp Plantagenet's view of Virginia, 1649 [161]

Almost thirty-five years now had passed since George Percy first sighted the coast of Virginia. Though the colony had grown significantly in those years, it had hardly become the kind of "nation" its original settlers had envisioned. Jamestown, the so-called capital and the only place that could call itself a town, consisted of a few, measly buildings with only a handful of residents.[162] Everyone else lived on widely dispersed ramshackle tobacco farms scattered across the tidelands surrounding Chesapeake Bay.

As a place to live, the colony that Berkeley first saw as he arrived in 1642 was not much to brag about. In essence, Virginia was a very typical boom town, filled with single poor men obsessed with growing tobacco for profit. Of these, most were recent arrivals, bonded servants working in the fields for anywhere from four to ten years in order to pay the cost of their transportation. For these new immigrants, of which approximately seventy percent were men, the death rate from disease remained merciless.[163]

Since money equaled tobacco and tobacco required land, the free settlers had spent the last twenty years scattering across the tidelands searching for unpossessed land they could claim for their own. They took with them the few women settlers, as well as their entourage of male indentured servants. The result was a colony dominated by single men, most of whom were considered nothing more than temporary slaves by their masters. One man who lived it even put the experience in verse.

> A canvas shirt and trouser then they gave,
> With a hop-sack frock in which I was to slave:
> No shoes nor stockings had I for to wear,
> Nor hat, nor cap, both head and feet were bare.
> Thus dress'd into the field I next must go,
> Amongst tobacco plants all day to hoe,
> At day break in the morn our work began,
> And so held to the setting of the sun.[164]

Not surprisingly, this widely scattered boom-town gold-rush culture of money-hungry men and unfree laborers had few parishes or churches. If one thinks of such places, one realizes that the last thing many of these tobacco prospectors wanted was a strong minister or clergyman questioning the morality of their actions. "They have 20 churches in Virginia," Berkeley himself admitted in a public relations piece he wrote in 1649.[165] Twenty churches, many of which did not have ministers, for a 1649 population that had grown to between fourteen to fifteen thousand people.[166]

The colony also had no schools and no requirements for instructing its children. Compare this to New England, where as early as 1642 it was required that all children be taught to read and write, and by 1647 every town was required to hire a schoolmaster.[167]

What made the situation even more miserable for the poorest inhabitants was that the colony's laws comprised of only a handful of poorly worked out rules, none of which dealt with the wide-spread abuse of indentured servitude by those with power. While the "custom of the country" said that an adult immigrant should work for approximately four years to pay his transportation costs, this term was hardly firm, and many servants who arrived without written contracts were quickly exploited.

Of these exploited servants, a very small but increasing number were the black slaves. Before the 1630s, these slaves generally came from either Dutch, Spanish, or Portuguese trades who would sometimes appear in Chesapeake Bay. Then the Dutch conquered Brazil from the Portuguese in 1630, Curacao from the Spanish in 1634, and finally the Portuguese slave-trading bases in Africa in 1637. For the next seventeen years, until 1654, the Netherlands had exclusive control of the trade, and was the source of all the slaves that ended up coming to Virginia.[168]

Even so, the number of black slaves in Virginia in the 1640s remained tiny. Of the colony's 1641 population of 7,000, probably no more than one to two hundred were blacks. Like the white immigrants, they were mostly servants working the fields for their white masters. Because of their unfamiliarity with British law, however, they were probably the most exploited of those servants. Many were expected to serve far longer terms of servitude than the whites, ranging anywhere from the customary four years to as long as thirty years. Many had no set term at all, and were essentially treated as slaves for life.[169]

A few, however, did finish their service and become free. We know that about this time, Anthony and Mary Johnson, formerly owned by Richard Bennett, were making their move to the Eastern Shore as free citizens. In the two decades since their arrival as slaves, they had married and had been blessed with four children, two boys and two girls. By 1645 Johnson was apparently on the Eastern Shore, working land for planter Philip Taylor. According to court records, Taylor and Johnson went out into the corn fields one day to settle their deal. As Anthony said,

> Now Mr. Taylor and I have divided our corn and I am very glad of it. Now I know mine own. He finds fault with me that I do not work but now, I know mine own ground, I will work when I please and play when I please.[170]

Like all freed British settlers without money or property, the Johnsons were struggling to make their way amid the difficulties of frontier life.

All in all, for most whites and blacks alike, Virginia was a wretched place to live. Colonists had little money, using tobacco instead as a form of currency. Uneducated, isolated, and unfree, most worked in an atmosphere of death and misery. Entertainment was limited to horse-racing, gambling, hunting, and drinking at the innumerable taverns that sprouted up across the bay.[171] In every way, the colony resembled the American wild west, with its saloons, its corruption, its violence, and its unruly exuberance.

To this scruffy place came Royalist William Berkeley. He carried with him the king's instructions, numbering thirty-one specific commands for running the colony, from such mundane matters as requiring each farmer to plant a specific quantity of corn, hemp, flax, and mulberry trees to the more significant issues of establishing Royalist control over the colony's churches and government.[172] Like any good Royalist who believed in top-down rule, Berkeley accepted the idea that the king, unfamiliar with Virginia and 3,000 miles away, could impose these requirements on Virginia.

Upon arrival, however, Berkeley did not immediately force the king's agenda upon the colony. Instead, he stepped lightly at his first assembly session in 1642, immediately establishing that he was not John Harvey reincarnated by joining with the planters to reaffirm their property rights and oppose the threatened re-establishment of the Virginia Company.[173] At the same time he elevated the House of Burgesses into a separate legislative body, giving its members greater status as well as a bigger voice in the managing of their government.[174]

Berkeley's light touch did not mean that he did not have Royalist friends on the Virginia Council or in the House of Burgesses. On the contrary, among the colony's most powerful men were many die-hard Royalists, as revealed by a review of the members of the Council and the House during Berkeley's first term. On the Council was Christopher Wormeley (tables 1, 2), who had been Governor of Tortuga and whom some said had lost that colony to the Spaniards due to his carelessness. When he became a Justice in Virginia the accusations of cruelty and oppression were so strong against him he had been forced to go to England to defend himself.[175] Wormeley's brother Ralph had himself come to Virginia in 1636 and had settled in Royalist York County. In 1649 he would join Berkeley in resisting Commonwealth rule, having been appointed to the council by the exiled Charles II.[176]

Henry Browne (tables 1-5, 7, 8) had refused to participate in the ouster of Governor Harvey in 1635, and was turned out of office for a time as a result. Upon his arrival in 1642, Berkeley immediately reappointed him to the council. When the Commonwealth took over the colony in 1652, he stood with his king and once again retired from public office, returning only when the monarchy was restored to power in 1660.[177]

Edmund Scarburgh (tables 2, 5, 6, 8) had been born in 1618, had been younger than twelve when his parents came to the colony. He had either come with them, or arrived at eighteen years old when his father died, before 1635.[178] Throughout the thirties and forties he had been aggressively building a large and thriving plantation on the Eastern Shore, and by the time of his first election to the House of Burgesses in 1643 (see table 2), he had already accumulated over 1,500 acres.[179] In future years he would defend his rank, his power, and the Anglican church to which he belonged with the bluntness and ferocity typical of a frontier-raised Virginian.

Richard Kemp (tables 2, 3, 7, 8) had been appointed by the king as secretary of state in 1634, and was able to keep his post despite his participation in the ouster of Harvey, mainly due to his close ties to Lord Baltimore and Secretary of State Sir William Windebanke. When Berkeley returned to England for a year in 1644, he was quite willing to entrust the governorship to this Royalist ally. Furthermore, Kemp joined the other Royalists in the colony in forming complicated family links: his wife was the daughter of Christopher Wormeley as well as the stepdaughter of William Brocas.[180]

Table 1–The Grand Assembly, April 1642

Previous Governor: Francis Wyatt
New Governor: William Berkeley

Assembly, including Council and House

Royalist:	Neutral:	Puritan:
*Hill, John	*Ludlow, George	+Bennett, Richard
+Browne, Henry	*Minifie, George	+Robins, Obedience
+Wormeley, Christopher	Dew, Thomas	
Hill, Edward, Sr.		Parliamentarian:
	Unknown:	+Mathews, Samuel, Sr.
Uncertain Royalist:	*Butler, William	
*Townsend, Richard	*Fallowes, Thomas	Uncertain Parliamentarian:
	*Fowler, Francis	*Pierce, William
	*Franklin, Ferdinand	+Harwood, Thomas
	*Hardy, George	
	*Johnson, Joseph	
	*Worleigh, George	
	+Bernard, Thomas	
	+Upton, John	
	Chiles, Mathew	
	Dacker, William	
	Gough, Nathaniel	
	Harrison, Benjamin	
	Pettus, Thomas	
	Weale, John	
	Windham, Edward	

	Present Slave-holders	Including future slave-holders	Raised in Virginia	Incumbents	Including those who served previously
Assembly	7 of 29	10 of 29	1 of 29	12 of 29	20 of 29

Legend for all tables:
*-incumbent
+-served previously
Bold names have purchased slaves or have claimed headrights on blacks.
<u>Underlined names will eventually buy slaves or claim headrights on blacks.</u>
Italicized names either were born in Virginia or immigrated as minors.

Source for all tables: See Appendix B

George Reade's brother was private secretary to Windebanke, who seems to have arranged with Governor Harvey and Richard Kemp that a secure position would be arranged for Reade once he landed in 1637. First they put him in "command of some forces sent upon a new plantation." Then, when Kemp went to England in 1640, he arranged to have the king appoint Reade secretary of state in his stead. Such patronage made Reade one of Harvey's strongest supporters, and in later years a reliably dedicated Royalist as well.[181]

Table 2-The Grand Assembly, March 1643

Current Governor:
William Berkeley

Council

Royalist:	Neutral:	Puritan:
*Browne, Henry	*Ludlow, George	*Bennett, Richard
*Wormeley, Christopher	+Higginson, Humphrey	
Kemp, Richard		Parliamentarian:
	Unknown:	*Mathews, Samuel, Sr.
Uncertain Royalist:	*Pettus, Thomas	
*Townsend, Richard	Bernard, William	

House

Royalist:	Neutral:	Parliamentarian:
*Carter, John, Sr.	+Chiles, Walter, Sr.	+Stegg, Thomas-SPEAKER
Scarburgh, Edmund	+Fludd, John	
	Lloyd, Cornelius	Uncertain Quaker:
Uncertain Royalist:	**Taylor, Phillip**	Chesman, John
*Hutchinson, Robert		
Filmer, Henry	Unknown:	Uncertain Puritan:
Taylor, William	*Aston, Walter	+Chew, John
	*Branch, John	
	*Gough, Mathew	
	*Windham, Edward	
	+Crew, Randall	
	+Flint, Thomas	
	+Jones, Anthony	
	Bayly, Arthur	
	Davis, William	
	Death, Richard	
	Hoddin, John	
	Llewellin, Dan	
	Sadler, Roland	
	Smith, Toby	
	Webb, Stephen	

	Present Slave-holders	Including future slave-holders	Raised in Virginia	Incumbents	Including those who served previously
Council	5 of 10	8 of 10	1 of 10		
House	5 of 27	8 of 27	1 of 27	6 of 27	13 of 27

    Rowland Burnham (tables 3, 4) was a Royalist who had come to Virginia a few years before Berkeley, fleeing the civil war. For most of the forties he seems to have maintained a residence in England, but with the execution of the king in 1649, he finally cut his ties there and brought his wife to Virginia. He lived in York and Lancaster counties where he served as a burgess, and like Kemp and Wormeley, would develop close family ties to other Royalists. Over the years he would build a 3400 acre estate, repeatedly claiming headrights on black slaves.[182]

    All these men would have agreed with Berkeley's belief in rule-from-above: the best and wisest government should descend from the king, and be administered by the educated aristocracy that owned the land.

Table 3-The Grand Assembly, March/June 1644

Current Governor:
William Berkeley

Council

Royalist:	Neutral:	Parliamentarian:
*Browne, Henry	*Higginson, Humphrey	Mathews, Samuel, Sr.
*Richard Kemp	*Ludlow, George	
+Brocas, William	+Minifie, George	Uncertain Parliamentarian:
+West, John		+Pierce, William
	Unknown:	
Uncertain Royalist:	*Bernard, William	
*Townsend, Richard	*Pettus, Thomas	
+Willoughby, Thomas		

House

Royalist:	Neutral:	Uncertain Parliamentarian:
Burnham, Rowland	*Lloyd, Cornelius	Zouch, John
Hill, Edward, Sr.-SPEAKER	Walker, John	
		Uncertain Puritan:
Uncertain Royalist:	Unknown:	*Chew, John
*Hutchinson, Robert	*Crew, Randall	
Calithrop, Christopher	*Death, Richard	
Sidney, John	*Gough, Mathew	
	*Hoddin, John	
	*Llewellin, Dan	
	*Smith, Toby	
	*Webb, Stephen	
	+Hardy, George	
	+Whittbye, William	
	Bishopp, John, Sr.	
	Brewster, Richard	
	Douglas, Edward	
	Hull, Peter	
	Loveing, Thomas	
	Poythers, Francis	
	Roper, William	
	Shepherd, John	
	Travis, Edward	
	Warren, Thomas	
	Westropp, John	
	Worleich, William	

	Present Slave-holders	Including future slave-holders	Raised in Virginia	Incumbents	Including those who served previously
Council	5 of 13	8 of 13	2 of 13		
House	4 of 30	8 of 30	2 of 30	10 of 30	12 of 30

Like England, however, Virginia also had a number of dissenting individuals for whom Berkeley could only view as rivals or opponents. In the 1640s this opposition generally divided into two factions, Puritans and what I label Parliamentarians. Among the Puritans there was Richard Bennett (tables 1, 2, 4, 5, 7) who I have already mentioned, and Obedience Robins. Bennett had now been a local justice and burgess for more than a decade, and a councilor since Francis Wyatt appointed him

Table 4-October 1644

Current Governor:
Deputy Governor Richard Kemp

Council

Royalist:	Neutral:	Puritan:
*Browne, Henry	*Higginson, Humphrey	*Bennett, Richard
+Brocas, William	*Ludlow, George	
	+Minifie, George	Parliamentarian:
		+Claiborne, William
Uncertain Royalist:	Unknown:	
*Townsend, Richard	*Bernard, William	Uncertain Parliamentarian:
		+Pierce, William

House-Held over without elections

Royalist:	Neutral:	Parliamentarian:
*Burnham, Rowland	*Lloyd, Cornelius	Hatcher, William
*Hill, Edward, Sr.-SPEAKER	*Walker, John	
Fantleroy, Moore		Puritan:
	Unknown:	+Robins, Obedience
Uncertain Royalist:	*Bishopp, John, Sr.	
*Hutchinson, Robert	*Brewster, Richard	Uncertain Parliamentarian:
*Calithrop, Christopher	*Crew, Randall	*Zouch, John
*Sidney, John	*Death, Richard	Wood, Abraham
	*Douglas, Edward	
	*Gough, Mathew	Uncertain Puritan:
	*Hardy, George	*Chew, John
	*Hoddin, John	
	*Hull, Peter	
	*Llewellin, Dan	
	*Loveing, Thomas	
	*Poythers, Francis	
	*Roper, William	
	*Shepherd, John	
	*Smith, Toby	
	*Travis, Edward	
	*Warren, Thomas	
	*Webb, Stephen	
	*Westropp, John	
	*Whittbye, William	
	*_Worleich, William_	
	+Bernard, Thomas	
	+Cocke, Richard	
	Heyrick, Henry	
	Jordan, George	

	Present Slave-holders	Including future slave-holders	Raised in Virginia Incumbents		Including those who served previously
Council	5 of 10	8 of 10	1 of 10		
House	6 of 38	9 of 38	3 of 38	30 of 38	35 of 38

in 1639. His brother Phillip was an ardent Puritan with ties to New England.[183]

Robins (tables 1, 4) had come to Virginia in 1621 as an apothecary's apprentice. Since then he had built up a sizable estate on the Eastern Shore, serving as justice and burgess throughout the thirties. In 1633 he had refused to issue warrants against a Puritan minister, and throughout his life he feuded repeatedly with Edmund Scarburgh, Royalist and fellow Eastern Shore planter.[184]

Table 5-The Grand Assembly, February 1645

Current Governor:
*Deputy Governor Richard Kemp

Council

Royalist:	Neutral:	Puritan:
*Browne, Henry	*Higginson, Humphrey	*Bennett, Richard
+West, John	*Ludlow, George	
+*Yardley, Argoll*		Parliamentarian:
	Unknown:	*Claiborne, William
Uncertain Royalist:	*Bernard, William	
*Townsend, Richard	+Pettus, Thomas	Uncertain Parliamentarian:
+*Willowby, Thomas*		*Pierce, William

House

Royalist:	Neutral:	Puritan:
*Fantleroy, Moore	**Charlton, Stephen**	Bennett, Phillip
*Hill, Edward, Sr.-SPEAKER		Lloyd, Edward
+*Scarburgh, Edmund*	Unknown:	Meares, Thomas
	*Bernard, Thomas	
Uncertain Royalist:	*Hardy, George	Uncertain Parliamentarian:
*Calithrop, Christopher	*Poythers, Francis	*Wood, Abraham
*Hutchinson, Robert	+Corker, John	+Harwood, Thomas
+Hoe, Rice	+Heyrick, Henry	**Burroughs, Christopher**
	Barrett, William	
	Baugh, John	
	Harmer, Ambrose	
	Price, Arthur	
	Prince, Edward	
	Ridley, Peter	
	Rogers, John	
	Smith, Arthur	
	Stevens, George	
	Yeo, Leonard	

	Present Slave-holders	Including future slave-holders	Raised in Virginia	Incumbents	Including those who served previously
Council	5 of 12	8 of 12	3 of 12	8 of 27	13 of 27
House	6 of 27	8 of 27	3 of 27		

The Parliamentarians in Virginia's government included men like Samuel Mathews, Sr., Thomas Stegg, and William Claiborne. Mathews, Sr. (tables 1, 3) had been in Virginia since the 1620s and been on the council since 1623. He had taken a leading position in ousting Governor John Harvey, and though he had some Puritan leanings, his links with Parliamentarian forces in England were much more extensive and powerful. When the civil war broke out, Parliament appointed Robert Rich, the second earl of Warwick, as its "Governor-in-Chief and High Lord Admiral of all" British colonies. Rich, who believed in self-rule and was willing to let Virginians "make election of such Governor as you shall conceive most fit," recommended Mathews for the post rather than

Table 6-The Grand Assembly, November 1645

Current Governor:
Governor William Berkeley

Council-no known record

House

Royalist:	Neutral:	Parliamentarian:
*Hill, Edward, Sr.	+Chiles, Walter, Sr.	+Hatcher, William
*Scarburgh, Edmund-SPEAKER	+Fludd, John	Johnson, Thomas
+Burnham, Rowland	+Lloyd, Cornelius	
	+Walker, John	Puritan:
Uncertain Royalist:	Mottrum, John	*Bennett, Phillip
*Calithrop, Christopher		Major, Edward
*Hoe, Rice	Unknown:	
Wyatt, Anthony	*Bernard, Thomas	Uncertain Parliamentarian:
	*Hardy, George	*Burroughs, Christopher
	*Harmer, Ambrose	*Wood, Abraham
	*Poythers, Francis	Swann, Thomas
	*Price, Arthur	
	*Prince, Edward	Uncertain Puritan:
	*Ridley, Peter	Wells, Richard
	*Yeo, Leonard	
	+Crew, Randall	
	+Upton, John	
	Barker, William	
	Chandler, John	
	Eppes, Francis, Sr.	
	Seward, John	
	Sparrow, Charles	
	Warne, Thomas	
	Wetherall, Robert	

	Present Slave-holders	Including future slave-holders	Raised in Virginia	Incumbents	Including those who served previously
House	9 of 36	13 of 36	4 of 36	15 of 36	23 of 36

the royally-appointed Berkeley.[185] Unfortunately for Mathews, Berkeley would not step down, nor would Parliament do anything to depose him.

Thomas Stegg (table 2) gave up a Council seat to become Speaker of the House when Berkeley arrived.[186] As the Civil War heated up, he obtained a commission from that same Robert Rich to ply the seas capturing Royalist ships for Parliament.[187] When Parliament finally decided to send a force to subdue its rebellious Royalist colony in 1652, Stegg, along with Bennett, was chosen to represent Parliament's interests.

William Claiborne (tables 4, 5), a diminutive man from an ordinary middle-class English family, was also part of that

Table 7-October 1646

Current Governor:
Governor William Berkeley

Council

Royalist:	Neutral:	Puritan:
*Browne, Henry	*Ludlow, George	*Bennett, Richard
*West, John		
+Brocas, William	Unknown:	
+Kemp, Richard	*Bernard, William	
	*Pettus, Thomas	
Uncertain Royalist:		
*Willowby, Thomas		

House

Uncertain Royalist:	Neutral:	Parliamentarian:
*Hoe, Rice	*Chiles, Walter, Sr.	*Johnson, Thomas
Robbins, John	*Walker, John	
		Puritan:
	Unknown:	*Major, Edward
	*Crew, Randall	+Meares, Thomas
	*Harmer, Ambrose-SPEAKER	+Lloyd, Edward
	+Barrett, William	
	+Douglas, Edward	Uncertain Parliamentarian:
	+Jordan, George	*Wood, Abraham
	+Llewellin, Dan	
	+Loveing, Thomas	Uncertain Quaker:
	Bagnall, James	Eyres, Robert
	Ball, Henry	
	Cocke, William	
	Fawdown, George	
	Gwin, Hugh	
	Luddington, William	
	Shepeard, Robert	
	Stoughton, Sam	
	Taylor, Thomas	

	Present Slave-holders	Including future slave-holders	Raised in Virginia	Incumbents		Including those who served previously
Council	4 of 9	6 of 9	1 of 9			
House	7 of 26	9 of 26	2 of 26	8 of 26		15 of 26

Commonwealth commission sent to subdue the colony. He had first come to Virginia in 1621, at the tender age of 21, as the official surveyor for the Virginia Company. By 1624 he had become Virginia's first secretary of state, and in the early 1630s he had established an Indian trading post on Kent Island, in the northern reaches of Chesapeake Bay. When the king gave that part of Virginia to the Catholics in 1633, Claiborne and his men were forcibly removed. For years, Claiborne tried anything and everything to reclaim Kent Island for himself, even using military force against the Maryland Catholics when available.[188]

These men, Royalists and dissenters alike, defined the extremes of the Assembly's political landscape. In between, however, was that vast

Table 8-The Grand Assembly, November 1647

Current Governor:
Governor William Berkeley

Council

Royalist:	Neutral:
*Brocas, William	*Ludlow, George
*Browne, Henry	
*Kemp, Richard	Unknown:
*West, John	*Bernard, William
	*Pettus, Thomas

House

Royalist:	Neutral:	Puritan:
+Fantleroy, Moore	+Charlton, Stephen	*Meares, Thomas
+Hill, Edward, Sr.	+Lloyd, Cornelius	
+*Scarburgh, Edmund*	Elliot, Anthony	Uncertain Parliamentarian:
Lee, Richard, Sr.		+Harwood, Thomas-SPEAKER
	Unknown:	
Uncertain Royalist:	*Crew, Randall	Uncertain Puritan:
+Hutchinson, Robert	*Jordan, George	+Wells, Richard
+Sidney, John	*Shepeard, Robert	
+Taylor, William	*Stoughton, Sam	Uncertain Quaker:
Woodhouse, Henry	+Chandler, John	*Eyres, Robert
	+Davis, William	
	+Flint, Thomas	
	+Freeman, Bridges	
	+Harris, Thomas	
	+Poythers, Francis	
	+Ridley, Peter	
	+Upton, John	
	George, John	
	Lambert, Thomas	
	Morgan, Francis	
	Poole, Henry	
	Presley, William, Sr.	

	Present Slave-holders	Including future slave-holders	Raised in Virginia	Incumbents	Including those who served previously
Council	3 of 7	5 of 7	0 of 7		
House	8 of 32	12 of 32	1 of 32	6 of 32	24 of 32

amorphous center of Virginians who did not assign much importance to the political and religious ideas that were at the very moment beginning to tear England apart. While many clearly favored one political faction over the other, they gave their first priority to staying alive and getting wealthy.

Merchants like Cornelius Lloyd and Walter Chiles, Sr., for instance, had numerous ties with both the Puritans and the Royalists. Lloyd in particular is an interesting case. His brother Edward was a fervent Puritan, but Cornelius was willing to put his name on a document defying the rule of Cromwell's Puritan parliament.[189]

Other self-interested planters were Richard Townsend (tables 1-5) and Abraham Wood (tables 4, 7). Both men had similar backgrounds, arriving in the New World as indentured servants and rising from common servant to landed gentry in a single lifetime, a practical impossibility in England. Politically, however, they lived their lives on opposite ends of the spectrum. Townsend made numerous trips to England, lived in Royalist-dominated York county, received favors from Berkeley, and was quite willing to buy slaves to increase his wealth,[190] while Wood never owned slaves, was raised on Samuel Mathews' plantation, spent much of his life exploring the frontier, and had numerous close links with the Indians.[191] Yet neither ever took a known firm stand on any political issue, and like Royalist Edmund Scarburgh, they both carried a life perspective almost entirely shaped by the harsh frontier life of Virginia: what came first was their survival, not philosophical ideas about government and religion.

Not surprisingly, this ethic of survival-above-all-else applied universally to all Virginia planters, even the more outspoken Royalists and dissenters mentioned above. These were self-interested men, focused not on moral and intellectual issues but in how they could increase their wealth, power, and prestige. As historian Bernard Bailyn said, "They were tough, unsentimental, quick-tempered, crudely ambitious men concerned with profits and increased landholdings, not the grace of life."[192] This is why they had grabbed Governor Harvey by the collar in 1635 and kicked him out: he had threatened the legal title to their land.[193]

It was also why, as the English Civil War was raging throughout the decade of the 1640s, Berkeley's first term as governor, little evidence of this conflict was noticeable in Virginian society during this time.

Berkeley, it seems, understood how self-centered the majority of Virginians could be. Moreover, with the king at war with his parliament in England, the governor could ill afford his own kind of fight in Virginia. The Virginia planters might not think much about issues of liberty of conscience, government rule, or religion, but they definitely cared about their property rights and their ability to make money. When he called his second Assembly in April of 1643, Berkeley very carefully negotiated with these men, imposing his will only with their tacit or overt support. As a result, he was able to consolidate his power and get much of what he wanted.

Berkeley's success and willingness to do this stems partly because he chose to treat all these men not as his inferiors, but as fellow members of an aristocracy living in Virginia. By strengthening the House of Burgesses and the property rights of its members, Berkeley was not so much establishing Parliamentary rule but strengthening the positions of these powerful Virginia planters. For Berkeley, concentrating power at the top was the best route for solving Virginia's problems.

These actions however also served Berkeley well with the Puritans and Parliamentarians. Men like Obedience Robins, Richard Bennett, and Thomas Stegg immediately benefited by the larger role

Berkeley seemed to be offering them. By reinforcing their positions of power as Burgesses, the Governor actually appeared to be adopting a more moderate Royalist position than his king, and thus gained the support of those who might have opposed him.

Berkeley's skill as a writer and poet gave him an additional advantage, allowing him to put his own personal stamp on all legislation in a colony where many could not read or write. The obvious improvement in style, detail, and depth in the writing of the colony's laws after 1642 indicates the unseen hand of Berkeley, former poet and scholar. Thomas Ludwell, writing in the 1660s, confirms this, saying that Berkeley was "the sole author of the most substantial parts of [the government], either for laws or other inferior institutions."[194]

Subtle diplomat or not, Berkeley now proceeded to carry out King Charles I's instructions. Loosely interpreted, more than half the laws passed in 1643 reflected the intent of these instructions.[195]

First, Berkeley moved to establish church government in Virginia. The king's instructions required Berkeley to take control of the colony's churches as well as obtain loyalty oaths from all citizens,[196] including the banning of any dissenting religious beliefs. So, twelve of the seventy-three acts passed by the 1643 Grand Assembly dealt with establishing the Anglican church in Virginia, with the very first act of 1643 specifically outlining the structure of this government-run religion, with the governor at its head. Other acts passed by this same Assembly barred Catholics from public office, and demanded that "for the preservation of the purity of doctrine and unity of the church . . . all non-conformists upon notice shall be compelled to depart the colony with all convenience."[197]

The make-up of this 1643 Assembly (Table 2) illustrates why Berkeley had little problem imposing these laws. Such a church was familiar to these British immigrants and intensely desired by most of them. Those with known dissenting Parliamentarian views, such as Stegg and Mathews, did not express much interest in religious issues, and since Berkeley's initial reforms left control of the local parish to "the commander and commanders of the county,"[198] strengthening the church strengthened their own power. Hence, these reforms were fine with them.

Only one individual in the 1643 Assembly, Richard Bennett, had links with a dissenting religious faction, and following a pattern that we will see repeated in later years, he took no public action that we know of to oppose this legislation. Possibly he felt that the law would not affect him or his Puritan friends, living as they did on isolated farms and plantations, many miles away from any Anglican minister. Possibly he thought that by leaving parish control with the local community leaders, he and his Puritan friends would still be free to practice their religion. Or possibly he did oppose this legislation, without success. What we do know is that his presence on the council did nothing to mitigate the nature of Berkeley's legislation.

Next Berkeley and the Assembly moved to codify the laws concerning indentured servitude, making official and legal the custom of using free British citizens for forced, unpaid labor. The 1643 legislation forbade the harboring of runaway servants, protected servants from ill-usage by their masters, established term lengths for servants arriving in the colony without indentured contracts, outlawed trade between free men and other men's servants, and required servants to get permission from their masters before marrying. Furthermore, the 1643 legislature also abolished the use of servitude as a form of punishment by the colonial government.[199]

While Berkeley's intentions might have been good, attempting to standardize these customs as well as prevent abuse, his actions also acted to legitimize the exploitation of poor immigrants as free labor, actions that almost certainly pleased the self-interested desires of the Virginia planters with whom he had to deal.

Finally, Berkeley and the 1643 Assembly turned to regulating many of the details of private and commercial life. Almost a third of the laws passed in 1643 dealt with this kind of governmental control, from regulating the size of fences to restricting the practice of law, from insisting that guns be carried to church to requiring that each man plant two acres of corn. Furthermore, the Assembly moved to regularize the colony's legal system, setting the terms, responsibilities, and salary for county sheriffs, clerks, and burgesses, as well as codifying the court and jury system.

All of these regulations were inspired by specific problems: guns in case of Indian attack, fences to keep cattle and pigs out of crops, and corn to guarantee sufficient food supplies. Yet all served to establish the members of the Assembly as the sole arbiters of public policy. Not only did they reserve the right to tell everyone else how high their fences should be, they also strengthened their ability to enforce those regulations by solidifying their power as justices on the local county courts.

In future years, the governmental approach that Berkeley established in 1643 would be reflected again and again in legislation, even when he wasn't governor. Berkeley's Royalist philosophy of birthright, church government, and rule-from-above, all for the good of society, became legislative custom: birthright by the increasingly severe servant laws; church government by the establishment of a state religion with penalties for dissent; and rule-from-above by the numerous and never-ending social and commercial regulations.[200]

For the rest of the forties, Berkeley acted to increase his popularity among the landowners of Virginia. Almost immediately upon arrival he began building his own plantation. Named Green Spring, the enormous brick mansion on his plantation announced to all and sundry Berkeley's willingness to live in the colony. On this manor he devoted extensive energies to the growing and development of a wide range of crops, including silk, rice, oranges, grapes, maize, barley, wheat, rye, and cotton.[201] In retrospect, we can see in Green Spring the blueprint and forebearer of every antebellum southern plantation that was to follow.

When the Powhatan Indians attacked the colonists in 1644, it was Berkeley's leadership that won the war. At first he and the March, 1644 Assembly put the militia under William Claiborne's command while arranging to send Berkeley back to England to try and obtain money and ammunition.[202] In England the governor found the civil war raging fiercely and Charles simply too pre-occupied to help Virginia. For a short while Berkeley fought with the king's army, and probably could have obtained a command comparable to his brother John's, who held an important post. Instead he decided to return to Virginia, where he found that in the ensuing year Claiborne had used the militia not only to fight the Indians but also to try to regain Kent Island from the Maryland colonists.[203] Berkeley took over command, and it was his own decisive action that led to the capture of Opechancanogh, chief of the Powhatans and main instigator of the 1644 attack. Upon learning of Opechancanogh's whereabouts Berkeley immediately took horse with some troops and rode to trap him, ending the war and making possible a peace treaty with the Powhatan tribe that lasted more than thirty years.[204]

Even as he was increasing his popularity, Berkeley nevertheless moved throughout the 1640s to strengthen his hold on the church government that he and Royalists favored. While he would consistently defend the personal and property rights of all planters, even those members of the Grand Assembly who had ties with the Puritans, such as men like Bennett and Robins, Berkeley steadily moved to squelch their religious rights.

In 1642 Philip Bennett, Richard Bennett's brother, brought to Virginia three Puritan clergyman from New England to teach religion. Berkeley's response was to oppose their preaching and drive them from the colony. Then, in 1644, when his own minister, Thomas Harrison, became a Puritan and refused to administer the rite of baptism and to read the book of Common Prayer as required by the Anglican Church, Berkeley immediately expelled him from his parish.[205]

None of these actions, nor the 1643 legislation, succeeded in eliminating religious dissent in Virginia. In February 1645, with Berkeley in England and Richard Kemp acting as governor, Puritans Phillip

Bennett, Edward Lloyd, and Thomas Meares were elected to the House of Burgesses (tables 5, 6, 7). The presence of these Puritans and the absence of Berkeley seemed to inspire the Assembly to devote seven of its twenty acts to religion,[206] most of which addressed the "scarcity of pastors, many ministers having charge of two cures."[207] To solve this lack by allowing the open preaching of Puritan ministers, however, would have been unacceptable to Royalists like Kemp, Browne, West, Yardley, and Scarburgh. Instead, it seems that this 1645 Assembly forged a compromise: while several acts required approved Anglican ministers to officiate at multiple parishes, another act allowed those parishes still unserved "to make use of any other minister as a lecturer to baptize or preach." What that lecturer or minister could preach was delicately avoided.[208]

By November 1647, however, the only certain Puritan attending the assembly was Thomas Meares. Berkeley was back from England, and Royalists like Scarburgh, Kemp, Browne, and Richard Lee, Sr. now dominated in the Assembly (table 8). Lee, the founder of that famous southern family, first appeared in Virginia records in the late 1630s as a witness in legal contracts. As the years passed he developed, like Richard Kemp, the typical tangled and close family ties with Royalists like Wormeley, Burnham, Brocas, and Kemp,[209] while also linking himself politically with Berkeley and the Royalists. First he became Clerk of the Quarter Court in 1640.[210] Then the king would appoint him attorney general in 1643, followed by secretary of state in 1649. In 1650, after the execution of Charles I, Berkeley would send Lee to Europe to find King Charles II in order to offer him sanctuary in Virginia as well as obtain for Berkeley a new commission.[211]

The 1647 Royalist controlled Assembly immediately passed a law requiring ministers to follow Anglican practices.

> Upon diverse information presented to this Assembly against several ministers for their neglect and refractory refusing after warning given them to read common prayer or divine service upon the Sabbath . . . all ministers in several cures throughout the colony do duly upon every Sabbath day read such prayers as are appointed and prescribed unto them by the said book of common prayer.[212]

Should a minister refuse to obey, his parishioners were permitted to withhold their parish taxes.

Yet, despite the Royalist dominance in positions of power, their belief in privilege, in deference to leaders, in the idea that religion and government should be administered by a ruling caste of aristocracy and that such affairs "doth pass [the] capacity"[213] of the lower members of society, the Virginia Royalists of the 1640s did not pass any laws sanctioning slavery. This in spite of their willingness to deal with issues of servitude, quickly codifying the labor customs in 1643, and in spite of the fact that more than 40 percent of the Council was made up of men who were already slave-holders (see figure 4.1). When we include future slave-owners, that percentage actually rises to anywhere from 62 to 80 percent of the Council at any time (see figure 4.2).

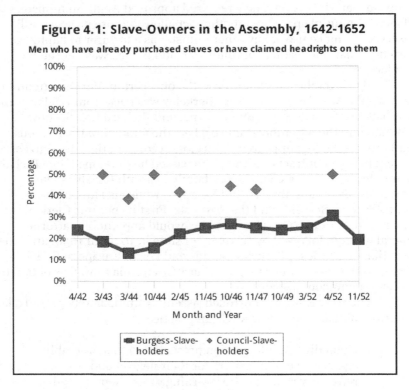

Furthermore, when we look at how these men were willing to use the law to enrich themselves, the lack of slave laws becomes truly inexplicable. Consider for example a comparison of the king's ninth instruction to Berkeley to the actual law eventually passed in 1643. In order to reimburse councilors for their work, instruction nine stated

> that they and ten servants for every councilor be exempted from all public charges and contributions assessed, and levied by the Grand Assembly (a war

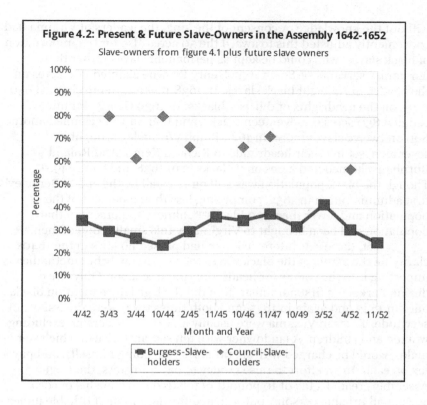

defensive, assistance towards the building of a town or churches and the ministers duties excepted).[214]

Other than levies assessed because of war, town or church construction, councilors were to be partly exempt from taxes. Compare this with the law passed by the 1643 Grand Assembly:

> It is enacted . . .that those of his majesty's Council for the colony shall according to his said majesty's instructions be freely exempted from all public charges and taxes, church duties only excepted.[215]

The law was clearly more favorable to the councilors than intended by the king's instructions. Obviously, the members of the Council had no problem fashioning the law for their own personal benefit.

An even better example is how they had subtly changed Sandys' headright system for their own benefit. Originally the headright was meant to encourage British immigration to the colony, awarding the 50 acres for the transportation of Englishmen only. By the 1630s, however, without the knowledge or consent of the king, the leaders of Virginia had conveniently adjusted this to award the 50 acres also for the importation of black slaves, who could be kept as permanent slaves rather than temporary servants.[216] So, in 1637 Henry Browne claimed 400 acres on the headrights of eight black slaves. In 1638 Richard Kemp claimed 650 acres on the headrights of thirteen blacks. In 1639 George Minifie claimed 850 acres on seventeen blacks, and then in 1642 claimed another 600 on twelve slaves more. In 1649 Ralph Wormeley brought in seventeen, selling their headrights to Richard Kemp. And Roland Burnham claimed 350 acres on 7 blacks, 2 in 1638 and 5 in 1649.[217] Though the black population was still quite small by the end of Berkeley's first administration in 1652, comprising less than 2 percent of the population and fewer than 300 blacks[218], almost 25 percent of that population had been brought to Virginia by this small group of men.[219]

Yet, these self-interested men had passed no laws in the 1640s clarifying the status of the black slaves they were slowly but methodically importing into the colony for headrights and free labor. Only twice during Berkeley's first administration did the legislature mention blacks, once in 1643 and again in 1645, and both times taxes were the issue, not servitude. Taxes in Virginia were usually assessed per person, excluding women and children. A landowner with ten servants, five of which were males, would be charged with six tithables (including himself), and pay a tax on each. In an effort to raise money to pay ministers, the 1643 Assembly created a tax of 10 pounds of tobacco and a bushel of corn against all tithable persons, but widened the definition of tithable under this particular tax to include "all youths sixteen years of age as upwards, as also for all negro women at the age of sixteen years."[220] While before black women were not tithable, like white women, now they were.

Then in 1645 the colonists needed to raise money to prosecute the war against the Powhatan Indians. With Berkeley still in England and Richard Kemp acting as governor, the Assembly passed a law that once again widened who was tithable, this time adding black men.[221]

Though neither law restricted the freedom of blacks, being intent instead on the collection of taxes, both laws did treat blacks differently from and inferior to whites, and hence reflected the Royalist notion of caste, foreshadowing later developments. Yet, because these two laws actually taxed the owners of black servants at a higher rate than white servants, they actually served to discourage the ownership and importation of slaves. From 1643 to 1649 (when these laws were changed) few patents on black slaves were claimed, with members of the Assembly claiming only 8 blacks for headrights, Bartholomew Hoskins patenting six, and George Reade and Phillip Bennett each patenting one.

Whether or not this definition of tithable was unfair to black slaves or to their owners, in 1649, with both Hoskins and Reade now freshman burgesses, the Assembly redefined tithable again, removing black women from the rolls and limiting taxation to

> all male servants imported hereafter into the colony of what age soever they be . . .. excepting . . .such as are natives of this colony [American Indians] and such as are imported free, either by their parents or otherwise, who are exempted from levies, being under the said age of sixteen years.[222]

Now only male servants, black or white, of whatever age, would be tithable. Skin color was no longer the issue.

Why didn't Berkeley, the Royalists, and the slave-holders in the Assembly move to legalize slavery? If such men dominated the government of Virginia, believed in the ideas of caste and rule by birthright, and were willing to own slaves at a much higher percentage than the rest of the population, why did they pass no laws institutionalizing the custom?

Unquestionably, the low numbers of blacks, never higher than two percent of the population, allowed the colony's leaders to ignore the whole issue. This benign neglect permitted each slave-holder to treat his black slaves as he wished. Some could therefore take advantage of the vagueness of the law and keep their black slaves enslaved for life. A burgess like Rowland Burham, who served in 1644, 1645, and 1649, could repeatedly import black slaves into the colony, claim headrights on them, and then, in his will, bequeath a "negro" each to his sons and daughters.[223] Others, such as Stephen Carleton, a devout Anglican with both Parliamentary as well as Royalist connections and who served as a burgess in both 1645 and 1647, would be free to take the moral route and treat his slaves more like white servants, freeing them after they had served a term of bonded servitude. In 1648 Carleton did just that, making freedom agreements with most of his black slaves so that they would eventually become free.[224]

That the percentage of slave-holders in the House of Burgesses never rose as high as in the Council probably contributed as well to the lack of slave codes (see figures 4.1 and 4.2). It would have been difficult at best for the majority of legislators, having no slaves and never having profited from them, to have legally endorsed the custom, especially considering the discomfort English citizens seemed to feel over the idea of slavery. These were almost all English-born and English-raised men, native-born Virginians never comprising more than sixteen percent of House membership and usually much, much less (see figure 4.3). As Englishmen they desired to recreate their image of an English community on North American soil, and as I have noted, England did not

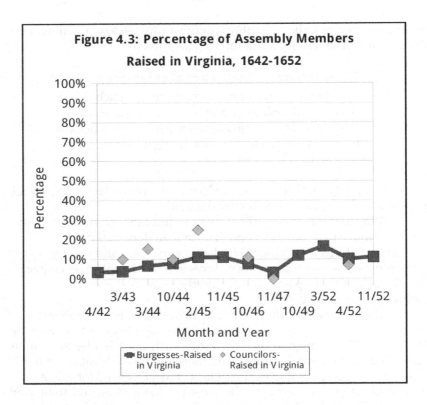

Figure 4.3: Percentage of Assembly Members Raised in Virginia, 1642-1652

have slaves. Even some Royalists, such as Edward Hill, Sr., the speaker of the House in 1644 and 1645 and a man who never owned or claimed headrights on slaves, would have probably felt uneasy about any law that institutionalized the custom of slavery.

Though the low number of both blacks and slave-holders in Virginia might partly explain the lack of slave codes in the 1640s, it does not provide a completely satisfactory explanation. Other considerations, such as religion and politics, were probably as important, and must be recognized.

Some legislators would have opposed these codes on religious grounds. Since, according to some interpretations of the Bible at this time, all Christians were equal before God, to pass laws that would treat blacks as inferior merely because their skin color and African background could only be viewed as immoral, and hence difficult if not impossible to consider. Such men as Richard and Phillip Bennett, Edward Lloyd, Obedience Robins, and Thomas Meares belong to this group. Richard Bennett, with his aristocratic background and powerful family connections in England, had freed a number of his slaves, including

Anthony and Mary Johnson. And Edward Lloyd, Obedience Robins, and Edward Major never owned slaves at all, despite ample opportunity.

Others, like Stegg, Mathews, and Claiborne, had links to Parliamentarian factions in England and would have probably opposed the institutionalization of slavery on moral and political grounds. Such ideas would have seemed odious to these English Parliamentarians transplanted to Virginian soil, being especially proud of Britain's belief in individual rights. They would have feared that once it was possible for slavery to be established for blacks, it could then be established for whites, and they would see their rights and privileges as free British citizens evaporate. In fact, the civil war in England was being fought at this very time for some of these very reasons.[225]

None of these dissenters, however, neither Puritans nor Parliamentarians, would have abolished slavery, and in fact, Bennett and Mathews owned slaves. The seventeenth-century world simply did not yet see the custom with total moral repugnance,[226] even if some rejected it for themselves. Yet, the dissenting philosophies of some clearly provided a brake and opposition to the most abusive policies of others. That men such as Bennett, Stegg, and Mathews were eventually pushed to the fringe of political power during Berkeley's first administration indicates their willingness to oppose him on some issues.

These political dynamics were about to change, however. All throughout Berkeley's first term as Virginia's governor the Civil War in England had raged, with events generally going against the king. As the decade wound down, the Royalists in England found themselves faced with the stark reality that their country was against them, and that they must flee its shores. Some went to France, some to Spain. Others, however, came to Virginia, and in that flight they helped forge the chain that brought slavery to America.

5. A Royalist Safe Haven

From late 1645 onward, the Royalist cause in England began to crumble.[227] In September 1645 King Charles' army lost the city of Bristol to the Parliamentarian army led by Sir Thomas Fairfax. Regrouping under the command of Lord Ralph Hopton, the king's army was pursued westward onto the narrow peninsula that forms the southwestern tip of the island of England. Even as the king's army retreated, its size shrank, from desertion and lack of discipline. By March 1646 Hopton's army had practically dissolved, and he was forced to sue for peace. The prince fled to the Scilly Islands, and Hopton's army disbanded.

Two months later, in May 1646, Charles sought sanctuary with the Scottish army at Southwell, thinking that he could fasten his cause to them. Unfortunately, the Scots had different plans. No longer did they see him as a divine ruler for whom they were obliged mindless obedience. Instead, they demanded that he accept their ideas about government and religion before they would go to war to restore his rule in England. Without the slightest regard for the nobility of the king, they took him prisoner, hoping that they could use their position of strength to negotiate an agreement with him. They wanted him to discard his ability to veto without override any action of Parliament, as well as to accept the Scottish desire for joining the government with Presbyterianism. Rather than leading a Scottish army back to England in triumph, Charles found himself guarded and imprisoned.

In June 1646 Henry Jermyn arrived on the Scilly Islands from France with letters from both the queen and the king begging the prince to abandon England for refuge in France. Despite the protests of Hyde, Hopton, and others, who felt that the king would never win back his country by seeking aid from foreign powers, the prince sided with Jermyn and, on June 26th, fled across the channel to France.

Having been moved north to Newcastle near the Scottish border, Charles continued to negotiate with the Scots, stalling for time in the hope he could maneuver them into helping him without agreeing to their terms. After more than half a year of talks, however, the Scots had had enough. By January 1647, they realized the impossibility of making Charles agree to Presbyterianism, and when the English Parliament offered them a 200,000 pound ransom for Charles, they agreed. On February 3, 1947 they transferred control of the king to the English

Parliament. Charles was now a prisoner of his sworn enemies, who took him south to Holdenby House in Northamptonshire.[228]

The king's cause continued to deteriorate. Having beaten the king, Parliament moved to disband their own military forces, the expense of which was becoming unbearable. Unfortunately, the Parliamentarian army refused to disband, and its most radical members, the Agitators, were becoming increasingly impatient with Parliament's willingness to negotiate a settlement with the king. By June, 1647, they decided to bypass Parliament entirely. Under Cromwell's orders, Cornet Joyce rode out with 500 troops to Holdenby House with orders to "either secure the person of the King from being removed by any other, or, if occasion were, to remove him to some place of better security for the prevention of the design of the aforesaid pretended traitorous party."[229]

On June 3 Joyce and his men surrounded Holdenby House, taking control from the Parliamentarian guards. At ten that evening Joyce faced the king and told him they wished to move him. After receiving assurance that his personal safety would be protected, that he would not be forced to do anything against his conscience, and that his servants could accompany him, the king agreed to go.

At six the next morning King Charles stepped outside to find himself face to face with Joyce and his 500 men. "What commission have you to secure my person?" the king now asked Joyce, who tried to evade the question.

Twice more the king demanded Joyce's written commission, refusing to accept the evasions. Finally, Joyce turned and pointed at his troops and said, "Here is my commission. It is behind me."

"It is a fair commission," the king replied. "And as well written as I have seen a commission written in my life."

The king was now under the control of the radical army, led by Cromwell. In the next two months that army moved to occupy London and to take control of the government entirely. In this military coup they based their actions on some remarkably modern democratic ideals. "The Agreement of the People," written in 1647 by some of the most fanatic Levelers in the New Model Army, became the basis for the Commonwealth government. As stated in its preamble:

> Having by our late labors and hazards made it appear to the world at how high a rate we value our just freedom . .. we do now hold ourselves bound in mutual duty to each other to take the best care we can for the future to avoid both the danger of returning into a slavish condition and the chargeable remedy of another war.[230]

That document mandated general elections every two years, limited the powers of Parliament, demanded religious tolerance, outlawed a military draft, and required equal treatment before the law for all citizens, regardless of rank.[231]

Despite their expressed desire to establish this democratic ideal, however, the army used as its commission not the sovereignty of the people, but that same commission of force used by Joyce. Parliament did not dissolve itself for new elections as promised, but instead used the arbitrary backing of the army to control society while personally benefiting themselves. The House ordered all theaters closed, charging the spectators with fines and the performers with floggings. It moved to remove the Vice-Chancellor of Oxford University, and to expel those members of the University who would not swear allegiance to Parliament. And its members eagerly used their newfound power to obtain bribes and payments in exchange for influence and votes.

Meanwhile, negotiations with Charles were going nowhere. The fanatic presence of the army prevented any easy compromise, though the king seemed to offer reasonable terms. Furthermore, even the moderate Parliamentarians were unwilling to trust him enough to put him back in power. Too often they had seen him make promises in negotiation which he broke as soon as he felt able.

Nor could they easily get rid of him. The monarchy was still popular among the English, representing as it did stable government and long tradition, neither of which were available under the chaos of the civil war.

For rest of 1647 the situation worsened, so that by the beginning of 1648 the mood against military rule was strong enough that riots and uprisings were occurring in scattered places throughout England. In league with the Scots, the Royalists attempted once again to raise an army and wrest control of the government from Parliament by force.

All to no avail. The king's supporters lost every battle during the summer of 1648. Hamilton surrendered in Uttoxeter, Preston capitulated in Colchester. By late 1648 all opposition to the army and Parliament had been squelched. In November the radicals in the army moved to remove from Parliament anyone who might disagree with them. On December 6th Colonel Thomas Pride, backed by army troops, stood guard in the lobby of the House with a list of those with whom the army would not permit entrance. Before the day was out, Pride's Purge produced a Parliament entirely hostile to monarchy and the king. To the men that remained, Charles was a traitor and a dictator who stood in the way of peace and just government, and if he couldn't be trusted to rule according to their rules, he should simply be disposed of, by any means necessary.

Parliament now moved to put Charles on trial, and after a short almost perfunctory proceeding, sentenced him to death for being "a tyrant, traitor, murderer, and public enemy to the good people of this nation."[232] On January 30, 1649 Charles was led to the scaffold outside the Banqueting Hall at Whitehall and beheaded.

When the news of the king's defeat and execution reached Virginia in the middle of 1649, Berkeley and his Royalist friends wasted little time in aggressively attacking the few religious and political dissenters living in Virginia. They immediately drove the Puritans from both Nansemund and Lower Norfolk counties. More than 300 Puritans fled Virginia, moving north to settle in Maryland with the more tolerant Catholics. Among these refugees were former burgesses Edward Lloyd, Thomas Meares, and Phillip Bennett, as well as Richard Bennett, member of the Virginia Council.[233] Furthermore, Parliamentarian Samuel Mathews fled to England, while Thomas Stegg went to sea to pillage Royalist ships.[234]

By 1650, it appeared Berkeley and his Royalists allies had purged Virginia of any significant Puritan or dissenting presence. Even those men who had not fled Virginia, such as Obedience Robins, Edward Major, and William Claiborne, had become inactive, no longer attending Assembly sessions (see tables 9 and 10).

Furthermore, Berkeley refused to acquiesce to Commonwealth rule, and made Virginia a haven for the Royalist refugees fleeing their defeat in England. In the three years following Charles' death the largest influx of future legislators, twenty-five in number, arrived in Virginia.[235] In 1649 alone the Royalists chartered seven ships for Virginia's shores,

Table 9-The Grand Assembly, October 1649

Current Governor:
Governor William Berkeley

Council-no known record

House

Royalist:	Neutral:	Parliamentarian:
*Hill, Edward, Sr.	+Chiles, Walter, Sr.	+Hatcher, William
+Burnham, Rowland	+Walker, John	
+Carter, John, Sr.		Uncertain Parliamentarian:
+Robbins, John	Unknown:	*Harwood, Thomas-SPEAKER
Hoskins, Bartholomew	*Chandler, John	+Swann, Thomas
Reade, George	*Lambert, Thomas	
Trussell, John	*Poythers, Francis	Uncertain Puritan:
Wormeley, Ralph, Sr.	+Barrett, William	Whittaker, William
	+Hardy, George	
	+Smith, Toby	Uncertain Quaker:
	+Sparrow, Charles	*Eyres, Robert
	+*Worleich, William*	
	Dunston, John	
	Pitt, Robert	

	Present Slave-holders	Including future slave-holders	Raised in Virginia	Incumbents	Including those who served previously
House	6 of 25	8 of 25	3 of 25	6 of 25	18 of 25

one of which carried Francis Moryson, Sir Thomas Lunsford, Henry Chickeley, and Manwaring Hammond, all of whom had held significant military posts fighting for the king. Moryson's father had been the secretary of state to King James, while Sir Thomas Lunsford had been Charles I's Lieutenant of the Tower.[236] Henry Chickeley had been imprisoned by Parliament for a short time,[237] while Manwaring Hammond was eventually enlisted by Berkeley to procure pardons from the restored Charles II for the colony's submission to commonwealth rule.[238] All these men were given sanctuary in the homes of Berkeley and Ralph Wormeley upon their arrival in the New World.[239]

Nor were these the only important Royalist refugees. In 1651 an additional 1,600 Royalist prisoners were granted permission to immigrate to Virginia,[240] providing Virginia with a cadre of men who in later years would form the nucleus of the Council and House, writing the laws and setting social policy.[241]

Consider, for example, refugees John Carter, Sr. and Nathaniel Bacon, Sr., both of whom arrived in Virginia in these years and would both become important allies to Berkeley. Bacon was the son of a Royalist minister, James Bacon, and a distant relation of Francis Bacon. One of the early settlers along the York River, Bacon helped make York County a stronghold for Royalist sympathizers, and for the next twenty-five years would be one of Berkeley's strongest allies.[242]

Table 10-The Grand Assembly, March 1651 to March 1652

Current Governor:
Governor William Berkeley

Council-no known record

House
Royalist:
*Hoskins, Bartholomew
Trussell, John
+Fantleroy, Moore
+Hill, John
Baldridge, Thomas
Lee, [Hugh]
Presley, William, Jr.
Speke, [Thomas]
Travers, Raleigh
Wilford, Thomas

Neutral:
+Lloyd, Cornelius

Uncertain Royalist:
+Sidney, John

	Present Slave-holders	Including future slave-holders	Raised in Virginia	Incumbents	Including those who served previously
House	3 of 12	5 of 12	2 of 12	2 of 12	6 of 12

Legend for all tables:
*-incumbent
+-served previously
Bold names have purchased slaves or have claimed headrights on blacks.
Underlined names will eventually buy slaves or claim headrights on blacks.
Italicized names either were born in Virginia or immigrated as minors.

Source for all tables: See Appendix B

John Carter was a Royalist who had already lived in Virginia for a short while, serving as a burgess for Nansemond County in 1642. In 1649 he moved to the colony permanently, and would be accused in 1658 of contempt against both the Commonwealth and Cromwell. At the Restoration in 1660, Berkeley would immediately appoint him councilor, and he would serve on that body as Berkeley's ally until his death in 1669.[243]

Other important Royalist refugees included Joseph Bridger, John Page, and Edward Digges. Bridger came in 1649, had been an officer in the king's army, and would be named by Berkeley to the Council in 1670. In 1676 he would be named a "wicked and pernicious councilor" by the rebels in Bacon's Rebellion and would lose heavily in the rebellion. Despite this the king would immediately reappoint him to the council in 1678.[244]

John Page came to Virginia in 1650, and by 1656 had claimed headrights on more than 3,000 acres in Royalist York county. Known for his dedicated loyalty to Berkeley, he was elected to the House in 1672, and stood by the governor during Bacon's Rebellion in 1676, despite suffering severe losses. In 1677 he served on the court martial that, under Berkeley's leadership, condemned many rebels to death, and would be appointed by Berkeley Sheriff of York county for this support.[245]

Edward Digges arrived in 1650. His father had been an important officer under both King James and Charles I, and had once been declared by Edwin Sandys to be "a papist and a Royalist." His father had also been known to have an independent spirit. Eventually he became the king's master of rolls in 1630, despite having once been imprisoned for expressing "matters derogatory to the king's honor." Digges' older brother Dudley was also a known Royalist who had written extensively in defense of Charles I.[246]

Using his family's wealth, Digges quickly accumulated over 4,000 acres by a combination of purchase and headright claims, and within a short time he had established close ties to both Berkeley and the other Royalists, eventually marrying John Page's daughter. By 1654 he used these ties to gain election to the council.[247]

The Royalist beliefs of these men is evident and clear. Also arriving in the three years following the king's execution were many other future legislators whose Royalist links, while expressed less plainly, can still be deduced by their actions and friends. Daniel Parke, Sr., for example, settled in Royalist York county and was appointed justice four years later in 1655. In 1659, on the eve of the Restoration of the king, he became sheriff, followed this in 1660 by his election as burgess for York county. Then, in 1670 Berkeley appointed him councilor, a solid indication of where Parke's loyalties lay.[248]

Other examples include William Moseley, Edward Dale, and Peter Ashton. Moseley was appointed a justice after the Restoration, and in the early 1660s served on the court that condemned the Quakers for defying the Royalist-controlled Restoration Assembly.[249] Dale moved to Lancaster county where in 1669-70 Berkeley appointed him first Justice, then Sheriff.[250] Ashton meanwhile became burgess for Charles county in 1656, was appointed sheriff in 1658, and in 1662, made justice by Berkeley's friend and secretary, Thomas Ludwell.[251]

Though none of these men left a direct record of their political beliefs, their quick rise during the Restoration years when the Royalists were in command, combined with their arrival in Virginia during its three year defiance of Parliament, are strong indicators of where they stood.

Finally, of the refugees that came to Virginia during this three year period were future Assemblymen for whom we know practically nothing: Robert Abrahall (1650), Francis Gray (before 1653), John Holmwood (before 1650), William Gooch (in 1650), George Collclough (in 1651), William Underwood, Sr. (in 1650), Valintine Peyton (in 1650), and George Mason (in 1651).[252] Their arrival in Virginia during its years of political defiance to Parliament, combined with their presence in the Royalist-leaning Assemblies of the future, offer us a slim indication of their political beliefs and how those beliefs influenced legislation.

All told, this Royalist tide of refugees had an enormous influence on Virginia's culture, law, and society. From among its members would eventually appear almost all of the important family names that would dominate southern history for the next two centuries.[253] More significant, however, was how they would change Virginia in the next two decades, bringing with them their belief in caste, privilege, and aristocracy.

In England the army and House of Commons continued to rule England, despite repeated failed attempts by Charles II and the Royalists to retake the kingdom by force. In 1649 Cromwell defeated a Royalist army in Ireland. In 1650 he did the same in Scotland. Then, in 1651 the Royalists raised a new army in Scotland under the leadership of the young king and attempted an invasion of England. Pursued by Cromwell and an overwhelming Parliamentary force, the Royalists were defeated soundly at Worcester in September 1651, forcing Charles to flee in disguise through the English countryside, eventually slipping across the channel to France.

During these same years, 1649 to 1652, Berkeley and his now Royalist-dominated Assembly continued their defiance of Parliament. When the Commonwealth government finally marshaled a force in 1651 to retake the colony under the leadership of Richard Bennett, William Claiborne and Thomas Stegg, Virginia's Royalist-controlled Grand

Assembly (see table 10) met to loudly proclaim its continuing loyalty to Charles II, declaring that

> these laws [of England] . . . .tell us that no power on earth can absolve or manumit us from our obedience to our prince.[254]

Berkeley himself stood and proudly avowed that

> By the Grace of God we will not so tamely part with our king, and all these blessings we enjoy under him, and if they oppose us, do but follow me, I will either lead you to victory, or lose a life which I cannot more gloriously sacrifice than for my loyalty and your security.[255]

Berkeley and his allies were quite willing to publicly proclaim their loyalty to their king, even though that king had been dead for three years and his party utterly defeated in England.

Berkeley had now been governor of Virginia for ten years. In that decade the colony had grown enormously, its population almost doubling to approximately 15,000.[256] Peace had been established with the Indians, and Virginia was prospering as it never had before. A new boom in tobacco had begun, and thousands of new acres were being settled throughout the Chesapeake.[257] And by making Virginia a haven for the well-bred Royalist refugees, Berkeley had brought a cultured air to the New World.

Yet, the colonists still lived in dispersed settlements, without towns, and were devoted almost exclusively to the growth of tobacco. And despite Berkeley and the Assembly's effort as well as the eager desire of most of the colony's citizens, the Anglican Church in the colony barely functioned, with most parishes still lacking a minister. For many reasons, not least of which was the Civil War in England tying up resources, the Anglican church in England could not provide sufficient clergymen.

A more important reason for this lack of new ministers, however, was the unwillingness of Berkeley and Virginia's men of power to cede any control to the church. In England a parish priest was appointed to a lifetime post within a well-established independent church hierarchy under the king's jurisdiction. In Virginia no such hierarchy existed, and instead the governor appointed ministers chosen by the local planters. Because such lifetime appointments gave the clergymen an independence that neither Berkeley nor anyone else in Virginia wanted them to have, Berkeley and the planters were always reluctant to make such appointments. Instead, clergymen were appointed yearly by the individual parishes, and were therefore always subject to the disapproval of their parishioners.[258] The result was that few high quality Anglican clergymen were willing to come to Virginia.

This lack of a working Anglican church, combined with Berkeley's aggressive effort to expel anyone who preached a dissenting message, meant that the colony was left with almost no moral leadership. Hence the mere "20 churches" noted by Berkeley in 1648.[259]

Furthermore, except for a small free school in Elizabeth County that had been bequeathed to the colony in the will of Benjamin Symms in 1643, the colony had no education system to speak of.[260] Additionally, under the headright system male immigrants still came to the colony in vastly larger numbers than women, skewing the sex ratio.[261] The death rate continued to be high, though there were signs it was beginning to diminish.[262] And the system of labor still depended almost entirely on the distorted form of bonded servitude that had been instituted by Edwin Sandys in 1619, and in fact, the laws passed by the Assemblies under Berkeley had helped legalize this labor system.

Neither the instructions of King Charles nor the legislative leadership of William Berkeley had addressed these chronic problems. The consequences to the children growing up in Virginia could only be profound, and quite likely terrible.

Simultaneously, a thriving and slowly growing free black community now existed within the colony, living alongside a growing number of black slaves, all the result of the ambiguous manner in which the government under Berkeley had dealt with the issue of slavery. Though Virginians in general were not buying slaves, nor were they showing any propensity to encourage the practice, the time was soon coming when the Virginia leadership would decide to face the issue of those black citizens, and deal with it in the worst possible manner.

6. Virginia Under the Commonwealth

After almost thirty years as a resident of Virginia, first as a slave, now as a free man, Anthony Johnson was finally ready to claim his own. On July 24th, 1651 he sailed across Chesapeake Bay to Jamestown, walked into the court building and submitted to Secretary of State Richard Lee the certificates for five headrights, entitling him to the ownership of 250 acres of unoccupied land. Johnson indicated his desire for property along the Pungoteague River across Chesapeake Bay on the Eastern Shore, and Lee made out the land patent for Governor Berkeley to sign. Then, with Berkeley's signature in hand, Johnson got back in his boat and sailed back across the Bay, now a proud landowner.[263]

That he was a black man and a former slave made no difference: he was a free resident of Virginia. His social standing would therefore depend more upon by his abilities and his personal achievements than on his skin color or former status. And since the law said those headright certificates were legally his, the land was legally his as well.

For the next four years Anthony Johnson and his family would appear repeatedly in both Northampton County records as well as the colonial patent books, almost always initiating action on their own, and almost always getting what they wanted. And these emboldened actions and success happen to also coincide almost exactly with the simultaneous rise of their former owner, Richard Bennett, to the governorship.

By 1651 the revolutionary Commonwealth government in England had finally gotten to the business of subduing its recalcitrant Royalist colony. Samuel Mathews and Thomas Stegg had been in England discussing the situation with Parliament, and on September 26, Stegg, along with Richard Bennett and William Claiborne, were appointed as Commissioners to retake Virginia. Under the leadership of Captain Robert Dennis, two armed frigates would sail from England, carrying orders "to use all acts of hostility that lies in your power to enforce" the colony's surrender. The Commissioners were to

> cause and see all the several Acts of Parliament against kingship and the House of Lords to be received and published; as also the Acts for abolishing the Book of Common Prayer, [and] . . . to administer an Oath to all the Inhabitants and planters there, to be true and faithful to the common-wealth of England as it is now established without a king or house of Lords.[264]

Stegg would sail with Dennis in the frigate *John*, while the *Guinea* would pick up Bennett and Claiborne on the Maryland coast. Both ships were armed with significant cannon, though neither carried troops.[265]

Sometime in December 1651 the *Guinea* arrived off of Chesapeake Bay, where both Claiborne and Bennett apparently boarded, and then waited for the arrival of the *John*. Weeks passed, during which Berkeley and the Royalists on shore were readying their resistance, mustering a force of approximately 1,000 men to fortify Jamestown.[266] Finally, Bennett and Claiborne could wait no longer. They opened negotiations with Berkeley and Secretary Lee, assuming (rightly as it turns out) that Dennis, Stegg, and the *John* were lost at sea.[267]

Without significant troops to back them up, Bennett and Claiborne could only offer Virginians the mildest of terms. Even Berkeley, who had worked to make Virginia a Royalist stronghold for a full decade, saw no reason to fight. Their offer left all land rights untouched, changed no basic laws or policies, and protected Virginians from taxation by Parliament. Furthermore, the personal estates of Berkeley and his Council were protected, and Berkeley was officially given a full year to leave the colony, though this banishment was never enforced and probably never intended to be enforced.

Bennett and Claiborne did not even demand that the Book of Common Prayer be abolished, despite Parliament's specific instructions. Instead, parishes were only required that "those things which relate to kingship or that government be not used publicly." Ministers were allowed to remain in place, and all their agreements and dues were left as before.[268]

So, on March 12, 1652, Virginia surrendered, and William Berkeley quietly retired to his home at Green Spring. New elections were called and by April the first House of Burgesses under Parliamentary rule met. Emulating the system used by the revolutionary government in England, all power was retained in the House, which would then choose both the Governor and the Council. Subsequently, Richard Bennett was elected Governor and William Claiborne Secretary of State. On the Council, Royalists Henry Browne and William Brocas retired, and were replaced by Claiborne and twenty-year-old Samuel Mathews, Jr. whose father was still in England. (see table 11).

The new governor, Richard Bennett, had previously been the highest ranking Puritan in Virginia's Assembly.[269] Though not much is known about his governing style, we can surmise from what we do know that strong leadership was not a part of that style. It was his brother Phillip who brought the Puritan ministers to Virginia in 1642, Bennett only offering aid and support. It was his Puritan friends who fled Virginia in 1649 to go to Maryland, Bennett merely following for a short while to

Table 11-The Grand Assembly-April 1652

Current Governor:
Governor Richard Bennett

Council

Royalist:	Neutral:	Parliamentarian:
*West, John	*Ludlow, George	+Claiborne, William
+*Yardley, Argoll*	+Higginson, Humphrey	+Mathews, Samuel, Sr.
Uncertain Royalist:	Unknown:	Uncertain Parliamentarian:
Littleton, Nathaniel	*Bernard, William	Harwood, Thomas
Taylor, William	*Pettus, Thomas	
	Eppes, Francis, Sr.	Uncertain Quaker:
	Freeman, Bridges	**Chesman, John**

House

Royalist:	Neutral:	Parliamentarian:
*Fantleroy, Moore	*Lloyd, Cornelius	+Hatcher, William
+Hill, Edward, Sr.	+Dew, Thomas	**+Johnson, Thomas**
+*Scarburgh, Edmund*	+Fludd, John	
Warner, Augustine, Sr.	+**Mottrum, John**	Puritan:
Willis, Francis	*Mathews, Samuel, Jr.*	+Major, Edward-SPEAKER
		+Robins, Obedience
Uncertain Royalist:	Unknown:	
+Woodhouse, Henry	+**Bishopp, John, Sr.**	Uncertain Parliamentarian:
Lee, Henry	+George, John	+Burroughs, C[hristopher]
Martin, John	+**Gwin, Hugh**	
<u>Soane, Henry</u>	+Hardy, George	Uncertain Puritan:
	+Lambert, Thomas	+**Moone, John**
	+Mansell, David	+Whittaker, William
	+Morgan, Francis	Jones, William
	+Pitt, Robert	
	+Sheppard, John	
	+Wetherall, Robert	
	+Whitbye, William	
	Fletcher, George	
	Hoskins, Anthony	
	Ransom, Peter	
	Stephens, George	
	Williamson, James	
	Woodlief, [John]	

	Present Slave-holders	Including future slave-holders	Raised in Virginia	Incumbents	Including those who served previously
Council	7 of 14	8 of 14	1 of 14		
House	12 of 39	12 of 39	4 of 39	2 of 39	26 of 39

Legend for all tables:
Source for all tables: See Appendix B
*-incumbent
+-served previously
Bold names have purchased slaves or have claimed headrights on blacks.
<u>Underlined names will eventually buy slaves or claim headrights on blacks.</u>
Italicized names either were born in Virginia or immigrated as minors.

offer aid and support. It was Bennett who seems to have been in charge of negotiating the colony's surrender in 1652, which means it was he who offered Berkeley and the Royalists such easy terms. And it was William Claiborne, not Bennett, who led the way north in 1654 in the attempt to reincorporate Maryland back into Virginia as part of the surrender. In all these disputes, Bennett seems to have always acted as the mediator and negotiator, not as the initiator of action and change.

A Royalist Safe Haven

Yet, we know he was a Puritan with many Parliamentarian contacts, and the laws passed in both Virginia and Maryland under his leadership at this time reflect this fact. As an example, in 1654 the Maryland Assembly, reconstituted now under Bennett and Claiborne's command, passed a revision of the laws of Maryland. One of those laws addressed the issue of free speech, stating that

> it is the mind of this Assembly that any free subject of the Commonwealth shall have free liberty not only to petition to seek redress of grievances but as also to propound things necessary for the public good.

Furthermore, the legal revision specified that the Maryland Assembly was subject to elections every three years.

Also included in the revision was a law requiring liberty of conscience, allowing that all Protestant dissenters

> such as profess faith in God by Jesus Christ (though differing in judgment from the doctrine of worship and discipline publicly held forth) shall not be restrained from but shall be protected in the profession of the faith and exercise of their religion.

Since this law was mainly written to establish religious tolerance for the Puritan refugees that had moved to Maryland from Virginia, it also stated

> that none who profess and exercise the Popish religion commonly known by the name of the Roman Catholic religion can be protected in this province by the laws of England . . .[and] are to be restrained from the exercise thereof.

Like the Puritans in England, Bennett and his Puritan allies did not fully or consistently interpret democratic principles as we understand them today: liberty of conscience did not yet apply to Catholics.[270]

In Virginia, however, to establish religious freedom and free speech even as limited as this was difficult if not impossible. Both the House and the Council still had its share of outspoken Royalists, including past Berkeley allies like Edmund Scarburgh and Edward Hill, Sr. as well as new Assembly members like Augustine Warner, Sr., and Francis Willis (see table 11).

Augustine Warner had come to Virginia in 1628 as a seventeen-year-old under the tutelage of Adam Thorowgood, Sr. He first settled in Royalist York county, becoming that county's representative in this first

Commonwealth Assembly. Later he would live in Gloucester county, serving as its burgess in 1659. In 1660 he was appointed to the council by Berkeley and the Restoration Assembly. Warner's Royalist loyalties would become especially evident in 1661 when he used his position as justice to have Quaker Thomas Bushrod arrested for "mutinous and slanderous" language against the local clergy.[271]

Francis Willis had also come to Virginia as a young man, and had served as the county clerk in Royalist York County for many years. He had been a vocal supporter of Governor Harvey, and had been disbarred from all offices because of it. When Berkeley arrived in 1642, he had restored Willis, who soon became a York county justice.[272]

Also attending Richard Bennett's first Assembly as governor were men for whom we have uncertain but plausible evidence of Royalist allegiances. As an example, Henry Soane had immigrated to Virginia just prior to 1651, when most Royalist refugees were arriving, was wealthy enough to pay the transport for both himself and his family, and would eventually be chosen Speaker in 1661 by one of the most Royalist-dominated assemblies ever (see table 21).[273] Similarly, Henry Lee immigrated in 1649 with his family, and as I have already noted, his brother Richard Lee Sr. was known to be a close ally of King Charles II.[274] Likewise, Henry Woodhouse had many Royalist family ties (he was directly related to Francis Bacon), his father had been appointed Governor of Bermuda by King James, and after the Restoration, Berkeley and the Council used his house to hold Council meetings, all striking indications of Woodhouse's Royalist leanings.[275]

Despite the presence of these men, however, the 1652 Grand Assembly was no longer dominated by Royalists. Unlike the last few Assemblies presided over by Berkeley (tables 9-10), the Assembly (table 11) now included a significant number of Puritan and Parliamentarian dissenters. Puritans Edward Major and Obedience Robins were re-elected to the House, with Major being chosen as House Speaker. Major had originally come to Virginia as a headright of Richard Bennett, probably worked on his plantation as an indentured servant, and was now one of his neighbors.[276]

Also re-elected was Thomas Johnson, who only days before had called a mass meeting in Northampton county demanding that there be "annual choice of magistrates," that "there be a free and general vote for governor wherein we shall elect Mr. Richard Bennett," and that Northampton County should be exempt from taxes because they had not been represented in the Assembly since 1647.[277]

Chosen as well for this Assembly were men whose Parliamentarian, Puritan, and even Quaker allegiances are uncertain but plausible. As an example, though we have very little direct knowledge about John Chesman's political or religious beliefs, we do know that he was elected a councilor by this Parliamentarian/Puritan leaning Assembly, and that his son Edmund would become an active Quaker in the coming years and would eventually die in prison for participating in Bacon's Rebellion in 1676.[278] In another case, Thomas Harwood, who was also elected to the Council, had close commercial links with Claiborne and Mathews, had taken a prominent position in deposing Governor Harvey in 1635, and had actually been arrested by Harvey in England. Harwood himself had said that "Governor Harvey had so carried himself in Virginia that if he returned he would be pistolled or shot."[279]

The Assembly began the Commonwealth era divided between the very factions that had fought the British civil war, and during Richard Bennett's entire three year term as governor, from 1652 to 1655, as well as Royalist Edward Digges' two year term that followed, from 1655 to 1657, the philosophical conflict between these factions was reflected in the laws of the colony. While the Royalists wanted to maintain top-down rule and church government, the Puritans wanted to establish their own form of church government, and the Parliamentarians wanted to institute governmental bottom-up rule. Simultaneously, the first Quakers appeared in Virginia in 1657, demanding liberty of conscience for all.

Despite the infusion of Puritan and Parliamentarian members in the Grand Assembly, however, the political and social dynamics still worked in favor of the Royalists. Numbers alone were against the dissenters. Beyond protecting their property rights, few Virginians cared about Parliamentarian ideas, and most of the Puritans who had been passionate about religion had left the colony in 1649. Nor did they seem interested in returning now, even with Puritan Richard Bennett as governor. The Royalists who had been chosen by Berkeley to run the local courts and church vestries were mostly still in office, and the large numbers of wealthy Royalist refugees who had arrived after 1649 were beginning to obtain tracts of land and exert their influence.

In the face of this political dynamic, the 1652 Assembly under Richard Bennett set out to revise all of Virginia's laws, to make them conform to Parliament's instructions.[280] Yet, even though there was a Puritan Governor and the Assembly had an increased number of Puritans and Parliamentarians, this legal revision included nothing in it resembling the Maryland Act on religious toleration cited above. Instead, much like the religious compromise of the 1645 Assembly, the 1652 legal revision delicately avoided the issue of church government, repealing all previous laws establishing the Anglican Church, but specifying nothing to replace it. The single act on religion in this revision merely maintained the colony's parishes and churches without requiring any definitive religious practice.[281] While the Puritans were now free to practice their religion in Virginia, and the Parliamentarians had successfully removed all specific religious prohibitions in the law, the Royalists had artfully retained control of their Anglican church by agreeing to remove any official mention of that church in the law. William Berkeley's weak church government still existed, controlled by the men whom Berkeley had picked in the previous decade.

That many Royalist doctrines continued to dominate the political discourse of Virginia during the Commonwealth era, despite the efforts of the invigorated Puritans and Parliamentarians, can also be seen by how little the 1652 legal revision under Richard Bennett actually changed the basic legal and social structure of Virginia. Rule-from-above, a basic Royalist ideology, is illustrated by the willingness of the 1652 Assembly to maintain the previous regulations on lawyers, millers, tavern-owners, surveyors, and doctors. The laws governing trade, business practices, and the treatment of servants were continued as well. Furthermore, the Assembly actually attempted to increase government control, as when it dictated that all profits were now limited to 50 percent, and restricted travel and settlement by requiring passes to leave the colony as well as repealing of the October 1644 law that had allowed people to "remove and dispose of themselves for their best advantage and convenience."[282] Though the specific intentions behind any of these laws might have been good and reasonable, the overall pattern is one of an authoritarian leadership believing itself responsible for managing the lives of the rest of the population.

We must remember that often the Puritans agreed with the Royalists about rule-from-above and church government, though they differed fundamentally with the Royalists on the details. That the Puritans in Maryland were willing to exclude Catholics from their vision of liberty of conscience illustrates this. In Virginia, the two factions might have actually teamed up to pass some of the legislation mentioned above (such as the law limiting profits).

Furthermore, many Virginians, regardless of their religious and political values, firmly believed in the concept of deference to leaders and to those in the aristocracy. The general population was expected to meekly accept the policies and decisions of those in power.[283] Hence, despite Thomas Johnson's vocal support of Bennett in the Northampton protest mentioned above, Johnson was fined 500 pounds of tobacco for instigating the protest, actions that "had been very mutinous and repugnant to the government . . . [and to] the Assembly, and scandalous to the place and persons."[284]

In this context we should not be surprised, therefore, that the April 1652 legal revision contained a law forbidding public dissent against "the government now established."[285] Unlike Maryland, which had been a haven for both Roman Catholic and Puritan dissenters, vocal opposition was not to be permitted in Virginia, home to Royalist

Table 12-The Grand Assembly, November 1652

Current Governor:
Governor Richard Bennett

Council-no known record

House

Royalist:	Neutral:	Puritan:
*Fantleroy, Moore	*Dew, Thomas-SPEAKER	*Robins, Obedience
*Willis, Francis	*Lloyd, **Cornelius**	
Pyland, James-EXPELLED	*_Mathews, Samuel, Jr._	Uncertain Puritan:
	+Charlton, Stephen	*Whittaker, William
Uncertain Royalist:		
*Soane, Henry	Unknown:	Uncertain Parliamentarian:
+_Calithrop, Christopher_	*Gwin, **Hugh**	*Burroughs, C[hristopher]
Hammond, John-EXPELLED	*Lambert, Thomas	+_Wood, Abraham_
	*Ransom, Peter	Fleet, Henry
	*Stephens, George	Montague, Peter
	*Wetherall, Robert	
	*Whittbye, William	
	*Williamson, James	
	+Llewellin, Dan	
	+Sparrow, Charles	
	Edwards, William II	
	Gill, Stephen	
	Gooch, William	
	Harris, William	
	Hone, Theophilus	
	Perry, **Henry**	
	Reynolds, C[hristopher]	
	Thomas, William	
	Underwood, William	
	Watson, Abraham	
	Woodfife, John	

	Present Slave-holders	Including future slave-holders	Raised in Virginia	Incumbents	Including those who served previously
House	7 of 36	9 of 36	4 of 36	16 of 36	21 of 36

Legend for all tables:
*-incumbent
+-served previously
Bold names have purchased slaves or have claimed headrights on blacks.
<u>Underlined names will eventually buy slaves or claim headrights on blacks.</u>
Italicized names either were born in Virginia or immigrated as minors.

Source for all tables: See Appendix B

resistance to the Commonwealth as well as Royalist doctrines of rule-from-above and deference. In November 1652, the Assembly expelled two members, John Hammond and James Pyland, for having outspoken Royalist opinions (see table 12), and in July 1653 the Assembly once more expelled a member, this time minister Robert Bracewell, possibly because he was of the Anglican church (see table 13). That same 1653 Assembly then used the law banning dissent to outlaw the publication of a paper in Northampton County initiated by Edmund Scarburgh, expelling Scarburgh from all offices as a result.[286]

Table 13-The Grand Assembly, July 1653

Current Governor:
Governor Richard Bennett

Council-no known record

House

Royalist:	Neutral:	Parliamentarian:
*Fantleroy, Moore	*Dew, Thomas-SPEAKER	+Johnson, Thomas
+Yardley, Argoll	*Lloyd, Cornelius	
	*Mathews, Samuel, Jr.	Puritan:
Uncertain Royalist:	+Chiles, Walter, [Sr. or Jr.]	+Major, Edward
*Calithrop, Christopher	-SPEAKER1	
*Soane, Henry	Melling, William	Quaker:
+Wyatt, Anthony		Horsey, Stephen
Boucher, Dan	Unknown:	
Yardly, Francis	*Edwards, William II	Uncertain Puritan:
	*Harris, William	*Whittaker, William
	*Watson, Abraham	Thornbury, Thomas
	*Whittbye, William-SPEAKER2	
	+Bishopp, John, Sr.	Uncertain Parliamentarian:
	+Butler, William	*Montague, Peter
	+Fawdown, George	
	+Fletcher, George	Uncertain Catholic:
	+Morgan, Francis	Broadhurst, Walter
	+Pitt, Robert	
	+Sheppard, John	
	Baldwin, John	
	Booth, Robert	
	Bracewell, Robert-EXPELLED	
	Hackett, Thomas	
	Hockaday, William	
	Iversonne, Abraham	
	Pate, Richard	

	Present Slave-holders	Including future slave-holders	Raised in Virginia	Incumbents	Including those who served previously
House	8 of 37	8 of 37	8 of 37	12 of 37	24 of 37

    Yet, despite the authoritarian similarities between the Puritans/Parliamentarians and the Royalists, there were also significant differences. Under Bennett's rule, the colony's official religion was eliminated, and some religious toleration appeared possible, as indicated by the later arrival of two Quaker leaders, followed by the growth in Quaker popularity during this time period in some parts of the colony.[287]

    Furthermore, the influence of the Parliamentarians can be seen in the Assembly's effort to expand voting rights during the Commonwealth years. In March 1655 the Assembly gave the right to vote to one householder per family, whether or not that householder owned or rented his land.[288] Then in March 1656 the Assembly expanded voting rights to include all freemen, "whereas we conceive it something hard and unagreeable to reason that any persons shall pay equal taxes and yet have no votes in elections."[289] Even the British Commonwealth had not yet approved such a radical form of suffrage.

The very design of Virginia's government during this period, with both the Governor and Council chosen by the elected burgesses instead of appointed from above, illustrates how unlike the Royalists was Commonwealth rule. Even if it meant they might lose an election, the dissenters instituted a governmental system that was inherently ruled from below.

Other differences between Berkeley's Royalist rule and the Commonwealth era can be seen by subtle membership changes in the House of Burgesses. Throughout the latter half of Berkeley's first administration the House saw a very slight increase of men willing to own slaves (see figures 6.1 and 6.2), with the numbers of present or future slave owners rising to between 30 to 40 percent. While many of these legislators did not possess slaves in the 1640s, such as Edmund Scarburgh and Richard Lee, their eventual willingness to do so illustrates their nonchalant acceptance of the idea of an ethnic caste system.

For the first few years of Commonwealth rule, however, the House saw a drop of potential slave-holders, declining about twenty points to just over 20 percent. Initially at least, Commonwealth rule attracted men who would never demonstrate an interest in profiting from slave labor. Furthermore, throughout the entire Commonwealth era the number of House members who actually held slaves at the time steadily dropped, eventually reaching its lowest level in 1659, the lowest since 1644. During this period of time, it seems, the men in power seemed as yet unwilling to buy slaves.

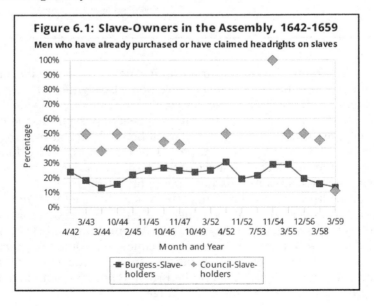

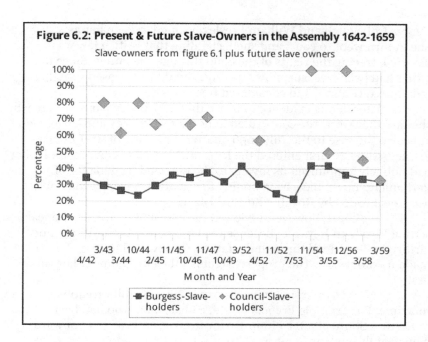

Finally, the distinctions between the Royalist rule of the 1640s and the Commonwealth rule of the 1650s can be seen by the other events outside the legislature. That small but thriving free black community had begun to develop in Virginia during the 1640s, and now many of them, including Anthony Johnson, were settled on the Eastern Shore of Virginia. We can learn much about the Commonwealth ideas about caste and ethnic privilege by how they treated these free blacks.

The Johnson family is the most demonstrative example. A year after Anthony had patented that first 250 acres, his son John patented an additional 550 acres of land adjacent to his father's estate. When Johnson's plantation was destroyed by a fire in 1653, he formally petitioned the local court for relief. The court not only exempted Mary Johnson from paying taxes, as would be the case if she were white, it also exempted the Johnsons' two daughters. In 1654 his other son Richard patented an additional 100 acres, also adjacent to the Johnson plantation.

Johnson, a free man and property owner, was not the only black to demonstrate the existence of such Parliamentarian ideas of liberty of conscience and individual rights at this time in Virginia. In 1656, during the latter part of the Commonwealth era, a mulatto by the name of Elizabeth Key sued for her freedom. Her mother had been a black slave, but her father a free white planter. As a child she had been indentured to John Mottrum. When Mottrum died in 1655, the overseers of his estate,

George Collclough and William Presley, wanted to keep her as a slave. She in turn went to court and sued, claiming that British law was paternal, tracing the rights of the child through the father. Since her father had been free, she, now an adult, should be released from her indentured contract and considered free as well.

After taking evidence to prove that Elizabeth's parents were who she said they were, the court ruled in her favor. Presley and Collclough appealed the case to the July 1656 quarterly meeting of the Council. Presley, who had first entered the legislature in that 1652 Assembly that had publicly announced its loyalty to the king and its defiance of Parliament, might have been hoping for favoritism from a council that was now headed by Royalist Edward Digges.

The court, however, reviewed the case and denied the appeal, setting Elizabeth Key free. Because the law said that inheritance and birthright descended through the father, the law had to rule. Skin color made no difference. The property rights of the citizen under law came first.[290]

Other free Virginian blacks experienced similar treatment during this time. For example Francis Payne, a black man who had been a slave to Eastern Shore planter Philip Taylor from as early as 1637 onward, managed through arduous work to purchase his freedom circa 1652, followed soon after by that of his wife and children. Throughout the 1650s he used the courts just as the Johnsons did, successfully suing for his rights in various cases. In 1651 he went to court to obtain 310 pounds of tobacco owed to him by white planter Joseph Edlow. In 1655 he did the same to recover rent from free black farmer John Gussall.[291]

Tony Longo and his wife were able to obtain headrights for five individuals and by 1655 he owned at least 250 acres. Longo would have repeated conflicts with the law, but his treatment by the court, both good and bad, was generally the same as that for whites. When he could no longer support his children, the court had them bonded out as indentured servants, in accordance with the same term and arrangements afforded any white children.[292]

Emmanual Driggus, his wife, and many children were originally owned by Francis Pott. When Pott died in 1658, his wife married William Kendall, an Eastern Shore planter who seemed to never be able to rid himself of his slaves. He would acquire some either through marriage or purchase, and then find ways to free them, either by sometimes expensive self-purchase or simple manumission. By 1660 Kendall had freed Driggus, to whom he eventually leased a 145 acre plot.[293]

And then there was the seeming tolerance to interracial marriage. Elizabeth Key ended up marrying her white lawyer, William Greenstad.[294] Emmanual Driggus also married a white woman, as did Francis Payne.[295] Payne's marriage, to a woman named Aymey, lasted at least seventeen years without any evidence of opposition from others. When Payne died in 1673 he gave "unto my loving wife . . . my whole estate."[296]

Consider: interracial sex was to become the south's most hated racial crime, and black men who committed it would almost certainly be subjected to quick and very violent mob justice. In seventeenth-century Virginia, however, the reaction to these marriages was hardly noticeable. Even more indicative were the circumstances surrounding free black Philip Mongon (also spelled Mongom), who by the 1650s was a somewhat successful tenant farmer. In the late 1650s he had sexual relations with a white woman, who gave birth to an illegitimate baby. The court condemned him for the "sin of adultery," fined him 500 pounds of tobacco, and made him responsible for the child's upbringing. The woman, Margery Tyler, was sentenced to ten lashes for "the filthy sin of fornication." The focus of both sentences was to punish the crime of sex outside marriage, not interracial marriage, and neither punishment was significantly different that what would have been administered to whites in the same circumstances. Furthermore, Mongon remained a free member of society, eventually leasing several hundred acres near the Virginia-Maryland border.[297]

Similarly, when John Johnson, Anthony Johnson's son, was caught having sex with a white servant of Edmund Scarburgh in the early 1660s, the court merely demanded that he "put in security to keep the parish harmless . . . and shall pay and satisfy all damages." Furthermore, Edmund Scarburgh had to post 1,000 pounds of tobacco to guarantee his servant's behavior.[298]

It would seem that, during the middle years of the seventeenth-century, and especially the Commonwealth era, Virginian society did not automatically treat blacks as inferior subjects merely because they were black. Instead, the British belief that every man must be allowed the freedom to improve his circumstances took precedence. Tied to this was the belief in the idea of liberty of conscience, which said that each human being is equal in the eyes of God, and should be given the opportunity and freedom to find God in his or her own way. If Bennett and his Puritan and Parliamentarian allies believed in these ideas, and the evidence indicates that they did, they would have had to treat free Christian blacks with the same respect given a free Christian white. The decisions involving both Elizabeth Key and Anthony Johnson confirm this, demonstrating that the Governor and courts would protect the rights of all citizens, regardless of skin color, ethnic background, or caste status.

We must not assume that only during the Commonwealth period could blacks get a fair deal in court. This is not true. Free blacks like Francis Payne and Philip Mongon were able to maintain their free status, and sometimes even increase their property holdings, as far into the future as the 1680s, using the courts to support the legality of their actions.

Yet, it was in the 1650s when the largest number of these early blacks of Virginia obtained their freedom. Ideas of individual rights and liberty of conscience seemed at this time to hold a more dominant place in the social order, allowing us to wonder whether Virginia society might have actually avoided the scourge of slavery and chosen a different route.

Until 1660, Virginia was little different from any other British community when it came to slavery: few Virginians owned slaves, fewer still wanted to buy them, and the men in power had made no effort to legalize the custom. Like every other North American British colony, slavery was a distasteful side issue percolating up from the Caribbean and Africa. While no law made slavery illegal, and the moral issues surrounding the custom were cloudy and in dispute, the vast majority of British immigrants to these colonies wanted nothing to do with slavery, trying instead to establish a legacy for their own families, with or without British servants.[299]

Like the other British colonies of this time, the black population of Virginia was still extremely small, probably less than 400 people out of a total population of over 16,000. Furthermore, the Assembly had seen a drop in its slave-owning members. Parliamentarian ideas seemed to have had an influence on Virginian politics in the 1650s, and for a time at least the idea of legalizing slavery as an institution seemed unlikely if not impossible.

And yet, the members of the Commonwealth Assemblies made no effort to prevent the exploitation of the blacks slowly percolating into the colony, leaving them in a legal limbo. Though British law still required freedom for Christian converts as well as those born of a free father, the growing custom in Virginia was to ignore these laws. And though many Virginians did free their slaves, they did so sporadically, and with no concern for the British laws they all so ostensibly prized. Only in cases such as Elizabeth Key's, where the challenge was made and the evidence was clear, would the law be followed scrupulously.

Moreover, the Commonwealth Assemblies under Bennett made no effort to reform the headright system. Instead, the law was left untouched, so that any newly patented land continued to be concentrated in the hands of the powerful and wealthy planters – Royalists, Puritans, and Parliamentarians alike. Poor immigrants continued to be mostly male, and continued to be treated as temporary chattel that could be bought and sold, at a whim.

Sadly, the mid-1650s is in many ways the crucial turning point to this story. At the same time that these free blacks were proving they could compete equally in the free, open British society that the British treasured so much, other events were occurring that clearly made manifest the financial benefits of slave-ownership, as well as demonstrating the need to legalize the custom in order to gain those benefits. In a colony driven almost entirely by financial gain with little moral foundation, such events carried an almost seductive emotional appeal.

Following the 1650s, the white Virginians in power were offered a choice: they could encourage the approach to justice epitomized by the treatment of Johnson, Key, and other free blacks. Or they could finally begin succumbing to slavery's financial appeal, and make it the law of the land.

The blacks of Virginia would not be treated with as much justice for another three hundred years.

7. A Royalist Resurgence

Sometime in the fall of 1654, "a negro called John Casor" beseeched several white planters to help him get free from his owner. Casor claimed that he was being held illegally as a slave, that he had entered Virginia as an indentured servant, and that his owner had been denying him his freedom for at least seven years. With the help of Robert Parker and his brother George, two Eastern Shore freeholders, Casor ran away from his master's farm to work for Robert Parker.[300]

John Casor's master was black landowner Anthony Johnson.

At first Johnson's family convinced Anthony not to try and reclaim Casor. In a formal statement, Johnson released "John Casor, negro, from all service, claim, and demands." By early 1655, however, Johnson had changed his mind, and was willing to go to court to recover his "property."[301] On March 8, 1655, only weeks before Richard Bennett was to step down as governor, the court ruled in Johnson's favor, declaring that John Casor was Anthony Johnson's slave. Furthermore, it ordered Robert Parker to pay all court costs. John Casor returned to the Johnson family, and remained their property until as late as 1672.[302]

The ironies of this story are profound. While once again the Parliamentary government of Richard Bennett demonstrated its willingness to protect the rights of its property-owning free citizens, regardless of ethnic background, it also joined with a black man to confirm the lower status of another black servant. For "John Casor, negro," lifetime slavery was an acceptable condition, even though such a condition would have been considered unacceptable for any white British citizen.

Three weeks later, Richard Bennett yielded the governorship, going to Maryland to join William Claiborne in his effort to merge that colony with Virginia so that Claiborne could recover his lost plantation on Kent Island. In Bennett's place the House chose for governor Edward Digges, Royalist (see tables 14 and 15).

We do not know why Bennett was replaced. Possibly he no longer wished to be governor. More likely the Royalists in the Assembly wanted a Royalist as governor, and voted him out. Either way, Digges' election signaled the beginning of the end of any Puritan or Parliamentary leadership in Virginia.

A Royalist Resurgence

Table 14-The Grand Assembly, November 1654

Current Governor:
Governor Richard Bennett

Council-No known record except for one new appointment:
Royalist:
Digges, Edward

House

Royalist:	Neutral:	Parliamentarian:
+Carter, John, Sr.	*Dew, Thomas	*Johnson, Thomas
+Hill, Edward, Sr.-SPEAKER	*_Mathews, Samuel, Jr._	
+Hoskins, Bartholomew		Puritan:
+_Trussell, John_	Unknown:	Bond, John
	*Booth, Robert	
Uncertain Royalist:	*Pitt, Robert	Uncertain Puritan:
***Soane, Henry**	*Sheppard, John	*Whittaker, William
Abrahall, Robert	*Watson, Abraham	+Moone, John
Mason, Lemuell	*Whittbye, William	_Waters, William_
	+Bagnall, James	
	+Cocke, Richard	Uncertain Parliamentarian:
	+Gooch, William	+_Wood, Abraham_
	+Perry, Henry	Godwin, Thomas
	+Stoughton, Sam	
	+_Worleich, William_	
	Batte, William	
	Baynham, Alexander	
	Breman, Thomas	
	Dipnall, Thomas	
	Hamelyn, Stephen	
	Haywood, John	
	Hobbs, Francis	
	Holland, John	
	Mason, James	
	Walker, Peter	
	Webb, Wingfield	

	Present Slave-holders	Including future slave-holders	Raised in Virginia	Incumbents	Including those who served previously
Known Council	1 of 1	1 of 1	0 of 1		
House	11 of 38	16 of 38	6 of 38	10 of 38	22 of 38

Legend for all tables:
*-incumbent
+-served previously
Bold names have purchased slaves or have claimed headrights on blacks.
Underlined names will eventually buy slaves or claim headrights on blacks.
Italicized names either were born in Virginia or immigrated as minors.

Source for all tables: See Appendix B

As noted in chapter five, Digges, whose family had many Royalist ties, had arrived in the Royalist immigrant wave after the beheading of Charles II in 1649 and by 1654 had accumulated 4,000 acres and had been appointed to the Council by the Assembly. Known as a friend of Berkeley, once established at his plantation he became like Berkeley in that he tried to innovate and diversify Virginia's industries. He even went so far as to transport two Armenian silk manufacturers to the colony to develop the silk trade. When he died in 1675, he owned 108 slaves and a substantial estate.[303]

Digges' quick rise from immigrant to governor under Parliamentary ideas of government showed the Royalists in Virginia that

Table 15-The Grand Assembly, March 1655

Current Governor:
Governor Edward Digges

Council

Royalist:	Neutral:	Parliamentarian:
*West, John	*Ludlow, George	*Claiborne, William
*Yardley, Argoll	+Higginson, Humphrey	*Mathews, Samuel, Sr.
	Dew, Thomas	
Uncertain Royalist:	Mathews, Samuel, Jr.	Puritan:
*Taylor, William		Robins, Obedience
	Unknown:	
	*Bernard, William	
	*Pettus, Thomas	
	*Eppes, Francis, Sr.	
	*Freeman, Bridges	
	Perry, Henry	
	Gooch, William	

House-Held over without elections

Royalist:	Unknown:	Puritan:
+Carter, John, Sr.	*Booth, Robert	Bond, John
+Hill, Edward, Sr.-SPEAKER	*Pitt, Robert	
+Hoskins, Bartholomew	*Sheppard, John	Uncertain Puritan:
+Trussell, John	*Watson, Abraham	*Whittaker, William
	*Whittbye, William	+Moone, John
	+Bagnall, James	<u>Waters, William</u>
Uncertain Royalist:	+Cocke, Richard	
*Soane, Henry	+Stoughton, Sam	Uncertain Parliamentarian:
<u>Abrahall, Robert</u>	**+Worleich, William**	*Wood, Abraham
	Batte, William	Godwin, Thomas
	Baynhan, Alexander	
	Breman, Thomas	
	Dipnall, Thomas	
	Hamelyn, Stephen	
	Haywood, John	
	Hobbs, Francis	
	Holland, John	
	Mason, James	
	Walker, Peter	
	Webb, Wingfield	

	Present Slave-holders	Including future slave-holders	Raised in Virginia	Incumbents	Including those who served previously
Council	8 of 16	8 of 16	2 of 16	House-held over without elections	
House	11 of 38	16 of 38	6 of 38		

if they wished to take over the colony, they only had to take advantage of the Parliamentarian rules to do so. The results of the December 1656 election revealed this possibility very clearly (see table 16). Besides the re-election of Edmund Scarburgh, Francis Willis, and George Reade, Royalist refugees Francis Moryson, Henry Chickeley, and John Page entered the legislature for the first time.

Also elected for the first time was Royalist Joseph Crowshaw, who had been born and raised in Virginia. His father had been one of the first settlers, a man who was highly respected by John Smith for his knowledge of the Indians. In England, his family had many religious and aristocratic ties (his uncle was a devout minister who had preached to Lord Delaware as he prepared to leave for Virginia). Crowshaw, a

Table 16-The Grand Assembly, December 1656

Current Governor:
Governor Edward Digges

Council-not fully known

Royalist:	Neutral:
Bacon, Nathaniel, Sr.	*Mathews, Samuel, Jr.
	+Walker, John
	Unknown:
	Thomas, William

House

Royalist:	Neutral:	Puritan:
*Hoskins, Bartholomew	+Fludd, John	*Bond, John
+Fantleroy, Moore		
+Reade, George		Uncertain Puritan:
+Scarburgh, Edmund	Unknown:	*Whittaker, William
+Willis, Francis	*Walker, Peter	
Chickley, Henry	+Harris, William	Uncertain Parliamentarian:
Crowshaw, Joseph	+Hone, Theophilus	*Wood, Abraham
Lyggon, Thomas	+Lambert, Thomas	Ellyson, Robert
Moryson, Francis-SPEAKER	+Llewellin, Dan	
Page, John	+Ramsey, Thomas	Uncertain Quaker:
	Beazley, [Job or Robert]	Davis, Thomas
Uncertain Royalist:	Blake, John	
*Mason, Lemuell	Foster, Richard	
+Sidney, John	Holmwood, John	
+Wyatt, Anthony	Holte, Robert	
Ashton, Peter	Langley, Ralph	
	Lobb, George	
	Pritchard, Thomas	
	Smith, Nicholas	
	Streeter, Edward	
	Wade, Armiger	
	Wilcox, John	

	Present Slave-holders	Including future slave-holders	Raised in Virginia	Incumbents	Including those who served previously
Known Council	2 of 4	4 of 4	1 of 4		
House	8 of 41	15 of 41	6 of 41	6 of 41	20 of 41

resident of Royalist York County, himself favored the king, so much so that he got himself suspended as a justice in 1659 for challenging Commonwealth rule.[304] On the Puritan/Parliamentary side, few were chosen. Only six burgesses can be named whose history indicates any willingness to stand against the Royalists, and of these, two, Abraham Wood and William Whittiker, had never exhibited clear political or religious beliefs. Like Wood, Whittiker is another example of someone who we can only call a dissenter based on circumstantial evidence. He only served as a Burgess during the Commonwealth years, and was related to Alexander Whittiker, a well known Puritan who had come to Virginia in 1611. While both facts indicate Puritan leanings, neither establishes Whittiker's affiliations very convincingly.[305]

For two more men, Thomas Davis and Robert Ellyson, we have better knowledge, but only a little better. Davis was probably a Quaker, though we have no statements from him to prove it. However, his wife's family was very active in the Virginia Quaker movement, and in 1660, simultaneous with his marriage, Davis was dismissed as a Justice. This took place at the same time the Virginia government was attempting to repress the Quaker religion.[306]

Ellyson's brother had been a member of Parliament in 1648 during the civil war. When Ellyson immigrated from England in 1642, he came as a barber/doctor, going first to Catholic Maryland. There he married the daughter of Thomas Gerrard (who had been one of the early Catholic settlers, a close friend of Lord Baltimore, and in 1641, had actually taken the keys and books from a Puritan chapel in an attempt to shut it down).[307] By the mid-forties Ellyson had moved to Virginia, taking up the career of defense lawyer (an avocation despised by the Royalists) in Royalist York County of all places. While this career choice and place might suggest he had alliances among the Royalists, only during the Commonwealth years did he finally win appointment to any official posts, first becoming James City High Sheriff, then Burgess.[308] Though we have no direct evidence, these actions indicate his links to Parliamentarianism as well as his willingness to place himself in opposition to those in power.

Of the burgesses elected in 1656, only John Bond ever demonstrated publicly his opposition to the Royalist majority. Bond was a known Puritan, and like other Puritans, only held office during the Commonwealth years. When the king was restored in 1660, the Royalists in the Assembly immediately dismissed him from all offices because of his Puritan allegiances.[309]

Everyone else in the 1656 Assembly was either a known Royalist ally of Berkeley or a complete cipher, someone for whom we today have no information at all.

In retrospect, we can look at the Virginia of the 1650's and conclude that, despite their best efforts, the ideals of Parliamentarians for democratic government never had much chance of success. While some Virginia planters might have given lip service to such religious and idealist principles, their first goal continued to be economic profit. And except for some half-hearted attempts to bring ministers to the colony, the legislature of Virginia made little effort to improve the dismal state of the colony's religious institutions. Despite a Puritan governor from 1652 to 1655, there was no Puritan revival in the colony. When the Quakers attempted their own revival in 1657, they were imprisoned, as we shall see.

Furthermore, none of the laws during the 1650s actually addressed the fundamental and chronic problems caused by the headright system: Virginia was still a miserable place for most of its inhabitants. Family life continued to be disrupted by high mortality and an unnatural male/female sex ratio, the labor system remained oppressive (still dependent on a single crop economy and the headright system), and communities continued to be isolated, dispersed in large plantations rather than small family-owned farms, and lacking any religious or educational institutions.

Worse, by attempting to reject Royalist doctrines of privilege and rule-from-above, the Parliamentarians only succeeded in destroying the Royalist code of honor which had previously enforced some self-restraint on those in power. Without this code of honor, the privileges and status that came with power could now be used for brazen personal gain. Now, men became shameless in using the force of government to better their own interests.

Some quick examples: George Fletcher obtained from the Assembly a fourteen year monopoly on the brewing of alcohol in wooden containers.[310] William Claiborne, Abraham Wood, and Henry Fleet obtained from the Assembly a fourteen year monopoly on the profits arising from any exploration to the west and south.[311] Claiborne also used the Assembly to form New Kent county, probably to help enhance his power in the House and in the local courts.[312]

The best brazen example, however, of using government power for personal gain involved the seizure in June 1652 of a ship owned by Walter Chiles because it had not submitted to Parliamentary regulations. In order to recover his losses, Chiles made a deal with governor Bennett, Secretary Claiborne, Samuel Mathews Jr., and several other councilors. The government, having seized the Dutch ship *Leopoldus of Dunkirk* for the same violation, would sell it to Chiles for 400 pounds sterling. That money would then be distributed to Claiborne, Mathews, and the others who had helped clinch the deal.[313]

None of this is to suggest that such pay-offs had not occurred in Berkeley's administration, for they had, usually in the form of taxes to pay for "expenses." What had changed was the undisguised and aggressive manner in which these deals were proposed, negotiated, and then approved by the Assembly. William Berkeley might have been willing to benefit from his role as Governor, but a direct pay-off for services rendered – in other words a bribe – would have probably appalled him during his first term, violating his Royalist code of noblesse oblige. And we know this is so because during his first term, such direct pay-offs never happened.

Nor should we conclude from these pay-offs that everyone in the Commonwealth government was greedy, power-hungry, and without honor. Note that governor Bennett received no money in the Chiles deal, though such a pay-off was probably in his power to obtain. In fact, Bennett might have actually tried to squelch the deal when he tried to convince the Assembly not to choose Chiles as its Speaker, noting

> that it is not so proper nor so convenient at this time
> to make choice of him for that there is something to be
> agitated in this Assembly concerning a ship lately
> arrived, in which Left. Col. Chiles has some interest.[314]

The Assembly, however, in a probable effort to placate Chiles against further court action, ignored Bennett's advice and picked him anyway. (Interestingly, Chiles turned the job down, and the Assembly chose William Whittbye instead, who in turn received some of Chile's money from the above deal. Possibly Chiles didn't want the job, or possibly he recognized that the obvious conflict of interest would come back to haunt him.)

Yet, the unprecedented and blatant use of the Assembly to clinch this deal indicates how Virginia society was beginning a slide into increasingly selfish and corrupt economic and social practice for the sake of personal gain, without regard for consequences or morality.

This slide can best be seen by the actions of Edmund Scarburgh during the Commonwealth years. A tried and true Royalist, what Scarburgh did in 1655 and 1656 forcefully demonstrates how Virginians were more and more willing to consider almost any conduct for the sake of obtaining wealth and power. Furthermore, his actions illustrate how easily Royalist ideas of caste and privilege were transformed into bigotry, white supremacy and institutionalized black slavery, especially as morality, honor, and noblesse oblige exerted less and less influence on the powerful members of Virginia society.

Of all the men from this period of whom we know anything about, Edmund Scarburgh best epitomizes the Royalist planter. A man who despised religious dissent, who believed in top-down rule, who aggressively used government for personal gain, and who demanded the freedom to make as much money as he could in whatever way that he could, Scarburgh never showed any hesitation in pursuing what he wanted.

As already noted, in July 1653 the Parliamentarian-dominated Assembly had expelled Scarburgh from holding office, and had actually threatened to arrest him.[315] As the owner of at least four ocean-going ships, Scarburgh entrusted his estate to the care of friends and fled the colony. First he traveled to England, returning in 1654 with goods to sell. Bennett and the Council ordered that his ship be searched for arms and ammunition and seized if any were found.[316]

No arms were found, and Scarburgh was able to sail away. Ever aggressive, he now journeyed to the Dutch colony of New Amsterdam and began negotiations for the purchase of a sizable number of black slaves.[317] If the government was going to force him to leave his tobacco plantation, he was going to make his banishment pay.

The black population of Virginia was still extremely small. Since 1637 the Dutch had controlled access to the slave trade from Africa,[318] and therefore slaves were generally only obtainable when Dutch slave traders were forced to dock in the Chesapeake for supplies.[319]

As a planter who had already obtained 4600 acres for himself[320] and an additional 3050 acres for his oldest son Charles by transporting 141 people into the colony (three of whom were blacks),[321] Scarburgh surely understood that, though black slaves were not always treated in an identical manner as white servants, and though they did not come to the colony freely or with the expectation of eventually earning their freedom as did white servants, they could still earn the fifty acre headright for whoever paid their transportation costs. This corruption of the original intent of the headright, to encourage settlement by British citizens, was conveniently allowed by all Virginia governors from George Yeardley to Edward Digges, with the single exception of John Harvey, and he was practically ridden out of town on a rail for doing so.[322] Throughout the thirties and forties we have already noted how Henry Browne, Richard Kemp, Ralph Wormeley Sr., and others had taken advantage of this loophole to significantly enlarge their estates. In the 1650s the importation of black slaves for land continued to trickle in: in 1650 William Worleich imported three black slaves for 150 acres; in 1651 Rowland Burham brought in ten blacks to claim 500 acres; in 1652 George Ludlow imported four blacks for 200 acres, while Scarburgh himself brought in three for his son Charles; in 1653 Henry Soane claimed fifty acres on one black; and in 1654 Humphrey Higginson imported thirteen black slaves to patent 650 acres.[323]

Scarburgh now took this patent process one step further. Instead of buying slaves from the foreign slave traders, Scarburgh used his own ship to bring them into the colony himself. This way he could import a large quantity instead of the small numbers previously available from the rarely seen Dutch slave-traders. Also, by doing it himself with his own ships he could save money on the transportation costs while directly negotiating a better price for himself. Scarburgh negotiated with Dutch authorities in New Amsterdam to buy a shipment of slaves, numbering at least fifty-five, and returned to Virginia in March 1655.[324]

Royalist Edward Digges was about to replace Richard Bennett as governor, and on March 26 Scarburgh was cleared of all charges.[325] The very next day, only weeks after Anthony Johnson had won his case in court, Edmund Scarburgh claimed headrights on the black slaves, first assigning the rights of forty-one to his two daughters, Tabitha and Matilda, in order to obtain a large land estate of 2050 acres for his children. The headrights to the other fourteen slaves he sold to neighbors.[326]

Unquestionably, this action by Scarburgh was noted and commented on throughout the upper echelons of Virginian society, both by those who found it offensive, such as Scarburgh's Puritan neighbor Obedience Robins, and by those who found it an intriguing way to increase one's wealth. This latter group included such Royalists as Nathaniel Bacon Sr., who had just been appointed to the Council and would claim headrights on nine blacks ten years later,[327] Joseph Bridger, who entered the House for the first time in 1657 and would patent headrights on seventeen black slaves himself in a decade,[328] and Henry Corbin, who would claim two headrights on blacks in 1658. Other Royalists who would eventually copy Scarburgh were John Carter, Sr., who would claim 21 black headrights in 1665, and Robert Smith, who would claim two blacks as headrights in 1667.

Also influenced by Scarburgh's action were first-generation Virginians like John Savage and John Powell. Savage married the daughter of Obedience Robins, a man who never owned slaves despite accruing immense wealth. Several years after his father-in-law's death, however, Savage would himself claim 1250 acres on the headrights of twenty-five black slaves.[329] Powell, for whom we know very little except that he was raised in Virginia and would serve on the Assembly after the king's restoration, claimed land for seven blacks in 1665.[330]

In addition, future burgesses, for whom we know even less, such as John Blake (1 in 1658 and then 7 in 1664), John Weir (14 in 1663), and Roger Williamson (4 in 1666) all jumped on the bandwagon, bringing slaves to Virginia in exchange for a few acres of land.[331]

All these present and future burgesses, many of whom were either Royalist refugees or men raised in Virginia, would come to dominate the Virginia government in the next decade, during the very moment that government began legalizing the institution of slavery.

Their dominance, as well as most of these slave purchases, however, took place in the 1660s. In the 1650s, slavery as a legal institution still did not yet exist, nor was there any movement in the House or the Council to codify it, whether controlled by the Royalists or their opponents. Blacks remained a tiny percentage of the population, and few Virginians seemed interested in buying them. Furthermore, as noted earlier, the first two Assemblies in the Commonwealth period had seen a drop in its slave-owning members (figures 6.1 and 6.2). For a time at least the idea of legalizing slavery as an institution remained unlikely if not impossible.

This state of affairs was well illustrated by the Royalist-dominated December 1656 Assembly. It passed no laws sanctioning or even mentioning slavery, despite Scarburgh's actions and the apparent general acceptance to which it was received. Instead, it once again recognized the lack of ministers and religious leadership in the colony and asserted its desire to encourage their importation. In March, the Assembly had ordered that all counties be laid out into parishes, and increased the tax from ten pounds to fifteen pounds of tobacco in order to build churches and hire ministers.[332] The December Assembly now offered a twenty pound sterling award for anyone who transported a minister into the colony.[333] With both laws, the words were kept vague enough to allow any religious leader, regardless of sect or denomination, to consider coming to the colony. While this vagueness might have allowed Puritans to practice their religion freely, it really functioned more to allow the Royalists now in control to quietly encourage their Anglican church and only their Anglican church while under so-called Commonwealth rule.

We know that this encouragement did not apply to dissenting religious faiths because of how these Royalists treated Quakers Josiah Coale and Thomas Thurston when they arrived in the colony to preach their religion in 1657. The Assembly had just admitted that "many congregations in this colony are destitute of ministers whereby religion and devotion cannot but suffer much impairment and decay,"[334] and the law's wording did imply that the arrival of any preachers, Quaker or not, would be welcomed. Yet, Coale and Thurston were quickly imprisoned until a ship could be found to take them from the colony, this despite evidence that their visit to Virginia marked the beginning of a Virginian Quaker movement that six years later William Berkeley felt compelled to suppress.[335]

Though Virginians, including many Royalists, might have wanted some form of religious guidance, and though the law did not specifically outlaw any specific religious practice, the prevalent Royalist ideologies of church government and uniformity of religious practice made such dissenting beliefs impermissible, even if it meant the colonists would have no religion at all.[336]

These repressive measures, however, appear to have hurt the Royalists come the next election. In the 1658 elections Scarburgh, Chickeley, Crowshaw, Moryson, Pages, Reade, and several other known Royalists were all turned out, leaving the 1658 Assembly with the fewest incumbent burgesses since 1644, most of whom were men for whom we know practically nothing (table 17). As a result, the political make-up of this Assembly is probably the most unclear of any throughout the middle part of the seventeenth-century.

We do know what this Assembly did, however. First, it immediately replaced governor Edward Digges with the twenty-eight-year-old son of Samuel Mathews. Not much is known of Sam Jr., other than his father left him in charge of the family estate while staying in

Table 17-The Grand Assembly, March 1658

Current Governor:
Governor Samuel Mathews, Jr.

House

Royalist:	Neutral:	Puritan:
+Carter, John, Sr.	+Elliot, Anthony	*Bond, John
Bridger, Joseph	+Melling, William	
Horsmanden, Warham	Revell, Randall	Uncertain Parliamentarian:
Smith, John [Dade, Francis]	Webster, Richard	+Montague, Peter
—SPEAKER		+*Swann, Thomas*
	Unknown:	Michell, William
Uncertain Royalist:	*Harris, William	Kendall, William
Mason, Lemuell	*Foster, Richard	**Knight, Peter**
*Sidney, John	*Lambert, Thomas	
+**Soane, Henry**	*Ramsey, Thomas	Uncertain Quaker:
	Wilcox, John	*Davis, Thomas
	+Butler, William	Taberer, Thomas
	+*Edwards, William II*	
	+Loveing, Thomas	
	+**Worleich, William**	
	Blackey, William	
	Borne, Robert	
	Brewer, John	
	Carter, Edward	
	Caufield, William	
	Corker, William	
	Francis, Thomas	
	Goodwin, James	
	Ham, Jeremy	
	Haney, John	
	Hay, William	
	Lucas, Thomas, Sr.	
	Powell, John	
	Warren, John	
	Webb, Giles	
	Wynne, Robert	

	Present Slave-holders	Including future slave-holders	Raised in Virginia	Incumbents	Including those who served previously
House	7 of 44	15 of 44	10 of 44	9 of 44	19 of 44

Legend for all tables:
*-incumbent
+-served previously
Bold names have purchased slaves or have claimed headrights on blacks.
Underlined names will eventually buy slaves or claim headrights on blacks.
Italicized names either were born in Virginia or immigrated as minors.

Source for all tables: See Appendix B

England as Virginia's agent to the Cromwell government. His election, and this link to Cromwell, suggests strongly that he was a Puritan/Parliamentarian like his father.

The Assembly next elected as its Speaker a man using the name John Smith. Like many of his fellow burgesses, we don't know much about Smith. Some historians think his actual name was Francis Dade, a probable Royalist who first appeared in Virginia in 1651 under an assumed name and then died on his return voyage to England after the Restoration.[337]

Then this Assembly of unknown men passed the third major revision of Virginia's laws, totaling 131 Acts. This omnibus bill, written by a committee picked during the Royalist-leaning December 1656 Assembly (chaired by royalist Francis Willis, with its other members Royalist George Reade, Abraham Wood, a Virginia-bred man of uncertain Parliamentarian leanings, and John Wilcox, for whom we have almost no information),[338] cobbled together the laws of the Commonwealth era with many of the laws from Berkeley's administration.

While making no significant changes to the headright system, the revision continued the Royalist pattern of once again dictating terms to its citizens: new laws said that no tobacco could be planted after July 10th each year, that hides, iron, wool, sheep, and mares could not be exported, and that citizens were required to participate in the pursuit of runaway servants. Reinstituted laws from Berkeley's first administration once again regulated steelyards, doctors, tavern-owners, and millers, required everyone to plant mulberry trees and two acres of corn. Additionally, several harsh new taxes and fees were imposed, including an export tax of two shillings per hogshead of tobacco and the re-institution of the 1645 definition of tithable, taxing black women the same as men,[339] a legal change that could be viewed as an attempt to discourage slave-holding, considering that it taxed slaves more than white servants, and had in the 1640s reduced the importation of slaves.

Once this work was done, governor Mathews and the Council tried to dissolve the House without permission. While we don't know much about the House's political make-up, the council membership seemed to lean in the direction of the Parliamentarians (see table 18). As an example, when the House wished to know the Council's opinion on its desire to outlaw lawyers, Secretary Claiborne wrote back saying

> The governor and Council will consent to this proposition so far as it shall be agreeable to Magna Carta.[340]

The council seemed to be taking a position that opposed such a ban, contrary to the will of the Assembly which eventually passed it into law.

Yet, on April 1st, the day the Council called on the Assembly to dissolve, the Council make-up had shifted to the Royalists. Bennett and Wood were absent, replaced by Royalists Nathaniel Bacon and Francis Willis.

Why these men called for this action is not entirely clear. It is possible they did not like the direction the House was going. What we do know is that on March 27th Councilman Francis Willis had left town in defiance of Governor Mathews' orders, and when he was brought back against his will, he was one of the men who on April 1st convinced the

Table 18-The Council, March/April 1658

Current Governor:
Governor Samuel Mathews, Jr.

Council-as of March 13th, the opening of the House

Royalist:	Neutral:	Puritan:
*West, John	*Dew, Thomas	*Robins, Obedience
Reade, George	*Walker, John	+Bennett, Richard
	Unknown:	Parliamentarian:
	*Bernard, William	*Claiborne, William
	*Perry, Henry	
	*Pettus, Thomas	Uncertain Parliamentarian:
		Wood, Abraham

Council-on April 1st these men call for House to dissolve

Royalist:	Neutral:	Puritan:
*Reade, George	*Walker, John	*Robins, Obedience
+Bacon, Nathaniel, Sr.		
Willis, Francis	Unknown:	Parliamentarian:
	*Bernard, William	*Claiborne, William
	*Perry, Henry	
	*Pettus, Thomas	

Council-as of April 3rd after House refused to dissolve and modifies Council

Royalist:	Neutral:	Puritan:
*Reade, George	*Dew, Thomas	*Bennett, Richard
+West, John	*Walker, John	*Robins, Obedience
Hill, Edward, Sr.	Elliot, Anthony	
Carter, John, Sr.		Parliamentarian:
Horsmanden, Warham	Unknown:	*Claiborne, William
	*Bernard, William	+Wood, Abraham
	*Perry, Henry	
	*Pettus, Thomas	

	Present Slave-holders	Future slave-holder	Raised in Virginia
Council-March 13	5 of 11	5 of 11	1 of 11
Council-April 1	3 of 9	4 of 9	0 of 9
Council-April 3	5 of 15	6 of 15	1 of 15

Council and governor to issue the order calling for the House to dissolve.[341]

Unfortunately for them, the House refused to obey the order. The two governmental bodies now became involved in a week-long power struggle, reminiscent of the political battles between King Charles and the Long Parliament in England in the 1640s. When the smoke cleared, the House had successfully reasserted its power, firmly declaring that the House was

> not dissolvable by any power now extant in Virginia, but the House of Burgesses . .. [and] that we have in ourselves the full power of the election and appointment of all officers in this country.[342]

Francis Willis and Nathaniel Bacon, the apparent instigators of the dissolution order, were removed from the Council, and William Claiborne, who had backed it, was forced to resign as the colony's secretary.343

Was this a victory for Parliamentary government? If it was, it was a Pyrrhic victory. By the last years of the 1650s, the intellectual ideals of the English Civil War had become extremely unpopular, especially in a Royalist colony like Virginia. They represented chaos and endless political infighting, and since no one wanted to repeat the bloodshed and instability of the Civil War, the conflict over dissolving the House implied a return to those unhappy days. Furthermore, this House had passed a massive set of new laws, making life more difficult and expensive for everyone, while doing nothing to address the headright system that was the cause of most of the problems that still plagued the colony.

The Royalists would win heavily in the House elections of 1659 (see table 19). Both Nathaniel Bacon and Francis Willis would be re-elected to the House and immediately returned to the council. Also re-elected to the House were Joseph Crowshaw, Edward Hill, Sr., Moore Fantleroy, Augustine Warner, Sr, and James Pyland (who only seven years before had been expelled from the Assembly for speaking out against the Commonwealth government). Freshmen Royalists included Henry Corbin and Thomas Fowke, both of whom had close ties with the Royalist cause. Corbin had helped Charles II escape England after the defeat of his father's forces in 1649,344 while Fowke had been a Royalist refugee in 1650 with his brother Gerald.345

Faced with only scattered opposition, this newly empowered Royalist majority, which further renewed its dominance in the 1660 elections, now moved to install its political and personal agenda.

Table 19-The Grand Assembly, March 1659

Current Governor:
Governor Samuel Mathews, Jr.

Council-only committee assignments known

Royalist:	Neutral:	Parliamentarian:
+Bacon, Nathaniel, Sr.	*Dew, Thomas	*Claiborne, William
*Carter, John, Sr.	***Walker, John**	
+Willis, Francis	Elliot, Anthony	
*Horsmanden, Warham		
	Unknown:	
	*Bernard, William	

House

Royalist:	Unknown:	Puritan:
+*Crowshaw, Joseph*	*Blackey, William	*Bond, John
+Fantleroy, Moore	*Carter, Edward	
+Hill, Edward, Sr.-SPEAKER	**Caufield, William*	Parliamentarian:
+Pyland, James	*Hay, William	+Hatcher, William
+*Warner, Augustine, Sr.*	*Warren, John	
Corbin, Henry	*Webb, Giles	Quaker:
Fowke, Thomas	+Batte, William	**Bushrod, Thomas**
	+Jordan, George	
Uncertain Royalist:	+Paine, Florentine	Uncertain Parliamentarian:
***Mason, Lemuell*	+Pitt, Robert	+Godwin, Thomas
*Sidney, John	+Warren, Thomas	
+*Chiles, Walter, Jr.*	**Collclough, George**	Uncertain Puritan:
Stringer, John	Edlowe, Mathew	+Jones, William
	English, John	+Whittaker, William
	Harlow, John	
	Weir, John	

	Present Slave-holders	Including future slave-holders	Raised in Virginia	Incumbents	Including those who served previously
Council	1 of 9	3 of 9	0 of 9		
House	5 of 37	12 of 37	5 of 37	11 of 37	28 of 37

Legend for all tables:
*-incumbent
+-served previously
Bold names have purchased slaves or have claimed headrights on blacks.
Underlined names will eventually buy slaves or claim headrights on blacks.
Italicized names either were born in Virginia or immigrated as minors.

Source for all tables: See Appendix B

A Royalist Resurgence

8. The Consequence of Family

When the newly elected Royalist dominated Assembly took power in March 1659, it chose as its speaker Edward Hill, Sr. This was the fourth time that Edward Hill had been picked by the Assembly to be its speaker, in a legislative service spanning almost twenty years.

Though Hill was tied closely to Berkeley and the Royalists, he seemed well liked by everyone. He served in every Assembly from 1643 to 1652, with the exception of 1646. In that year Hill became governor of Maryland, commissioned by Catholic Leonard Calvert to quell a Parliamentarian rebellion against Lord Baltimore's rule. The revolt had deteriorated into violence and robbery, with mercenary soldiers plundering and terrorizing the local settlers.[346] Hill also received the endorsement of Anglican William Berkeley and the Assembly, commissioning him to go to Maryland and bring back several Virginians who had fled there without permission.[347]

Hill took over as governor, put down the mercenaries, and called an assembly to restore order. At this point, the Maryland Catholics decided to refuse his Anglican governorship. Unable to hold Maryland for the Royalists, Hill graciously returned to Virginia.

During the Commonwealth era, things did not go as well for Hill. He served only twice as a burgess, though in 1654 the Assembly again chose him as speaker as well as defending him against William Hatcher's accusations that Hill was "an atheist and blasphemer."[348] In 1656, however, the Royalist-dominated Assembly suspended him from office for leading a failed attack against the Indians.[349]

From what we can surmise, Edward Hill, Sr. tried to live his life like an old-fashioned Royalist from the era prior to the Civil War. He never owned slaves, nor obtained land by claiming headrights on them.[350] Though he desired power and assumed that such power could be freely wielded by him as he saw fit, the constraints of honor, noblesse oblige, mutual respect for the rights of others, and the needs of society seem to have always dictated his actions. When he took over in Maryland, he understood his responsibilities very clearly, using his power to stop the mercenary soldiers from plundering and to restore law and order. Though he probably tried to impose Anglican rule on this Catholic colony, when this seemed impossible he retreated to Virginia rather than cause more bloodshed, demanding only that the Marylanders pay him the salary he was due.[351]

In all, it seems that Edward Hill, Sr. was a very respected man. When the Royalists regained power in 1659, he was re-elected as the Charles City County burgess for the ninth time, and the Assembly immediately chose him as their speaker for the fourth time.

It would be the last time. In 1660, he would be appointed to the council by the restored Governor Berkeley, thereby barring him from House office. Soon after, sometime before 1663, Edward Hill, Sr. would die, leaving to his son, Edward Jr., an estate of over 6000 acres.[352] How Edward Jr. would use that wealth would contrast starkly with his father's life, and exemplify the effect Virginian culture was having on its children.

It was now more than fifty years since George Percy and that first band of adventurers had arrived on the shores of Virginia. Practically no one still lived from then, though a few had left descendants. On the day that Edward Hill, Sr. died, Virginia was something vastly different from what George Percy had imagined in 1607 it would become.

Though the population of the colony had now blossomed to over 25,000 people,[353] there were still no towns or villages. Jamestown, Virginia's largest "city," comprised no more than a single street with fewer than two dozen buildings. The entire population of the colony, which now included between 5,000 to 10,000 indentured servants and approximately 900 blacks (both free and enslaved),[354] still lived on scattered farms and plantations, now extending as far north as the Potomac River, and west and south along all the rivers almost as far as the fall line (where the land started its first rise into the mountains) and the border with North Carolina.[355] On these farms men now grew several million pounds of tobacco a year.[356]

In those fifty years the quality of life had not improved significantly. Most colonists still lived in ramshackle one room huts, made of simple plaster and mud with thatched roofs. Some had moved up to wood frame cottages with a brick chimney. And only since Berkeley's arrival had a few been able to build the large mansions of brick, stone, and wood, such as Berkeley's Green Spring, that in later centuries would become emblematic of southern culture.[357]

In those fifty years, immigration to the colony under Edwin Sandys' headright system had continued unabated, averaging somewhere between 500 to 1,500 new settlers a year, of which from one-half to two-thirds were indentured servants.[358] Despite this high immigration rate, the colony's population had not grown as fast as it should, probably because, first, most immigrants were still men, at a ratio of approximately three to one,[359] and second, the continuing high rate of death for all new settlers. Even after fifty years, the "seasoning" of new arrivals was still expected, at least "one in nine [to] die imported," as Beauchamp Plantagenet noted in 1648.[360]

Because of this high death rate, more than one third of all children could still expect to lose at least one parent before their thirteenth birthday, and be orphaned by the time they were eighteen. In Middlesex County alone, for example, sixty percent of all children born between 1650 and 1689 would lose one parent by the age of thirteen.[361]

After forty-plus years of boom, the price of tobacco was about to begin a twenty year downward spiral, showing only a few momentary and very temporary recoveries.[362] Furthermore, the methods of production had reached maximum efficiency, and a planter could no longer maintain his profit margin merely by increasing the production of his laborers.[363] The squeeze was on for this single crop economy, and Virginia planters, rather than diversifying, would increasingly propose a variety of unwieldy political schemes in their attempts to re-establish the boom in tobacco.

In fact, much of the legislation passed in Virginia during this century were shallow attempts to deal with Virginia's single crop economy. The laws requiring the planting of mulberry trees during Governor Digges' term were his attempt to encourage the silk industry. The desire to form towns resulted in repeated laws requiring market days in specific places as well as encouraging the formation of town corporations. The law restricting tobacco planting after July 10th, and charging two shillings per hogshead of tobacco were intended to limit tobacco production, both to encourage diversification as well as decrease tobacco supplies and thereby increase its price.[364]

Unfortunately, these superficial laws did nothing to reform the headright system responsible for shaping the immigration patterns of the colony and its single crop economy. Immigrants were still mostly single, uneducated men forced to work for years as unpaid laborers for wealthy owners of large plantations.

And the majority of Virginia's wealthy plantation owners continued to focus solely on their own self-interest, refusing to be weaned from tobacco, especially if such weaning meant that their potential earnings might be even slightly reduced. Everyone wanted to make a killing, and tobacco was still the way to do it. Let someone else, like Governors Berkeley and Digges with their government stipends, lose money and cultivate silkworms.

The result was that none of these laws were able to achieve what they were designed to do. Either everyone ignored them, they were used by some for corrupt advantage and pay-off, or the legislature was forced to continually tinker with them, repealing, reviving, revising, repealing, only to revive them again.

When the Royalist dominated Assembly met in 1659, it immediately proceeded to do exactly that. The two shilling tax on tobacco hogsheads was repealed. So was the prohibition on the export of hides and iron, as well as the requirement to plant mulberry trees. At the same time, this Assembly re-instituted the licensing of tavern-keepers, which had been passed in 1646 and repealed in 1652. And it once again changed the regulations on attorneys, which almost every legislature had tinkered with since 1643.[365] And as before, none of these changes did anything to actually change Virginia's fundamental problems. Single men continued to immigrate to the colony, living as indentured servants on scattered isolated farms growing tobacco, their 50-acre headright going not to them but to the planter who paid their transport.

No historians dispute the terrible disruption that took place in Virginian family life because of these conditions.[366] As historian Lorena Walsh has said, "Immigrants to the Chesapeake experienced an immediate and profound disruption in the patterns of family life, first in the selection of their mates, then in family politics, and finally in relationships with their children."[367] By not immigrating to the colony as family groups, as the immigrants did in New England and Pennsylvania,[368] and then being faced with high mortality, Virginians found it difficult if not impossible to develop a stable family life for their children. The long term ramifications to each succeeding generation could only be negative. As noted by historian Daniel Blake Smith in 1987, "In the absence of parental influence and kin contact, young people matured under little of the close supervision and control Englishmen clearly desired in their families."[369]

First of all, the unusually high death rate in Virginia clearly taught children that life was fleeting, that to expend much emotion on others would be a waste of energy. As William Byrd mused, "Before I was ten years old . . . I looked upon this life here as but going to an inn, no permanent being by God's will . . . therefore am always prepared for my certain dissolution, which can't be persuaded to prolong by a wish."[370] This cold approach to life was applied to family, friends, and acquaintances. As caring and hospitable southern colonists might have been, that feeling could only be superficial, knowing that in a moment family, friend, or acquaintance could be dead.

This terrible situation was further worsened by the lack of educational or religious institutions for providing support to Virginia's struggling families and education for their children.[371] Until 1642, Virginia had no schools, as noted in 1662 by an anonymous Anglican minister who had lived many years in Virginia,

> Their almost general want of schools for the education of their children . . .. renders a very numerous
>
> generation of Christian children born in Virginia . . .. unserviceable for any great employments either in church or state.[372]

While the wealthiest Virginians could send their children elsewhere to be educated (Puritan Richard Bennett sent his son to Harvard in Boston[373] and Royalist Richard Lee, Sr. sent his son to Oxford, England[374]), most Virginians could not afford such schooling for their children. The result was a pitiful literacy rate, especially among the poorer members of society, almost all of whom could not read. This was far different from either England or New England, which were both experiencing rising literacy among all levels of society, even the poor.[375]

Nor was this lack of schools helped by the colony's continuing and desperate lack of Anglican clergyman, as noted by that same minister: "Many parishes as yet want both churches and glebes [land owned by the parish for producing church income], and I think not above a fifth part of them are supplied with ministers."[376] Furthermore, whenever anyone else arrived who could have possibly offered some form of ethical instruction, such as the Puritans in the forties and the Quakers in the late fifties, the Royalists who controlled the courts and legislature took aggressive action to force them out. And since the clergymen already in Virginia were under the direct control of the secular government, they could never question the actions of that government.[377] By 1659, Virginia had been a colony without any system of independent ethical education for more than fifty years.

Children are essentially selfish creatures. A fundamental requirement of child-rearing is to teach children how to live together peaceably and to treat others with respect and decency.[378] While the first generation of Englishmen, even if they were hard-core Royalists, might have found the outright ownership of another odious, their children, raised poorly with little moral supervision in a society dominated by Royalist ideas of caste and rank and filled with unpaid indentured labor, would have had fewer scruples about such an idea.

Social research has repeatedly demonstrated what common sense has always known: there are terrible consequences to children growing up in broken homes within communities ill-equipped to compensate for parental failures.[379] While the loss of a parent might not prevent a child from learning the subtle difference between right and wrong, when that loss is repeated again and again, is accompanied by little or no education, and is placed in a social context that encourages wrong behavior, the consequences on the child as well as society are profound. Crime and violence increase, and the ethical rules of society break down, no longer respected nor obeyed by its citizens. In seventeenth century Virginia we can see an aspect of this process by noting that the colony's rate of illegitimacy was two to three times higher than England's, despite the small number of women in the colony.[380] The ethics of the parents had not been passed to the children, who now grew up knowing of such ideas only through the distorted view of a foggy translucent window.

As sociologist and Democratic politician Daniel Patrick Moynihan noted in 1986, "a community that allows a large number of young men to grow up in broken families . . . never acquiring any stable rational expectations about the future – that community asks for and gets chaos."[381] Raised within an insecure family structure, their parents dying frequently and without warning, within a society with no schools or churches, the first generation of children born in Virginia had no one to teach them that Virginia's labor system of forced unpaid workers was either unusual, temporary, or extreme. In fact, Virginia's children instead saw men like Edmund Scarburgh demonstrate how the aggressive use of this labor system was the surest path to success. And similar to the modern cycle of abused children becoming abusing parents, this cycle accelerated from generation to generation, until the Royalist idea of a justly ruled society administered by an educated aristocracy became a society of castes defined by racial bigotry. Ideas of law and mutual respect, of honor, of religion and equality before God and the law, became increasingly less important. Since looking out for "No. 1" had always been the primary goal of Virginian society,[382] and since each generation was taught less and less about the ethical limitations that must be placed on such economic goals, later generations were more willing to do almost anything to anyone to obtain greater wealth.

Edward Hill's son illustrates these sad changes, being a very different man from his father. In 1646 Edward Jr. married Tabitha Scarburgh and so inherited the many slaves her father had brought to the colony. Joining the Charles City county courts as a justice in 1659, by the 1670's he had become quite powerful there.[383] Unlike his father, however, he could not get elected to the Assembly, despite elections in Charles City county to replace deceased burgesses in 1665 and 1672. He was finally elected to the Assembly in 1676 during the height of Bacon's Rebellion, and the Assembly immediately expelled him from all offices for using his various positions of "public magistrates, officers or ministers for [his] private ends."[384]

Considered a loyalist to Berkeley, he was accused of seriously abusing his position as Charles City justice in the years prior to the Rebellion. The accusations asserted that he had imposed heavy taxes on the community, had used these public monies for his own personal use, had impressed men into military service without right or cause, and had extorted money from others in exchange for pardons already issued by the Crown's commission.[385]

In his lengthy defense statement, Edward Hill Jr. described himself as "a naked, unlearned, and unskilled Virginian, born and bred," who had "not the dress and learning of schools, nor have I the skill to clothe vice like virtue."[386] From his own words we can see the consequence of Virginia's shattered social fabric.

Nor is the Hill family the only example of this. Obedience Robins was known on the Eastern Shore as a religious man, a possible Puritan dissenter, and a rival to Edmund Scarburgh. Involved with parish duties throughout his life, he helped bring minister Thomas Teakle to Virginia in 1652.[387] Though his estate exceeded 4,000 acres, he had never claimed headrights on black slaves, and owned none when he died in 1663.[388] During his lifetime he served in the Assembly twice, the last time during Richard Bennett's first Assembly in April 1652. On March 12, 1655, just before Bennett stepped down as governor, Robins was appointed to the council, on which he remained until his death in 1662.[389]

In 1666, four years after his death, his son John patented 1,150 acres on the headrights of twenty-three persons, eight of whom were black slaves.[390] Furthermore, Obedience Robin's daughter had married John Savage, a man born and bred in Virginia who, as I have already noted, imported twenty-five black slaves into the colony only two years after Robin's death.[391]

William Ball, Sr. had lost his English estates because of his support of Charles in the war. Left with only a single ship, he fled to Virginia in 1650 with his wife and three children to become a tobacco merchant. When he died in 1680 he owned no slaves, nor is there any record of him ever obtaining headrights with the importation of slaves. His oldest son, William Jr., who had come to Virginia as a boy of nine, would be elected burgess in 1668 after the Restoration, and when he died in 1694 his estate included 23 black slaves.[392]

William Edwards I was born in England, and in the late 1630s worked for Adam Thorowgood as a land manager until Edwards' death in 1640. There is no record of him ever owning slaves. His son, William II, began patenting land in 1643, accumulating 3,370 acres from land purchases and claimed headrights, none of which included black slaves. He served in the Assembly as a burgess for Surry County during the Commonwealth years, the last time in 1658.[393]

William Edwards III was born in Virginia. When he began patenting land soon after his father's death in 1673 he quickly claimed 50 acres for the transportation of one black slave. Over the next twenty years he would claim headrights on at least 13 black slaves, building an estate of more than 6000 acres.[394]

Rice Hoe was born in 1599 in England and immigrated to Virginia in 1618.[395] In 1635 he returned to England and was certified to conform to "orders and discipline of the Church of England."[396] With this in hand he came back to Virginia with his wife and began building an estate.[397] In 1641 he and others petitioned for the right to explore west and south of the colony, and in 1643 the Assembly granted this right.[398]

The Consequence of Family

By 1643 he had accumulated almost 3000 acres in Charles City County, and was elected burgess for that county three straight times from 1645 to 1646.

Once, in 1643, he claimed four black slaves as headrights, of whom one was a slave named John. When Hoe was approaching death in 1653, he wrote out his will, which required "his heirs and assigns to set my Negro John free at the expiration of eleven years."[399]

This is most of what we know of Rice Hoe. From this slim data, the man apparently had some religious training in the Anglican church and was therefore probably aligned with the Royalist party. Additionally, he seemed to have mixed feelings about the ownership of slaves, and to have had the courage to explore and settle the frontier wilderness of Virginia.

Eleven years passed, and in 1665, Jack (as he was now called) had to go to court to sue for his freedom because Hoe's son, Rice Jr., refused to release him. As the court stated, "And now young Rice Hoe claims this your petitioner [Jack] for his lifetime, not withstanding his father's paper under his hand to the contrary." The son saw nothing wrong with enslaving another human being, despite his own father's written desires.[400]

William Presley, Sr. was a burgess in November 1647. Other than the 1,250 acres he patented in 1649 and 1650 for himself and his family from the headrights of twenty-five British immigrants, not much else is known about him.[401] He does not appear, however, to have ever owned slaves.

His son, however, is more well known, especially for his pithy and sometimes witty commentary during Bacon's Rebellion. "The Governor would have hanged half the country, if they had let him alone."[402] A member of the Assembly of 1651-2 that declared its defiance of Parliament, this Royalist ally of Berkeley returned to the Assembly after the Restoration, serving for the entire Long Assembly when the first slave codes were written and passed. And as a slave-holder himself, William Jr. was the man who refused to free mulatto Elizabeth Key in 1656.

Peter Montague owned no slaves when he died in 1655. He had lived in Virginia since 1621, when he had come to the colony as an employee of Samuel Mathews, Sr.[403] Over time he had been able to assemble a small estate of about 800 acres,[404] and subsequently served in the Commonwealth Assemblies in November 1652, July 1653, and March, 1658. A summary of what we know of him would indicate an uncertain alliance with Parliamentarian ideas. His son, however, had no problem owning slaves, and when he died in 1695 he was quite willing to bequeath these human beings to his heirs.[405]

Henry Woodhouse, whose father had been Governor of Bermuda, was burgess for Lower Norfolk County in 1647 and 1652. He had come to the colony in 1637 with his wife and daughter, owned an estate of just under a thousand acres, was known as a lawyer, a planter, a Justice, and a member of the county vestry.[406] When he died in 1655, he owned no slaves.[407] His son, however, owned several slaves, and when he died in 1686 his will ordered that they "be divided up" among his heirs.[408]

This is only a sampling. Though these second generation Virginian slave-owners never comprised more than a small percentage of Virginia's total population before 1680, their influence was strongly felt in the legislature. From 1658 to 1662 the percentage of Virginia-bred burgesses in the Assembly rose dramatically (see figure 8.1), reaching almost 35 percent of the House. Nor would these numbers ever drop significantly again. Furthermore, figures 8.2 and 8.3 show a simultaneous and precipitous rise in slave-owners in the Assembly. By 1666 more than half the House would be made up of men who owned slaves or had claimed headrights on them. In the Council slave-holding was even more prevalent, with almost every Council session comprised of a large majority of slave-holders.

I do not want to suggest that the majority of the immigrants, born and raised in Britain, avoided slave-ownership solely because of the ethical teachings of their parents in England. Clearly, for many economics made the decision: they could not afford a slave's higher

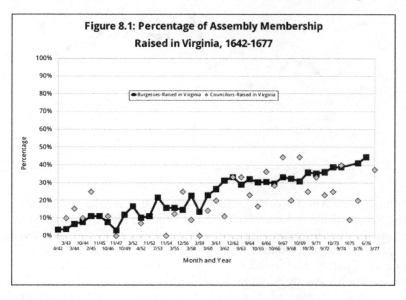

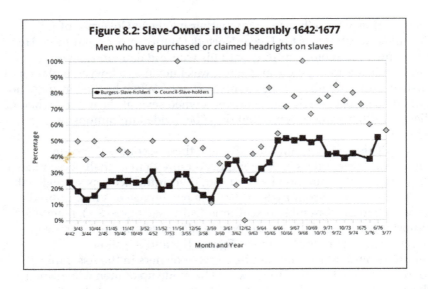

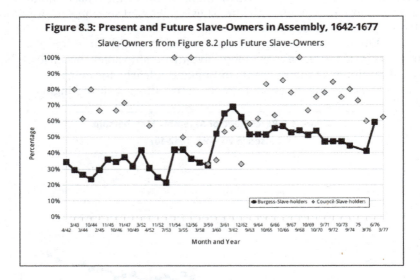

purchase price. Yet, for others who could afford that price, including many Royalists like Edward Hill, Sr. and William Ball, Sr., the idea of treating another human being as chattel must have seemed, at the very least, distasteful. Raised in an English society that prided itself in its concepts of justice and individual rights, they could not so quickly abandon such ideas by denying these rights to others.

    The first generation of native Virginians, however, brutally cold-blooded on issues of money and poorly educated in matters of ethics, had no such ideas to abandon. As larger numbers of these men began to appear in the legislature as well as society in the 1660s and 1670s, their influence was soon felt in legislation. Though most burgesses might have sincerely wished to change the colony's unhealthy social climate, fewer and fewer were willing to make the personal sacrifices necessary to actually bring about these changes. As the years passed, they would eventually give up with superficial legislative tinkering and actually legalize the worst aspects of that unhealthy social climate for their own benefit.

On September 3, 1658, Oliver Cromwell died and, though his son Richard had been declared successor, there were serious doubts about what would happen next in England, rumors flying that Charles II would soon be restored to the throne.[409] The 1659 Assembly, strongly Royalist and unsure of its ground, stalled, drawing up a committee to "address to his Highness [Richard Cromwell] for confirmation of their present privileges."[410] Up to now the Commonwealth government in England had left the colonists pretty much alone, demanding only that Virginia's government be by legislature and show no allegiance to royalty or church government. Since the Royalists now controlled that legislature, they did not want the new Protector to poke too deeply into how they were running their affairs.

The burgesses completed their tinkering in March 1659, having passed twenty-five new laws or amendments, and went home, still not knowing what would happen in England. In May of 1659 Richard Cromwell abdicated, leaving England without a ruler, Charles II still in exile on the continent. Then, in January 1660 Governor Samuel Mathews Jr. suddenly died, and Virginia was also bereft of its political leader.

In this uncertain situation, the colonists elected a new Assembly in March 1660, and these elections clearly reflected the sentiments in both England and Virginia, the desire by the majority of eligible voters for strong, authoritative leadership. As we can see from table 20, the Royalists won handily, and in fact, the March 1660 Assembly was probably the most heavily dominated Royalist Assembly since before the surrender in 1652. Besides the re-election of Royalists Bacon, Reade, Moryson, Warner, Carter, Crowshaw, and Willis, we see the re-appearance of Henry Browne, now willing to come out of retirement with the pending restoration of the king, Richard Lee, who had been out of office since the day he negotiated the surrender in 1652, and Edmund Scarburgh, of whom we have already discussed at length. Also elected for the first time was Royalist refugee Manwaring Hammond, who would

Table 20-The Restoration Assembly, 1660

Current Governor:
Governor William Berkeley

Council-March 1660

Royalist:	Neutral:	Puritan:
+Hill, Edward, Sr.	*Dew, Thomas	+Bennett, Richard
+Reade, George	*Walker, John	+Robins, Obedience
Moryson, Francis		
Warner, Augustine, Sr.	Unknown:	Parliamentarian:
	*Bernard, William	*Claiborne, William
	+Perry, Henry	
	+Pettus, Thomas	Uncertain Parliamentarian:
	Carter, Edward	+*Wood, Abraham*

Council-October 1660

Royalist:	Neutral:	Puritan:
*Hill, Edward, Sr.	*Dew, Thomas	*Robins, Obedience
*Moryson, Francis	*Walker, John	
*Reade, George		Uncertain Parliamentarian:
Warner, Augustine, Sr.	Unknown:	*Wood, Abraham*
+Browne, Henry	*Bernard, William	
Lee, Richard, Sr.	*Carter, Edward	
Molesworth, Guy		

House-March 1660

Royalist:	Unknown:	Puritan:
*Bacon, Nathaniel, Sr.	*Caufield, William	*Bond, John
*Carter, John, Sr.	*Pitt, Robert	
*Corbin, Henry	*Warren, John	Quaker:
*Crowshaw, Joseph	*Webb, Giles	*Bushrod, Thomas
*Fantleroy, Moore	*Weir, John	
*Fowke, Thomas	+Hamelyn, Stephen	Uncertain Puritan:
*Willis, Francis	+*Powell, John*	+Waters, William
+Scarburgh, Edmund	+Smith, Nicholas	Bland, Theoderick-SPEAKER
Hammond, Manwaring	+Sparrow, Charles	
Jenings, Peter	+*Worleich, William*	Uncertain Parliamentarian:
	+Wynne, Robert	+Ellyson, Robert
Uncertain Royalist:	Baldry, Robert	+Knight, Peter
Chiles, Walter, Jr.	Browne, William	
Mason, Lemuel	Cary, Miles	
*Sidney, John	Catchmaie, George	
*Stringer, John	Curtis, John	
+Abrahall, Robert	Denson, William	
+Ashton, Peter	Griffith, Edward	
+*Calithrop, Christopher*	F*farrer, William*	
+Soane, Henry	Ford, Richard	
Claiborne, William, Jr.	Hill, [Nicholas]	
Cant, David	Morley, William	

	Present Slave-holders	Including future slave-holders	Raised in Virginia	Incumbents	Including those who served previously
Council-March	5 of 14	5 of 14	2 of 14		
Council-October	5 of 13	5 of 13	2 of 13		
House	12 of 48	25 of 48	11 of 48	18 of 48	32 of 48

Legend for all tables:
*-incumbent
+-served previously
Bold names have purchased slaves or have claimed headrights on blacks.
Underlined names will eventually buy slaves or claim headrights on blacks.
Italicized names either were born in Virginia or immigrated as minors.

Source for all tables: See Appendix B

soon be sent to England with Guy Molesworth to procure for Berkeley and the Royalists the king's pardon for submitting to Parliamentarian rule, as well as Peter Jenings, who had served in the king's army in the civil war and in 1670 would join the council and become Virginia's attorney general.[411]

The members of the new Assembly immediately recognized that the death of Governor Mathews put them in a difficult position. According to the Parliamentary rules of the moment, they had to choose a replacement acceptable to the Commonwealth. Yet as Royalists they could not pick a Parliamentarian or religious dissenter, especially because such a choice might also offend Charles II, who was quite possibly going to be the future ruler of England. Nor was anyone among them willing to step forward and put his head in the noose. In the end, the only candidate who seemed qualified, willing, and electable was fifty-five year old William Berkeley, still living peaceably in his home at Green Spring.

Negotiations were begun. Berkeley wanted to insure that his position as Governor would be stronger than Mathews had been, that he would not be at the beck and call of the Assembly, and that he would be able to tell them what to do. He also wanted to protect himself against accusations of disloyalty to Charles II for participating in a Commonwealth government. "I . . . have pressing fears . . . and am seriously afraid to offend him [Charles], who by all Englishmen is confessed to be in a natural politic capacity of being a Supreme power."[412]

A deal was struck, and Berkeley agreed to return as Governor. Officially, the deal seemed to reaffirm many of the Parliamentary powers of the Assembly, such as the ability to approve Berkeley's appointments to the Council and Secretary, the freedom not to be dissolved without its approval, and the guarantee of House elections every two years.

In truth, Berkeley was their man, and they would have done whatever he wanted. When Berkeley insisted, they gave him "free liberty of treating with them." More importantly, the Royalists who dominated the Assembly were of a like mind with Berkeley, knowing he would initiate policies that they themselves would endorse. These Restoration Royalists – reflecting the generational decay of Virginia's ethical culture – wanted to protect and guarantee their financial and political power, were unscrupulous about how that should be done, and knew William Berkeley would gladly work with them to do it.

So, on March 22, 1660, after several days of careful maneuvering, William Berkeley, cavalier, poet, soldier, and former Governor of Virginia, returned to Jamestown, taking his place once again as Governor of the colony.[413]

The Consequence of Family

9. The Restoration Assembly

Sir William Berkeley had now lived in Virginia almost twenty years. He had built Green Spring, fought a war against the Indians, gotten married, lost a wife (whose name is lost in history), grown tobacco and silk, and essentially made a commitment to the New World. Unlike almost all colonial governors ever appointed by the Crown, Berkeley had never treated the job as a temporary assignment in a strange and foreign land. Once he came to Virginia, he made it his home, and he devoted his life to building the colony for himself, his country, and his king.

When he wrote his play *A Lost Lady* in 1637, his hero, having discovered the murderers of his lady love, vows horrible revenge.

> I will prosecute till I have made
> All that were guilty of my loss of peace,
> Wash their impiety in their guilty blood,
> All places where I meet them shall be altars
> On which I'll sacrifice the murderers
> To appease the spirit of my injured Mistress.[414]

Though Berkeley was strongly committed to honor and justice within the context of his Royalist beliefs, it appears that he considered loyalty as its highest value. While the beheading of King Charles in 1649 must have seemed to Berkeley a terrible betrayal, his own removal from office in 1652 must have struck him as a betrayal several magnitudes worse, especially considering his commitment to Virginia and his dedication to helping the colony grow and prosper. After eight years of enforced exile from government, his beloved colony ruled by the confused anarchy of Parliamentarian ideals, Berkeley now returned to power with a vindictiveness quite unlike his more careful, politically astute maneuverings of the 1640's.

In March 1660 Berkeley did not yet know if the king would be returning to power in England. Yet, he and the now Royalist-dominated Assembly (see table 20) wasted little time in re-establishing Berkeley's brand of Royalist control over Virginia. Immediately they repealed the House's right to approve Council appointments, enacted in 1659 during Mathews' administration.[415] Also repealed was the requirement that House elections be held every two years.[416] The Assembly also outlawed dissent or criticism of "the present government," and banned the Quakers from the colony, outlawing Quaker gatherings or the publication of their ideas. Furthermore, William Claiborne, member of Parliament's commission in 1652, was removed as Secretary of State.[417]

Then Berkeley and the Assembly did an immediate and cursory review of all the laws passed during the Commonwealth period. This review, quickly and sloppily done, first voided all previous laws and then only haphazardly and inconsistently confirmed some or repealed others. As an example, while confirming the fourteenth act of 1658 forbidding marriage by servants without permission, acts fifteen and sixteen on the hiring of runaway servants and their capture were both forgotten. Other basic issues, such as the appointment and term for sheriffs and undersheriffs (act twenty, March 1658), probate administration by the county courts (act thirty, March 1658), and the definition of what is tithable (act forty-six, March 1658), all slipped through the cracks, voided but not redefined in any way.[418]

What this review did accomplish, however, was establish beyond a shadow of a doubt that Berkeley, in cooperation with his Royalist allies who controlled the Assembly, was now in charge. To emphasize the point, the two shilling tax on tobacco exports was renewed, and the appointment of the tax collectors was put directly under Berkeley's control. Other regulations, such as the licensing of tavern-owners and the establishment of a court of admiralty, were also renewed and given to the Governor and Council to administer.

Furthermore, the blatant self-interest and pay-offs that had begun to appear during the Commonwealth period were now adopted with a vengeance by the Royalists, both those British-born as well as raised in Virginia. And for Berkeley himself, whatever code of honor he had once held on government rule seemed to dissolve in a desire to get his own back after eight years of exile. Berkeley was paid 700 pounds sterling out of the two shilling tax, as well as 50,000 pounds of tobacco from government funds. Furthermore, all customs from Dutch trade as well as all port charges were assigned directly to him.[419]

This was only the beginning. In May 1660 King Charles II was finally restored to the throne of England, and by the end of July a new commission appointing Berkeley Royal Governor of Virginia had been issued.[420] When the news arrived in Virginia in September, a good number of colonists in York county celebrated with fireworks, drinking, and the firing of guns and cannons.[421] Berkeley immediately re-called the Assembly (table 20), and in October they met in Jamestown to continue their work, knowing now that with the king back in power there was no longer any doubt about what direction they could take Virginia. The first act of October 1660 stated that

> Whereas by reason of the summons directed to the particular members of this Assembly convened

> *without a new election,* some doubt and scruples may seem to arise whether their power be legal or not, be it enacted and declared by the Governor, Council, and burgesses that . . . the authority of this present Grand Assembly is . . . warrantable and legal . . . and who ever shall presume to question the power thereof shall be adjudged a seditious person, and be liable to such censure as the law in such case doth inflict [italics added].422

"Without a new election." Since Berkeley's arrival in the colony in 1642, elections for the House of Burgesses had been held almost every single year. Now, having returned without holding new elections, Berkeley and this Restoration Assembly seemed to feel the need to underline the legitimacy of their continuing rule. This precedent, that elections were not necessary each time the Assembly reconvened, would become the rule for the next sixteen years.

The review of the laws went on, as sloppily and as haphazardly as before. All references to the Protectorate were removed, replaced with the "king's Majesty." The two year term for the Governor (act one, 1659) was repealed. The Governor was given sole power to issue passes for leaving the colony.423

The Assembly continued the empty practice of tinkering with the laws to regulate business. Blacksmiths were now regulated; the export of hides, iron and wool, repealed in March, was renewed; the price freeze on wines and liquors, voided in March, was also renewed.424

Furthermore, the use of government for personal pay-offs continued. Berkeley's salary was doubled. A tax of one bushel of corn per person was to be paid directly to him. Large amounts of tobacco were paid to Royalists Manwaring Hammond, Guy Molesworth, Francis Moryson, and Henry Woodhouse. Hammond was also exempted from taxes for the year. Jenkin Price, who had sheltered many Royalist refugees when they had been abandoned on the Eastern Shore in 1650, was rewarded for that service.425

And act twenty-two created a one-time tax of 450,000 pounds of tobacco "to be paid to several persons authorized to receive the same." The list of men to receive this generous pay-off is not completely known, but we do know that Lower Norfolk county alone reserved its share of this tax to Berkeley, William Claiborne, and Royalist Warham Horsmanden.426

The Restoration Assembly

Table 21-The Restoration Assembly, 1661

Current Governor:
Governor William Berkeley

Council-March 1661

Royalist:	Neutral:	Puritan:
*Browne, Henry	*Walker, John	*Bennett, Richard
*Hill, Edward, Sr.		
*Moryson, Francis	Unknown:	Uncertain Parliamentarian:
*Reade, George	*Bernard, William	*_Wood, Abraham_
*_Warner, Augustine, Sr._	*Carter, Edward	Swann, Thomas
+Carter, John, Sr.	*Perry, Henry	
Ludwell, Thomas	*Pettus, Thomas	

Council-June 1661

Royalist:	Unknown:	Uncertain Parliamentarian:
*Browne, Henry	*Bernard, William	*Swann, Thomas
*Hill, Edward, Sr.	*Pettus, Thomas	*_Wood, Abraham_
*_Ludwell, Thomas_		
*Moryson, Francis		
*Reade, George		
+Bacon, Nathaniel, Sr.		
Hammond, Manwaring		

House-March 1661-Held over without general elections

Royalist:	Unknown:	Puritan:
*Fantleroy, Moore	*Browne, William	*Bond, John-EXPELLED
*Fowke, Thomas	*Cary, Miles	
*_Scarburgh, Edmund_	*Catchmaie, George	Uncertain Puritan:
*Willis, Francis	*Caufield, William	*Bland, Theoderick-SPEAKER1
+_Presley, William, Jr._	*Denson, William	*_Waters, William_
+Travers, Raleigh	*_Ffarrer, William_	
	*Hamelyn, Stephen	Uncertain Parliamentarian:
Uncertain Royalist:	*Hill, [Richard]	*Ellyson, Robert
*Abrahall, Robert	*Pitt, Robert	*Knight, Peter
*Cant, David	*_Powell, John_	
*Chiles, Walter, Jr.	*Smith, Nicholas	
*_Claiborne, William, Jr._	*Warren, John	
*Soane, Henry-SPEAKER1	*Webb, Giles	
*Stringer, John	*Weir, John	
	*_Worleich, William_	
	*Wynne, Robert-SPEAKER2	
	Knowles, John	

	Present Slave-holders	Including future slave-holders	Raised in Virginia	Incumbents	Including those who served previously
Council-March	6 of 15	8 of 15	3 of 15		
Council-June	2 of 11	4 of 11	3 of 11		
House	12 of 34	22 of 34	9 of 34	31 of 34	33 of 34

Legend for all tables:
*-incumbent
+-served previously
Bold names have purchased slaves or have claimed headrights on blacks.
<u>Underlined names will eventually buy slaves or claim headrights on blacks.</u>
Italicized names either were born in Virginia or immigrated as minors.

Source for all tables: See Appendix B

The Assembly adjourned for the winter, reconvening for a third time without elections in March 1661 (table 21). Recognizing that certain of the king's policies were causing problems in Virginia (such the Navigation Acts, which required the colonists to sell their goods to England only, and the Northern Neck patent, which had literally given an already settled portion of the colony to several of the king's closest supporters, putting all land rights there in jeopardy), the Assembly asked Berkeley to go to England and lobby the king in their interest. The Assembly reserved 200,000 pounds of tobacco out of government funds to pay for the trip, and Berkeley agreed to go.[427]

Before leaving, however, Berkeley and the Assembly passed another forty acts, many of which continued the superficial social tinkering of prior legislatures that ignored the more basic problems of the colony. Flax seed was to be purchased and distributed with public funds in order to encourage a cloth industry. Tobacco planting after June 20th was forbidden in an attempt to limit production and therefore keep the price high. Certain imported luxury goods, such as silver and gold lace or thread and silk fabric, were banned from the colony.[428]

For the first time, however, the Assembly seemed to recognize the colony's desperate need for educational institutions. Act twenty included provision for a college, "for the advance of learning, education of the youth, supply of the ministry, and promotion of piety."[429] Land would be bought on which the school would be built "with as much speed as may be convenient." However, no money was allocated for the school, nor was anyone assigned the task of building it. The only reference to fund-raising for this project was act thirty-five, which asked Berkeley to solicit for charity and ministers while he was in England.[430] Such a half-hearted commitment explains why thirty years passed before the college was actually built.

To Berkeley and the Assembly, building a school was hardly the most important item on their agenda. Of greater necessity were the fundamental political changes they instituted this session, strengthening their own position of power and eliminating any opposition. Act seven limited to two the number of burgesses permitted from each county. Act nine limited the number of county court Justices to eight, to be picked by seniority with the approval of a committee set up by Berkeley and the Council. This committee included such Royalists as Nathanial Bacon, Sr., Edmund Scarburgh, Francis Moryson, and Francis Willis.[431] The justices that this committee would pick would wield enormous power, especially since one of the Commonwealth era laws that this Restoration Assembly did *not* repeal was the 1658 act that gave them the ability to follow the spirit of the law and not its letter in their judicial rulings.[432]

Act twenty-one limited the size of vestries to twelve members, all of whom had to take the oath of allegiance and supremacy to serve.[433] These vestries managed and ran the local parish churches, appointing ministers. Though they continued to refuse ministers lifetime appointments, this act also gave vestry members life terms as well as the power to choose their own replacements. Since the vestries also administered church lands, appointed local highway surveyors, assessed taxes for church purposes, and handed out punishments for sinful behavior, they held a great deal of social power.434

All these actions were surely used to push out dissenters and any Royalist opposition, and to keep new faces from entering the government unless they toed the party line.

To demonstrate that they meant business, John Bond, Puritan and burgess from Isle of Wight County since 1654, was expelled from all public offices because of his "factious and schismatical demeanors," typical seventeenth-century code words for anyone who opposed the leadership in power.435

The power of taxation was also moved upwards towards the top of Virginia's hierarchy. The Governor and Council ordered that an export tax of five shillings per barrel of provisions be imposed, and the Assembly confirmed it. The Governor and Council were also given the unilateral right for the next three years to impose an additional head tax of up to twenty pounds of tobacco. Control of tithable lists was moved upwards from the Sheriffs to the local Justices, who now had the power to impose double or triple taxes for any discovered fraud or error.436

It is a mistake to think that these actions focused power solely into William Berkeley's hands, thereby making him a tyrant. Instead, they distributed power among that small group of Royalist legislators, led by Berkeley, who already controlled the Council, House of Burgesses, and the local courts. Expressing the Royalist belief that rule should come from the privileged caste at the top of the social order, these laws focused power accordingly.437

Not content merely to cement their position of power, these men continued to expand their use of government for their own as well as their friends' profit. Act fourteen set the salary of the burgesses at 150 pounds of tobacco per day, plus all travel expenses. Clerks and Sheriffs were given greater power to sue for their fees. Edmund Scarburgh was rewarded 10,000 pounds of tobacco for starting a salt factory on the Eastern Shore. Henry Soane, the speaker of the House, was awarded a 6000 pound honorarium. And of course, there was that enormous payment of 200,000 pounds of tobacco to Berkeley for his trip to England.438

Finally, in a recognition that the hodge-podge of laws passed in 1660 had left Virginia's legal system a mess, "that the people knew not well what to obey nor the judge what to punish,"439 Francis Moryson, with the Clerk of the Assembly to help him, was assigned the job of once again reviewing the colony's laws, and to "present a draft of them with such alterations and amendments as they shall find necessary to the next Assembly."440 This would be the third major revision of Virginia's laws in less than ten years, and the first thorough revision under the Royalist banner following the king's restoration.

William Berkeley left for England in the spring of 1662, leaving Francis Moryson behind as acting governor. The Assembly adjourned, only to reconvene without elections a fourth time in March 1662 (table 22). Except for dealing with some private minor disputes and an Indian conflict in Westmoreland county, this Assembly's main action was to pass Moryson's complete revision of the laws of Virginia, replacing the laws passed during the Commonwealth era.

Table 22-The Restoration Assembly, October 1661 to March 1662

Current Governor:
Deputy Governor Francis Moryson

Council-October 1661

Royalist:	Unknown:	
*Bacon, Nathaniel, Sr.	*Bernard, William	
*Browne, Henry	*Pettus, Thomas	
*Hammond, Manwaring		
*<u>Ludwell, Thomas</u>		
*Moryson, Francis		
*Reade, George		
+Carter, John, Sr.		

House-March 1662-Held over without general elections

Royalist:	Unknown:	Uncertain Puritan:
*Fantleroy, Moore-EXPELLED	*Browne, William	*Bland, Theoderick-SPEAKER1
*Fowke, Thomas	*Cary, Miles	*<u>Waters, William</u>
*_Presley, William, Jr._	*Catchmaie, George	
*_Scarburgh, Edmund_	*Caufield, William	Uncertain Parliamentarian:
*Travers, Raleigh	*Denson, William	*Ellyson, Robert
*Willis, Francis	*<u>Ffarrer, William</u>	*Knight, Peter
	*Hamelyn, Stephen	
Uncertain Royalist:	*Hill, [Richard]	
*<u>Abrahall, Robert</u>	*Knowles, John	
*Cant, David	*Pitt, Robert	
*_Chiles, Walter, Jr._	*<u>Powell, John</u>	
*_Claiborne, William, Jr._	*<u>Smith, Nicholas</u>	
*Stringer, John	*<u>Webb, Giles</u>	
+Mason, Lemuell	*Weir, John	
	*_Worleich, William_	
	*Wynne, Robert-SPEAKER2	

	Present Slave-holders	Including future slave-holders	Raised in Virginia	Incumbents	Including those who served previously
Council	2 of 9	5 of 9	1 of 9		
House	12 of 32	22 of 32	10 of 32	31 of 32	32 of 32

Legend for all tables:
*-incumbent
+-served previously
Bold names have purchased slaves or have claimed headrights on blacks.
<u>Underlined names will eventually buy slaves or claim headrights on blacks.</u>
Italicized names either were born in Virginia or immigrated as minors.

Source for all tables: See Appendix B

This revision accomplished several purposes. Most obviously, it acted to organize all the laws into one book, so that everyone would be able to know what those laws were. This also served to straighten out the problems that had been created by the badly disorganized legal review of 1660. Laws that had been inadvertently dropped, such as the definition of tithable, were re-instituted, most of which without any significant changes.

To the Royalists the revision also re-established in writing church government under the Anglican church. The first fourteen acts re-instituted the 1643 religious laws passed during Berkeley's first administration, with power centralized under the Governor and the Council.[441] Because the revision voided all previous laws but did not include the March 1660 act for suppressing the Quakers (passed when Berkeley first took power), merely fining dissenters 200 pounds tobacco for "assembling in unlawful Assemblies and conventicles,"[442] it actually gave these religious dissenters a small window of opportunity for preaching their message. For the next year or so there seems to have been an increase in Quaker gatherings, so much so that in January 1662 George Wilson, Quaker preacher, was imprisoned, eventually dying three months later in what he called "a Jamestown dungeon." While he was alive he had repeatedly written to Francis Moryson and the Assembly trying to discourage them from passing laws oppressing the Quakers.[443] Instead, in April 1662 (about the time of his death) a major legal "proceeding against the Quakers" took place before the Council, led by Moryson, including the cruel punishment of two traveling Quaker woman, Alice Ambrose and Mary Thompkins. Their property was confiscated, they were whipped 32 times apiece, pilloried, and then expelled from the colony.[444]

This legal revision also confirmed and cemented the centralization of Virginia's government around the Royalist elite. Virginia was going to be ruled from above, power concentrated within the upper echelons of society. Confirming the political restrictions passed in October, the revision went on to add more. The Governor would now induct ministers, give final approval of all marriage licenses, approve all justice appointments, and have veto power on sheriff appointments from among the justices. The Governor and the Council had the right to tax each person up to twenty pounds of tobacco for the next three years, to sit on local courts, and only the Governor or Council could put a stay on court decisions. The Governor, Council, and House were to pick tax collectors, and were given full control over the spending of all fines imposed in the colony. The burgesses confirmed their pay at 150 pounds of tobacco a day plus traveling expenses, as well as their privilege from arrest during and up to ten days after each Assembly session.

Finally, since the legislators believed in the Royalist idea of rule-from-above, they confirmed almost all of the superficial business regulations from the previous two decades. Physicians, millers, tavern-owners, and surveyors were regulated; the export of iron, wool, sheep, mares, and hides was forbidden; ship-building and flax cultivation were subsidized; the planting of mulberry trees, two acres of corn, and the building of tanhouses was required; all trade with the Indians, unless commissioned by the Governor, was forbidden; and tobacco planting was outlawed after July 10th. Other acts prohibited the export of imported English goods, controlled the importation of rum, and regulated the treatment of ship passengers.[445]

And as during the Commonwealth era, these regulations did little to change the headright system that continued to shape the colony's oppressive labor system. While some of these Royalist legislators, both British and Virginian-bred, might have sincerely wished to make the colony a better and more stable place to live, none of them seemed willing to make the personal sacrifices necessary for this to really happen. They did not reduce their dependence on forced, unpaid servant labor, continuing the aggressive immigration of male laborers for the cultivation of tobacco. And while they had given lip service to their desire for a college, they did not build it. Nor did they act to give their clergymen some independence, and in fact they took steps to limit both education and religious dissent.

While these events were going on in Virginia, Berkeley was unsuccessfully lobbying his king in England. Despite his best efforts, Charles II would not cancel the Northern Neck patent, nor would he repeal the Navigation Acts,[446] despite Berkeley's plea "that forty thousand people should be impoverished to enrich little more than forty merchants, who being the only buyers of our tobacco, give us what they please for it, and after . . . sell it how they please."[447]

Furthermore, Berkeley failed entirely in obtaining any funds for the college, nor could he get the universities of England to commit any of their graduates for Virginia. He returned in the fall of 1662, carrying new instructions from the king that required little else but the building of towns as well as a call for new elections within a month of his return.[448]

With Berkeley's return we move into a new period of political rule in Virginia. From this point on, the worst aspects of Royalist ideology take control of the political life of the colony. Both the British Royalists and the second generation Virginians moved aggressively to solidify their positions of privilege and rank in an atmosphere of revenge, retribution, and an unrestrained hunger for money and wealth. Virginian society now moved in directions that no pre-Civil War Royalist would have ever wanted or dreamed.

Yet, it must be emphasized that Berkeley, Moryson, the Council, and the House still viewed themselves as conscientious and responsible leaders, trying to do the best they could for the colony. And we do have ample examples of Berkeley, the Council, and the House making decisions that were directly harmful to allies or friends who did wrong. Moore Fantleroy, a man who had stood with Berkeley in 1652 in defying the rule of Parliament, was expelled from office for a period of time for dealing dishonestly with the Indians. Other allies, such as Gerald Fowke, George Mason, Giles Brent, were also suspended and fined for attacking the Indians in Westmoreland county.449

Furthermore, Berkeley and the Assembly sincerely believed that Berkeley's trip to England was in the best interests of everyone in the colony. The price of tobacco had been plunging since the late fifties, and would not stop its free fall until almost the end of the sixties.450 The Navigation Acts were certainly a contributing factor. Dependent on tobacco for their income and forbidden from trading it on the open market, the Acts helped depress the price more than necessary.

Despite their sincere efforts, however, Berkeley and the Assembly were still following the oft-repeated Virginian pattern of rule-from-the-top, for the benefit of those on the top. And after fifty years, that pattern was now asserting itself in ways that were more authoritarian and selfish than ever before. Whatever influence the more moderate Royalist, Parliamentarian, or dissenting religious factions might have had in past years now ended with the Restoration.

10. The Choice of the Electorate

The results of the autumn 1662 general elections spoke eloquently of Virginia's future. Up until 1660, elections in Virginia had been regular and frequent, and in the sixteen previous elections between 1643 and 1660, an average of 75 percent of all burgesses had been replaced in each election (figure 10.1). Furthermore, the law still allowed all freemen to vote, regardless of their land status.[451] With no elections for almost three years, changes in the House of Burgesses would have seemed likely.

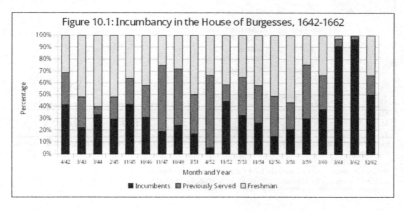

Yet, only 50 percent of the incumbents were replaced, the fewest ever. Furthermore, the political make-up of the House hardly changed, remaining firmly in Royalist control (see table 23). Though not much is known about many of these burgesses, a good number seemed allied with Royalist sentiments, especially those burgesses who had been raised in Virginia. Both Raleigh Travers and William Presley had signed the 1652 Declaration against Parliament[452] and neither served in the House again until after the Restoration. Joseph Bridger and Lemuell Mason were both accused of oppressive actions in the grievances filed by their respective counties in 1676.[453] William Claiborne, Jr, unlike his father, seemed a strong supporter of Berkeley. Raised in Virginia, issues of Royalism probably mattered less to him than maintaining his position as a powerful member of the colony's aristocracy. He served for the entire Long Assembly, backed Berkeley during Bacon's Rebellion, and was named in 1676 as a "wicked and pernicious councilor" by the rebels. In 1677 he served on the court martial that condemned to death many

Table 23-The Long Assembly, October to December 1662

Current Governor:
Governor William Berkeley

Council-October 1662

Royalist:	Neutral:	Uncertain Parliamentarian:
*Carter, John, Sr.	Stegg, Thomas, Jr.	+Swann, Thomas
*Ludwell, Thomas		
*Moryson, Francis	Unknown:	
	*Bernard, William	

House-December 1662

Royalist:	Unknown:	Quaker:
***Presley, William, Jr.**	*Ffarrer, William	Porter, John, Sr.
*Travers, Raleigh	*Hamelyn, Stephen	
+Bridger, Joseph	*Hill, [Richard]	Uncertain Parliamentarian:
	*Powell, John	*Ellyson, Robert
Uncertain Royalist:	*Weir, John	
*Chiles, Walter, Jr.	*Wynne, Robert-SPEAKER	
*Claiborne, William, Jr.	+Griffith, Edward	
*Mason, Lemuell	+Warren, Thomas	
Ramsey, Edward	+Yeo, Leonard	
	Barber, William	
	Cockeram, William	
	Gwillen, George	
	Walker, Thomas	
	Wallings, George	
	Williamson, Roger	

	Present Slave-holders	Including future slave-holders	Raised in Virginia	Incumbents	Including those who served previously
Council	0 of 6	2 of 6	2 of 6		
House	6 of 24	15 of 24	8 of 24	12 of 24	16 of 24

Legend for all tables:
*-incumbent
+-served previously
Bold names have purchased slaves or have claimed headrights on blacks.
<u>Underlined names will eventually buy slaves or claim headrights on blacks.</u>
Italicized names either were born in Virginia or immigrated as minors.

Source for all tables: See Appendix B

rebels, and for this service received a letter from Berkeley certifying his loyalty.454 Likewise, Virginia-bred burgess, Edward Ramsey served on that same court martial, and would also serve in the House for the entire Long Assembly.455

Of the burgesses attending the December 1662 Assembly, only one, John Porter, had ever demonstrated any links with a dissenting group, and the Assembly would soon deal with him.

Undoubtedly, the actions of Virginia's leaders in the previous three years to consolidate their power had made dissent difficult and impractical. The eight Justices of each county court were now appointed by the Governor's commission, and from these eight Justices the job of sheriff was rotated annually, the Governor having an additional veto power over any particular sheriff. Furthermore, Berkeley would soon begin appointing undersheriffs in each county, ignoring the rights of the sheriff to appoint his own assistant.456 Since the sheriff supervised the elections and counted the votes, these actions helped control the electorate process. Sheriffs could restrict electorate choices by limiting notice of elections and by intimidating voters, especially because voting was usually by show of hands. It would have involved enormous risk to stand up and defy the wishes of those in power.

Still, it would be a distortion of the situation to see this merely as the upper classes oppressing the lower. Evidence indicates that there was probably little objection from the electorate about how Berkeley and the Royalists were running things in the early years of the Restoration. Most of the colonists were now uneducated single men driven to seek power and wealth just like the Royalists. For two generations Virginia had been a place where raw self-interest and greed had dominated the cultural atmosphere, with few people allowed or able to express alternative ethical choices. It had now become the defining principle of life for the majority of the colonists, and when they were given the chance to vote in December 1662, the population chose men of like principles.

Probably the most depressing fact of this very depressing tale is that, while William Berkeley, Edmund Scarburgh, and other Royalists in the 1660s might have consciously chosen to abuse the power of government and to legalize the enslavement of blacks for their own ends, they were able to do so because they had the meek acquiescence of the British citizens living in the colony. In most of the matters that counted to the leadership of Virginia, the citizenry of Virginia was quite willing to defer to their rulers, to allow them the complete freedom to set the agenda as they wished.

This acquiescence by Virginians to Royalist rule mostly took place because a large percentage of Virginia's voting population likely favored Royalist policies. Witness the celebrations in York County at the restoration of the king.[457] Since many Virginians were Royalist refugees from the Civil War and had numerous contacts with England, they likely saw the chaos and confusion of the Civil War and the Commonwealth years as something now ending with the return of the king. To let Berkeley and the Assembly run things must have seemed natural and ideal: strong, authoritative leaders whose central goals were to increase everyone's financial profits.

Furthermore, the electorate's easy acceptance of Royalist rule happened because Virginian society was dominated by the Royalist idea of deference to authority. If everyone believes in the idea of rule-from-above, than everyone must defer to the decisions of those in charge, no matter what their beliefs. Even the Parliamentarians during the Commonwealth period had demanded and gotten deference when they passed laws to regulate business, trade, religion, and speech. Now, under the Royalists, William Berkeley deferred to the king when he insisted that towns and forts be built in ludicrous places,[458] the Assembly deferred to Berkeley when he dictated that the forts must be built to a certain size and made of bricks, and the public was expected to defer to the local justices who were placed in charge of construction and tax assessment.[459] In such an atmosphere, everyone waits for the leader to set the agenda, and when that leader says jump, everyone else tries to figure out how high.[460]

Unfortunately, democracy and deference do not go well together. Because democracy is built on the principle that rule comes from the bottom, that sovereignty resides not with the king, the Lords, or the Parliament, but with the citizenry that choose them, deference to those in authority creates a dangerous power vacuum. While the Royalist code of honor might have set rules of behavior for a ruling caste of aristocracy, democracy instead depends on a vigilant electorate to prevent the abuse of power. By deferring to its elected leaders, either due to fear or a reluctance to question, the citizenry gives those leaders a blank check, permitting them to abuse their positions in whatever way they wish.

Finally, the Royalists took such total control because, for much of the colonial period Virginia's electorate did not seem very interested in what their leaders were doing, and merely wanted to get on with their lives. The little evidence we have indicates that, by the end of the Commonwealth period, the numbers of men participating in each election could sometimes be appallingly low, sometimes less than a dozen in some counties. And even when all freemen were given the vote, as in Virginia between 1656-70, it was not unusual for less than half to participate in most elections.[461]

Yet, not all Virginians at this time were apathetic non-voters, deferential yes-men, or enthusiastic Royalists. In Lower Norfolk County a small but dedicated community of Quakers was still active and holding religious meetings, including several in November and December of 1662, at which a good number were arrested and fined. John Porter, Sr., county Justice and former sheriff, attended these meetings, listening to the sermons of his younger brother, John Jr., a popular and well known Quaker speaker.[462] When the December elections were held, it was John Porter, Sr. whom the Quakers in Lower Norfolk county elected to the House of Burgesses.

That winter the Assembly was sparsely attended. Only twenty-four burgesses showed up in Jamestown, with only six of sixteen counties sending both of their representatives. None at all came from Northampton county across Chesapeake Bay. Even the Council was poorly attended, with only six councilors present.

This small attendance might have acted as a modifying influence, as the session's legislation was strangely mixed. One act recognized that "the country poor men are most likely to suffer if they shall be forced to pay" equally the reward for killing wolves. Instead the act placed the tax on the owners of horses, since they usually had the money to pay it.[463]

Another act certified that the clothing and luggage of English indentured servants was actually owned by them, and could not be sold by their masters.[464] A third act was, in a strange way, restrictive to the power of masters. Previously, if an unmarried woman servant became pregnant, she was punished with two years additional service.[465] Since this would actually reward her master, it acted to encourage masters to impregnate their female servants. The new law maintained the two years of punishment, but transferred its administration to the local parish in order to deny the master its benefit.[466]

In other cases, however, the Assembly continued to strengthen its hold on power and wealth. Witnesses were now required to testify, even against themselves, and were to be imprisoned if they refused.[467] The definition of tithable was broadened, now including all women who worked in the fields.[468] Craftsmen, who did not grow tobacco and had been made exempt from taxes in order to encourage their immigration and ease their tax burden, were now required to pay taxes like everyone else.[469]

Another law imposed a fine of 2000 pounds of tobacco on anyone who refused to have their children baptized.[470] Since Quakers customarily did not baptize their children, this law was a direct attack on them. John Porter probably protested in House debates, albeit futilely.

Finally, act twelve stated that

> WHEREAS some doubts have arisen whether children got by any Englishman upon a negro woman should be a slave or free, be it therefore enacted and declared by this present Grand Assembly, that all children born in this country shall be held bond or free only according to the condition of the mother, and that if any Christian shall commit fornication with a negro man or woman, he or she so offending shall pay double the fines imposed by the former act.[471]

From now on, the children of female black slaves would automatically be classified as slaves as well, regardless of their father's status. William Presley, who six years earlier had lost ownership of Elizabeth Key because her father had been white, was a member of this Assembly, and was almost certainly involved in the passage of this law.

At the time the act was probably seen as a minor revision to the laws of servitude and slavery. The black population in Virginia in 1662 had increased only slightly in the last twelve years, hovering somewhere around 4 percent of the population, now approximately 1000 blacks out of close to 30,000 people (see figure 3.1).[472] Few men owned slaves, and many blacks lived as free men. Hence, few would be affected by such legislation.

Yet, as we know, this law was to have far-reaching effects, being a fundamental change to basic British law. While whites might still trace their freedom status, property rights and rank through their father, blacks could not. The children of black slave-women were now guaranteed a second class status in Virginia society.

Hence, this law clearly admitted that racial heritage, not slave status, determined an inferior position for blacks. It was as if Royalists like William Berkeley, William Presley, and Francis Moryson, having recognized the possibilities inherent in the Elizabeth Key decision, whereby mulatto children could not only be free but also inherit the estates of their white masters, acted in horror to stave off this possibility.[473] Previously, love and marriage between a free man and a servant woman would have elevated the woman and her children. The only distinction had been bondage, which was only a situation of circumstance and could therefore always be altered.

This law changed all that. Since the only people in Virginia ever relegated to slavery were blacks,[474] making their unborn children slaves clearly re-defined slavery from mere circumstance to racial heritage. The very language of the law indicated a desire to re-define the status of blacks to that of inferiority. To allow black children the same rights as white children was impossible, and from now on, the children of black slaves would begin their lives as slaves.

Considering the dominance in Virginia of the Royalist belief in caste within an unchanging Chain of Being, we shouldn't be surprised that such ideas degenerated into racial stereotyping and bigoted oppression, especially considering the lack of education or moral training within the colony. From the Royalist point of view, blacks clearly could not be part of the ruling British aristocracy determined by birthright, and hence any offspring born from love between the races had to be considered inferior as well.

Previous generations of British Virginians, while possibly willing to condone slavery in individual cases, would have been very reluctant to *legislate* inferiority on the children of a specific group of people. To some, such as Puritans Obedience Robins, Edward Lloyd, Thomas Meares, and Richard Bennett, such legislation would have seemed shameful. Even some Royalists, such as Edward Hill, Sr., would have probably questioned such an action.

However, Obedience Robins and Edward Hill, Sr. had both recently died, Edward Lloyd and Thomas Meares and the Puritans had fled to Maryland in 1649, and councilor Richard Bennett did not even attend the December session of the Assembly. A new generation was moving into power, and that made all the difference.

Despite the increased regulations against them, the Quakers in Lower Norfolk county continued to congregate. We know that in December and in May they held meetings, resulting in both arrests and fines.[475]

When the full Assembly reconvened without elections (see table 24) in September of 1663, the issue of suppressing this Quaker activity was number one on the agenda. First, they solved the problem of John Porter by expelling him from the House.[476] Then the Assembly passed an act specifically prohibiting the unlawful assembly of Quakers, stiffening the penalties significantly by adding banishment and imposing fines of 5,000 pounds of tobacco for bringing any Quakers into the colony. According to members of this Assembly, it was now a crime

> if any person or persons commonly called Quakers, or any other separatists whatsoever in this colony shall at any time . . . in the several respective counties . . . assemble themselves to the number of five or more . . . at any one time in any place under pretense of joining in a religious worship not authorized by the laws of England nor this country."

In addition, to allow Quakers "to teach or preach" in one's home was to be liable for a 5,000 pound fine.[477] Any informer who brought a Quaker to justice would be rewarded with half the fine.

Typical of his aggressive take-charge approach to everything, Edmund Scarburgh wasted no time in gathering a troop of forty men, and less than one month later he was leading an expedition across the disputed Eastern Shore border with Maryland to read this act and to arrest any Quakers he could find.[478] Not only would Quakers be unwelcome in Virginia, their presence in neighboring colonies would not be tolerated either. Since many had fled Virginia to Maryland in the previous three years, Scarburgh had no trouble tracking some down.

Scarburgh's description of Quaker Stephen Horsey is telling. Horsey was an

> ignorant, yet insolent officer . . . elected a burgess by the common crowd and thrown out by the Assembly for a factious and tumultuous person, a man repugnant to all government, of all sects yet professedly none, constant in nothing but opposing church government, his children at great ages unchristianed.[479]

Scarburgh here articulates the typical Royalist beliefs that now controlled Virginia. The offense of opposing "church government" (that is, the

Table 24-The Long Assembly, 1663

Current Governor:
Governor William Berkeley

Council-March 1663

Royalist:	Neutral:	Puritan:
*Carter, John, Sr.	*Stegg, Thomas, Jr.	+Bennett, Richard
*Ludwell, Thomas		
+Bacon, Nathaniel, Sr.	Unknown:	Uncertain Puritan:
+Lee, Richard, Sr.	*Bernard, William	Bland, Theoderick
Corbin, Henry	+Carter, Edward	
		Uncertain Parliamentarian:
		*Swann, Thomas
		+Wood, Abraham

Council-November 1663

Royalist:	Unknown:
*Bacon, Nathaniel, Sr.	Cary, Miles
*Corbin, Henry	
*Ludwell, Thomas	
+Reade, George	
Uncertain Royalist:	
Smith, Robert	

House-September 1663-Held over without general elections

Royalist:	Unknown:	Quaker:
*Bridger, Joseph	*Barber, William	*Porter, John, Sr.-EXPELLED
*Presley, William, Jr.	*Cockeram, William	
*Travers, Raleigh	*Ffarrer, William	Uncertain Parliamentarian:
+Jenings, Peter	*Griffith, Edward	*Ellyson, Robert
Fowke, Gerald-EXPELLED	*Gwillen, George	+Kendall, William
	*Hamelyn, Stephen	Andrews, William, Jr.
Uncertain Royalist:	*Hill, [Richard]	
*Chiles, Walter, Jr.	*Powell, John	
*Claiborne, William, Jr.	*Walker, Thomas	
*Ramsey, Edward	*Wallings, George	
	*Warren, Thomas	
	*Weir, John	
	*Williamson, Roger	
	*Wynne, Robert-SPEAKER	
	*Yeo, Leonard	
	+Knowles, John	
	Gray, Francis	
	Lucas, Thomas [Sr. or Jr.]	
	Peyton, Valentine	

	Present Slave-holders	Including future slave-holders	Raised in Virginia	Incumbents	Including those who served previously
Council-March	5 of 12	7 of 12	4 of 12		
Council-November	4 of 6	6 of 6	1 of 6		
House	8 of 31	16 of 31	9 of 31	23 of 31	26 of 31

Legend for all tables:
*-incumbent
+-served previously
Bold names have purchased slaves or have claimed headrights on blacks.
<u>Underlined names will eventually buy slaves or claim headrights on blacks.</u>
Italicized names either were born in Virginia or immigrated as minors.

Source for all tables: See Appendix B

Anglican church) is defined by Horsey's refusal to baptize his children, a Quaker custom. Worse was Horsey's election as a burgess in 1653 (when the Commonwealth controlled Virginia) by "the common crowd."

In England, this strong desire to restrict religious practice to the Anglican Church would have only forced the general population into a well-organized, sophisticated religious system run by the king and his party. While such a church might have magnified the power of the king, it would have also been able to provide moral guidance to the citizenry.

In Virginia this was not possible. Unlike Britain, Virginia did not have the luxury of picking and choosing the religion it wished. While Berkeley and the other Royalists might have wanted to impose Anglicanism on the colony, and while the general populace of poorer farmers and indentured servants probably wouldn't have minded in the least this choice, Anglican clergymen could not be convinced to immigrate to the colony. As noted by historian William Seiler, "By 1662 there were only ten clergyman to serve forty-five to forty-eight parishes."[480] Though Berkeley and the Royalists were able to successfully eliminate any serious dissent from the Anglican church, they were unable to establish a thriving Anglican church to replace these other religious movements. Essentially, Berkeley and the Royalists under him would have rather had no church than a church that he, as the king's representative, could not control.

The result was that this September 1663 law and its strict enforcement by men like Scarburgh succeeded at last in making the Virginia colony a homogeneous community: uneducated men seeking land, power, privilege and wealth and maneuvering for positions on the Council or in the House of Burgesses in order to make it happen. The colony was now a one party state, without opposition to its leaders. Those individuals who might have opposed such a community, believing such behavior unseemly or immoral, were finally gone, having been expelled or deciding on their own that it was better to simply move elsewhere.[481]

And since it was almost impossible to become a burgess without first being appointed a Justice or sheriff,[482] and since the Royalists had moved in to control these appointments,[483] anyone who wanted to obtain power either had to curry favor with those in power, or move away.

There would be no general elections in Virginia for thirteen years.

The Choice of the Electorate

11. The Long Assembly

What furies govern man, we hazard all
Our lives and fortunes to gain hated memories,
And in search of virtue, tremble in shadows.
—William Berkeley, from *A Lost Lady*.[484]

Sometime in the early 1660s, almost at the exact moment that Edmund Scarburgh was riding north to arrest Stephen Horsey and his Quaker friends, African-born Anthony Johnson, former slave and landowner, sold his Virginia farm and moved north to Maryland. Unlike Scarburgh, however, Johnson took his family with him, and by 1665 the whole Johnson clan had abandoned Virginia. Johnson leased a farm on the Maryland Eastern Shore from, of all people, Stephen Horsey, naming his new estate "Tonies Vineyard."[485]

Johnson's reasons for leaving Virginia are not known. Yet, that this move corresponded with the Restoration and the consolidation of power by the Royalists is very telling.

With the December 1662 elections began the reign of what historians now call the Long Assembly (tables 23 through 42. Because each session's membership changed so little, only tables 23, 24, 28, 29, 41-44 are placed in the text. Tables 25-27, 30-40 can be found in Appendix A). For thirteen years William Berkeley would call no general elections. Each year he and the Assembly would meet, pass some laws, settle some court cases, adjourn for six months to a year, and then reconvene again to do the whole thing over. Only when a burgess died or left the colony would a special election be held.

Though many of the members of the Assembly were replaced during these thirteen years due to death or retirement, figure 11.1 shows how completely different a situation this was from previous Assemblies, when the turnover rate for burgesses was fast and free-flowing. Now the turnover rate was slow and limited. It was as if Berkeley and his Royalist allies were satisfied with the overall Royalist make-up of the House and acted to freeze the situation, slamming the door on new-comers who might not have toed the party line.

As I have already described, this Long Assembly began its governance in much the same way as the Restoration Assembly that had preceded it, restricting dissent and strengthening the personal wealth and power of the ruling elite. Furthermore, these legislators, under Berkeley, continued the legal but petty tinkering of past Assemblies in an

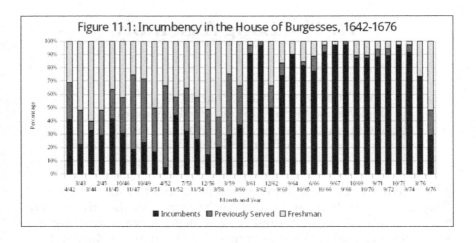

Figure 11.1: Incumbency in the House of Burgesses, 1642-1676

attempt to influence the social and economic structure of the colony. In December 1662 the Assembly authorized awards for the building of ships.[486] In September 1663 the Assembly prohibited the exportation of deer and calf skins.[487] In 1665 it also forbade the exportation of shoes.[488] Berkeley also made repeated but unsuccessful efforts to negotiate a joint cessation of tobacco planting with Maryland in a desperate attempt to prop up the falling price of tobacco.[489]

Yet, this kind of economic regulation was never the focus of the Long Assembly, and in fact by the late sixties and early seventies the Assembly moved to ease some of the trade regulations of past Assemblies.[490] In September 1663 the requirement to plant mulberry trees was partly lifted, and then repealed in 1666.[491] In 1664 the regulation of rum was repealed.[492] In 1668 the exportation of horses and mares was once again allowed.[493] In 1671 all laws prohibiting the exportation of hides, skins, iron, and wool were repealed.[494]

This is not to say that the ruling elite of Virginia had become supporters of laissez-faire and free-market policies. Royalist top-down-rule was still their guiding principle. The Long Assembly mandated the local counties to build looms and tanneries, to be run by the government.[495] Other regulations, such as those on millers, surveyors, and tavern-owners were never repealed (though some fees were lowered), while restrictions on the export of food and tobacco were constantly tinkered with throughout the Long Assembly's thirteen year rule.

Overall, however, they simply regulated less. Instead, the members of the Long Assembly focused on issues of power and caste: having extended and consolidated their political power in the Restoration Assembly, increasingly during its thirteen year reign the Long Assembly's legislation worked to enlarge the master's power over his servants and slaves. Under the leadership of an aging William Berkeley, faced with no election challenges, the threat of removal non-existent, the Royalist code of noblesse oblige practically forgotten in an uneducated society and the lingering anger from the English Civil War, this Royalist Assembly began to rework the law to institutionalize and make permanent their status as a superior caste defined by birthright.

It is with the character of William Berkeley upon which much of this story hinges. By 1665, he was sixty years old, a very old man for that time and place. He was not in good health, and in many ways Berkeley had become a carbon-copy of Charles I: stubborn, hostile to compromise, and arrogant.

It seems that Berkeley had become a truly bitter man after his 1663 trip to England. First he had been abandoned by his Virginian friends during the commonwealth years. Now he had discovered that his own king, the son of the executed Charles I whom Berkeley had stood steadfastly behind during the Civil War, had abandoned him as well. Berkeley returned to Virginia having gotten less than nothing from his king. Not only had Charles II refused to cancel the Northern Neck patents and the Navigation Acts, he had ignored Berkeley's advice and ordered him to build useless forts while also rebuking Berkeley for the colony's dependence on tobacco and its lack of prosperity. Charles II was even willing to praise the *Puritans* of New England, citing them as a model for Berkeley to copy, "who have in a few years raised that colony to breed wealth, reputation, and security."[496]

In a well-known story, Berkeley was visited in 1672 by a Quaker minister, William Edmundson. When Edmundson paid a call on Richard Bennett the next day, he mentioned that Berkeley had seemed to be in an ill-temper. Bennett asked, "Did he call you 'a dog, rogue, etc?'"

"No," said Edmundson.

"Then you took him in his best humor, those being his usual terms, when he is angry, for he is an enemy to every appearance of good."[497]

That anger of Berkeley was to find its expression in the laws passed by the Long Assembly in the coming years. Though Berkeley's commitment and love for Virginia was unquestioned, by the 1660s he appears to have cared little for the typical colonial farmer's problems. As Berkeley had noted when he was in England in 1663, "Another great imputation lies on the country that none but those of the meanest quality and corruptest lives go thither."[498] It had been these common folk that had killed his king. Not only did he now use the government to improve his power and wealth, as I have already outlined, he was now willing to work with his Royalist allies on the Council and in the House, men such as Nathaniel Bacon, Sr., Thomas Ludwell, Edmund Scarburgh, and

William Presley, to guarantee power and wealth for themselves, and to oppress that rank multitude who had killed their king.

Of the new faces entering the House in the early years of the Long Assembly (see tables 23 and 24), several deserve special mention. Thomas Ludwell, for example, had been close to Berkeley since his first arrival in the colony, having been apprenticed to Berkeley as a young boy and accompanying the governor when he first arrived in 1642. Though Ludwell was never elected to the House, Berkeley had him appointed by the king to the council in 1661, making him Secretary of State that year as well to replace William Claiborne, Sr. For the next sixteen years, Ludwell would be the governor's closest ally and confidant, working aggressively to defend Berkeley from attack both within the colony and from England.[499]

John Washington, whose grandson would be George Washington, had come to Virginia in 1657 as the mate of a merchant ship. His father had been an Anglican clergyman who had suffered badly in the civil war. By 1660 he was a settled slave-owner in Westmoreland County, eventually patenting more than 3,500 acres. He would become a justice in 1662, and join the House for the first time in 1665.[500]

John Leare came to Virginia in 1656 and eventually settled in Nansemond county. There he and his brother David would hold multiple offices, including county clerk, escheat master, notary public, surveyor, and sheriff superior, a fact of which the citizens of that county would complain loudly of in their grievances of 1676. In 1666, however, he was elected burgess, and as a member of the Long Assembly he would become a close ally of Berkeley, sticking with him throughout Bacon's Rebellion and losing heavily because of it.[501]

Nicolas Spencer, cousin of Lord Culpepper, arrived in Virginia in 1659. In 1666 he would be elected to Long Assembly for the county of Westmoreland, and in 1672 Berkeley would appoint him to the council, expressing in a letter how he and Berkeley thought well of each other. He, like Leare, would fight hard for Berkeley in Bacon's Rebellion, suffering significant property loss. Afterward, he would become a close ally to future governors, acting frequently as deputy governor in their absence.[502]

These men joined with Berkeley and his older allies to form the central Royalist core that was to control the Long Assembly for the next thirteen years.

And for those years, the oppression of servants and slaves in the Virginia colony became the social legislative policy of Berkeley and these men. As just one illustration we can trace the legal history which established the term length for indentured servitude. Many poor immigrants came to Virginia without a written contract, the ship captain paying their transportation costs, which was then repaid by a wealthy planter in Virginia. Some had been kidnapped, others had made oral agreements that could not be proved.[503] Because of this, their term of servitude was not set forth clearly, and could be and was often abused by their new master in Virginia under the headright system.

In 1643, during the second Assembly under William Berkeley's governorship, he and the Assembly made the first legal attempt to address this type of mistreatment, while doing nothing to change the headright system itself. Their solution was to limit the term for unwritten indentured contracts to four years for adults, five years for children between 12 and 20 years old, and seven years for children under 12.[504] This law, however, created several problems. Depending on a servant's declared age, his or her term of servitude could range from four to seven years, and under its vague terms it was actually possible for children as young as eight to be legally considered free.

In April 1652 the legal revision under Richard Bennett recognized these problems and simplified the law. The term was left four years for adults over 16 years old, while anyone younger would remain a servant until they were 21.[505] This law still had problems, since it encouraged servants to lie about their age and claim to be over 16, since anyone under 16 would have to serve at least five years, while someone over 16 could not serve longer than four years. Still, the new law was simple and easy to administer, and was therefore renewed unchanged in the 1658 revision under Samuel Mathews.[506]

Hence, for the first half century of the colony's existence, until the Restoration in 1660, a term of four years for any indentured servant over 16 years old was the custom of the country.

As odious as this policy might have been, with the Restoration in 1660 and William Berkeley's re-appointment as Governor, it was no longer sufficient. In March 1660, the first session of the Restoration Assembly under Berkeley decided to amend this law. While children under 16 years old would still remain servants until they were 21 (a minimum of five years), the term for adults was increased to five years.[507] While this seemed to clarify the contradiction in the law, it did so by giving the owners of servants an additional year of free labor.

In March 1662, this same Restoration Assembly amended the law again when it passed Francis Moryson's legal revision. While leaving the term for adults at five years, Moryson increased the term for children under 16 years old so that they would remain servants until they were 24, adding three years to their servitude for a new minimum of eight years.[508] This change, which took advantage of the helplessness of servant children, gave their owners an additional three extra years of free labor.

In October 1666, following Berkeley's return from England, the Long Assembly re-addressed this law, recognizing that the March 1662 amendment had reinstituted the earlier discrepancies that favored certain servants, depending on their age. While the law required a 15 year old to remain a servant until 24, it mandated only a five year term for a 16 year old, meaning they were free when they were 21, three years earlier. The Long Assembly's solution: raise to 19 the age requiring servitude until 24.[509] Under this set-up, every indentured servant in Virginia was now forced to work for free until they were at least 24 years old. Once again, the legislators had acted to increase the servitude of servants.

In less than six years, the Restoration and Long Assemblies under William Berkeley had moved to double the servitude, from four to eight years, for a 16-year-old boy arriving in Virginia without an indentured contract. None of these laws awarded the servant with any corresponding payment for his or her additional service.

While financial considerations might have been a major motivating force for these legal actions, the legislation more clearly illustrates the Royalists' inclination to delineate the social caste distinctions between property owner and servant as well as their willingness to use force and the rule of law to establish these distinctions, and to do so to the advantage of the owner. It is therefore not surprising that tensions between masters and servants increased during the 1660s, and owners were in constant fear of servant revolt.

In September of 1663 events took place that confirmed these fears for the Royalists. On Sunday, September 6th, a group of nine indentured servants, led by a man named John Gunter, gathered at the servant house of burgess Peter Knight in Gloucester County to plot a revolt. They would meet at midnight in the woods the next Sunday, September 13th, seize the arms of councilor Francis Willis and march from house to house, raising an army of rebellion to march to Green Spring and demand that the Governor either give them their freedom or allow them to leave Virginia.[510]

Unfortunately for these men, an indentured servant known only by the name of Berkenhead betrayed them, exposing the revolt to the authorities, who captured the rebels and squashed the rebellion. The members of the Long Assembly considered the threat so serious and its suppression so important that they awarded Berkenhead a reward of 5000 pounds of tobacco and declared September 13th a holy day.[511] Furthermore, this same session of the Long Assembly then passed laws giving the local County courts the power to supervise and administer the pursuit of runaways while also outlawing both the "meeting of servants" as well as the travel of any servant without permission.[512]

The anger, fear, and contempt that Berkeley and the Royalists in the Long Assembly felt for "those of the meanest quality and corruptest lives"[513] was repeatedly expressed in the laws passed during the next decade. In 1666 the Assembly established fines for anyone who harbored "loitering runaway servants for two or three days or more."[514] In 1668, the Assembly gave masters the right to administer corporal punishment on their runaway servants as an additional punishment over and above the additional time added to their term of servitude.[515]

In 1669 and 1670 the Assembly established a reward for anyone who captured a runaway servant, and declared that the runaway was to pay for this reward with a longer term.[516] Furthermore, the 1670 Assembly required that second time offenders would have their hair cut, and that the owner would pay half the reward if the servant ran away a third time. The runaway was to be passed from sheriff to sheriff until he was returned to his master, each sheriff "enjoined by virtue of this act . . . to whip them severely, and then to convey him to the next constable . . . who is to give him like correction."[517]

In 1670 the Assembly reduced the right to vote, limiting it to only those freemen owning or renting land, saying that freemen without land "have little interest in the country [and] do often make tumults at the election to the disturbance of his majesty's peace."[518]

In 1672 the Assembly declared that because of the increase of "vagabonds idle and desolute persons," the justices of the local county courts were authorized to place in apprenticeship "all children, whose parents are not able to bring them up apprentices to tradesmen, the males till one and twenty years of age, and the females . . . till eighteen years of age."[519]

All these laws were written on the Royalist principle that those in charge, being higher on the Chain of Being, were therefore inherently better than the rest of the population, and as such they had the right to dictate how everyone else should live their lives. Furthermore, the language of these laws, probably written by Berkeley himself, clearly expresses an increasing contempt for and fear of "those of the meanest quality and corruptest lives."[520]

Yet, though these Royalist legislators worked hard to legalize their social status while simultaneously improving their own financial wealth, it was politically impossible for them to make their status and birthright legally permanent, to create a caste of nobles as existed in France or Spain, and to establish a legally inferior status for other British citizens. As even Berkeley himself noted in 1663, all British citizens were endowed with the right to make their lives better.

> Those that come from [England] with those ungoverned manners and affections, change them for sober and thrifty passions and desires . . . and those that will either experimentally or morally weigh the nature and conditions of men, shall find . . . that if we

> will be provident and industrious for a year or two, we
> may provide for our posterity for many ages; the
> manifest knowledge of this makes men industrious
> and vigilant with us.[521]

To make a law that would forbid the possibility of advancement for some British citizens was unacceptable for many political reasons, not the least of which was that the promise of such advancement was central to British cultural life. Though it might be permissible to make difficult a British citizen's move up in status, it could not be made impossible.[522]

If Berkeley and the members of the Long Assembly could not make themselves permanent rulers over their white British servants, who else could they do it to? The answer is obvious.

Beginning in December 1662, when the Long Assembly enslaved the children of black slave women, these legislators began to legalize a caste system in Virginia, divided by race. If they could not create a lower caste of white, British, indentured servants, they could instead give that inferior status to the colony's few black slaves.

In September 1667 the Long Assembly passed a law stating that:

> Whereas some doubts have risen whether children
> that are slaves by birth, and by the charity and piety of
> their owners made partakers of the blessed
> sacraments of baptism, should by virtue of their
> baptism be made free; it is enacted . . . that the
> conferring of baptism doth not alter the condition of
> the person as to his bondage or freedom.[523]

This law was as important as the 1662 law, because it removed from blacks the one route open to them for obtaining their freedom by their own actions. By ending baptism as a route to freedom, it made it possible for slave-owners as well as traders to keep their "goods" forever. No longer could a slave-owner in Virginia lose his slave to conversion. No longer would it be possible for a slave to claim his freedom by converting to Christianity.

In 1668, the Assembly permanently confirmed the lower status of black women by stating that "negro women, though permitted to enjoy their freedom yet ought not in all respects to be admitted to a full fruition of the exemptions and impunities of the English . . . are still liable to payment of taxes."[524] In plain English, even if free, black women would be taxed, while free white women would remain exempt.

In 1669, the Assembly passed another act: "If any slave resist his master . . . and by the extremity of the correction should chance to die, that his death shall not be accompted felony, but the master . . . be acquit from molestation."[525]

In 1670, the Assembly declared that "no negro or Indian though baptized and [free] shall be capable of any such purchase of Christians, but yet not debarred from buying any of their own nation."[526] This same year the Assembly firmly legalized slavery for life for all non-Christians imported by ships (meaning blacks), and set a twelve year term for non-Christian adults imported by land (meaning Indians).[527]

In 1671, the Assembly clarified the status of black slaves belonging to orphans whose parents had died without a will. The county courts were authorized "to cause such negroes to be duly appraised, sold at an outcry, or preserved in kind, as [the courts] find it most expedient for preservation, improvement or advancement of the estate and the interest of such orphans."[528] It is interesting that in the preamble of this act, blacks were equated with the sheep, horses, and cattle of such estates.

In 1672, the Assembly once again declared that "it shall and may be lawful for any person who shall endeavor to take them, upon the resistance of such negro, mulatto, Indian slave, or servant for life to kill and wound him or them so resisting . . . and the person who shall kill or wound . . . any such for resisting . . . shall not be questioned for the same."[529] The Assembly also clarified the tithable law, stating that all black women over sixteen years old were taxable.[530]

As I have already discussed, these laws allowed the leaders of Virginia to take advantage of the headright system to increase their power as well as their wealth. As Edmund Scarburgh had so clearly demonstrated in 1655, if you imported slaves rather than servants you not only got a servant for life, you earned fifty acres of land on which you could make him work. Among the wealthy elites of Virginia this was now becoming an acceptable business practice, and throughout the decade of the sixties the members of the Long Assembly gained power and land by using this to their advantage. Councillor and Secretary of State Thomas Ludwell imported two blacks in 1663, and then two more in 1671.[531] Other councilors who did this included Robert Smith (who imported two blacks in 1667),[532] Nathaniel Bacon, Sr. (nine blacks in 1666),[533] and John Carter, Sr. (twenty-one blacks in 1665).[534]

In the House, burgesses like James Bridger (seventeen black slaves in 1666), John Savage (twenty-five blacks in 1664), John Powell (seven blacks in 1665), Roger Williamson (four blacks in 1666), John Weir (fourteen blacks in 1664), and William Kendall (two blacks in 1666) all took advantage of the very laws that they themselves passed.[535]

Others, such as councilors George Reade, Henry Corbin, and Augustine Warner, Sr., and burgesses Edmund Scarburgh, William Presley, John Washington, Adam Thorowgood, Jr., Daniel Parke, Sr.,

Raleigh Travers, Leonard Yeo, and Nicholas Hill, had either claimed headrights in previous decades or already owned slaves by the 1660s.[536] Others, such as John Leare, Nicholas Hill, and Thomas Warren,[557] would buy slaves in the coming years. By passing the slave laws, they were solidifying the "property rights" to the slaves they owned or wished to own, increasing their wealth while simultaneously establishing their position of privilege in a society of castes.

All of these men were members of the Assembly between 1666 and 1670 when the majority of the most oppressive servant and slave laws were passed (see tables 28 and 29). In the Council their number made slave-owners the majority at almost every Council meeting. In the House, the sixteen burgesses listed above comprised almost half the membership. With the added support of Governor Berkeley, it is not surprising that they had no problems passing laws that oppressed their "inferiors."

Moreover, the members of the Long Assembly were far more likely to buy and sell slaves than members of previous Assemblies (see figure 8.2). Prior to 1666, men who actually owned slaves or had claimed headrights on them in the House of Burgesses never exceeded 35 percent, and was usually much lower. After this date the number of slave-holders in the House consistently hovered just below 50 percent. In the Council the percentage of slave-holders after the Restoration sky-rocketed compared to earlier years, sometimes reaching 100 percent.

When we include those men who did not yet own slaves but would eventually do so (see figure 8.3), we can see that it is with the Restoration that the number of legislators willing to use slavery as a tool for profit and power increased significantly. While during the Commonwealth era present or future slave-holders never exceeded 36 percent, and during Richard Bennett's term as Governor was actually less than 20 percent, with the Restoration the percentage of such men rose to between 46 to 56 percent. The Long Assembly began with 50 percent, hovered between 40 to 50 percent for most of its existence, and only dropped below 40 percent with its last session.

This increase in slave-ownership closely paralleled the first significant appearance of Virginia-raised legislators in the House of Burgesses (see figure 8.1). While prior to 1660 such men usually made up from 3 to 17 percent of the House, after 1660 Virginia-bred legislators always comprised between 24 to 29 percent of the House, and after 1669 that number rose to above 35 percent. Hence, the perspective of men like Thomas Ludwell, who had first come to Virginia with William Berkeley when he was about thirteen years old,[538] and John Savage, who had been born in Virginia and whose father Thomas had been an early adventurer,[539] first became a significant influence on legislative policy during the reign of the Long Assembly. Unfortunately, that influence

Table 28-The Long Assembly, October/November 1666

Current Governor:
Governor William Berkeley

Council-October/November 1666

Royalist:	Unknown:	Uncertain Parliamentarian:
*Ludwell, Thomas	*Cary, Miles	*Swann, Thomas
+Bacon, Nathaniel, Sr.		
+Corbin, Henry		
+Reade, George		

Uncertain Royalist:
*Smith, Robert

House-October 1666-Held over without general elections

Royalist:	Unknown:	Uncertain Parliamentarian:
*Bridger, Joseph	*Barber, William	*Carver, William
*Jenings, Peter	*Blackey, William	*Kendall, William
*Leare, John	*Blake, John	Ballard, Thomas
*Meese, Henry	*Ffarrer, William	
*Presley, William, Jr.	*Griffith, Edward	
*Spencer, Nicholas	*Hill, [Richard]	
*Thorowgood, Adam, Jr.	*Holte, Robert	
*Travers, Raleigh	*Lucas, Thomas, [Sr. or Jr.]	
*Washington, John	*Powell, John	
+Scarburgh, Edmund	*Southcoat, Thomas	
	*Walker, Thomas	
Uncertain Royalist:	*Warren, Thomas	
*Baker, Lawrence	*Weir, John	
*Claiborne, William, Jr.	*Williamson, Roger	
*Filmer, Henry	*Wynne, Robert-SPEAKER	
*Parke, Daniel, Sr.	*Yeo, Hugh	
*Ramsey, Edward	*Yeo, Leonard	
*Savage, John	+Hone, Theophilus	

	Present Slave-holders	Including future slave-holders	Raised in Virginia	Incumbents	Including those who served previously
Council	5 of 7	6 of 7	2 of 7		
House	19 of 37	21 of 37	11 of 37	34 of 37	36 of 37

Legend for all tables:
*-incumbent
+-served previously
Bold names have purchased slaves or have claimed headrights on blacks.
<u>Underlined names will eventually buy slaves or claim headrights on blacks.</u>
Italicized names either were born in Virginia or immigrated as minors.

Source for all tables: See Appendix B

most expressed itself in the institutionalization of slavery for personal privilege and financial gain.

The 1660s were bad times for Virginia. The British and Dutch were at war, and in June 1667 the Dutch fleet entered Chesapeake Bay, destroying one warship and five merchant ships while capturing thirteen others. In August of that year, a terrible hurricane ravaged the colony, destroying homes and crops throughout the Chesapeake.[540] Furthermore, the price of tobacco continued its downward spiral, fed by such an enormous surplus of crop that even the hurricane could not significantly reduce it.[541] Additionally, this decade saw the first dramatic decline in

Table 29-The Long Assembly, 1667

Current Governor:
Governor William Berkeley

Council-June 1667

Royalist:	Neutral:	Puritan:
*Corbin, Henry	+Stegg, Thomas, Jr.	+Bennett, Richard
*Ludwell, Thomas		
*Reade, George		Uncertain Parliamentarian:
+Carter, John, Sr.		+Wood, Abraham
+Warner, Augustine, Sr.		

Uncertain Royalist:
*Smith, Robert

Council-September 1667

Royalist:	Neutral:	Puritan:
*Corbin, Henry	*Stegg, Thomas, Jr.	*Bennett, Richard
*Ludwell, Thomas		
*Reade, George	Unknown:	Uncertain Parliamentarian:
*Warner, Augustine, Sr.	+Carter, Edward	+Beale, Thomas
+Bacon, Nathaniel, Sr.		+Swann, Thomas
+Willis, Francis		

Uncertain Royalist:
*Smith, Robert

House-September 1667-Held over without general elections

Royalist:	Unknown:	Uncertain Parliamentarian:
*Bridger, Joseph	*Barber, William	*Carver, William
*Jenings, Peter	*Blackey, William	*Kendall, William
*Leare, John	*Blake, John	
*Meese, Henry	*Ffarrer, William	
*Presley, William, Jr.	*Griffith, Edward	
*Scarburgh, Edmund	*Hill, [Richard]	
*Thorowgood, Adam, Jr.	*Holte, Robert	
*Travers, Raleigh	*Hone, Theophilus	
*Washington, John	*Powell, John	
Allerton, Isaac(a)	*Lucas, Thomas, [Sr. or Jr.]	
	*Southcoat, Thomas	
Uncertain Royalist:	*Walker, Thomas	
*Baker, Lawrence	*Warren, Thomas	
*Claiborne, William, Jr.	*Weir, John	
*Filmer, Henry	*Williamson, Roger	
*Parke, Daniel, Sr.	*Wynne, Robert-SPEAKER	
*Ramsey, Edward	*Yeo, Hugh	
*Savage, John	*Yeo, Leonard	

	Present Slave-holders	Including future slave-holders	Raised in Virginia	Incumbents	Including those who served previously
Council-June	7 of 9	7 of 9	4 of 9		
Council-Sept	7 of 12	7 of 12	4 of 12		
House	18 of 36	19 of 36	12 of 36	35 of 36	35 of 36

immigration from England, the population only growing from 25,000 in 1662 to just over 31,000 in 1674.[542] The planters were being squeezed from all sides.

    Yet, while these difficult circumstances might have motivated the members of the Assembly to institute harsh slave and servant codes in order to ameliorate their problems, such hardships hardly justified the ruthless nature of the solution. Nor do the slave codes and the move to popularize slavery explain sufficiently why the planters of Virginia would choose racial oppression and forced labor as the solution to their problems.543

These men passed these laws under the leadership of William Berkeley and the Royalist banner that firmly believed that birthright and status made some human beings better than others; that knowledge, wealth, and power gave them the right to tell others how to live their lives; that with that power came the right to use it for their own personal benefit, regardless of the harm caused to others not of their caste. When Berkeley was asked in 1670 what provisions the colony had made for religious instruction, he indirectly expressed the above Royalist attitudes toward religious liberty, free expression, and the common folk of society:

> I thank God, there are no free schools nor printing [in Virginia], and I hope we shall not have these hundred years, for learning has brought disobedience, and heresy, and sects into the world, and printing has divulged them, and libels against the best government. God keep us from both!544

If you believe that some were fated to rule because of their birth, it is easy and natural to also believe that others are fated to slavery for the same reasons.

And there was no one in Virginia who could effectively question or defy the institutionalization of slavery. Lacking all forms of moral education, with no schools or religious institutions and the family structure continually shattered by death, Royalist doctrines of aristocratic rule and noblesse oblige had now deteriorated into cold-blooded personal greed and white supremacy, with such ideas dictating social and legal policy. And tragically, the existence of the laws now sanctioned the activity: from 1667 there is a marked increase in the importation of slaves.545

Yet, if the leaders of Virginia wanted to buy slaves and increase their wealth, this did not mean that all Virginians gladly participated in the trade. On the contrary, this growth in slavery proceeded slowly, and even as late as the 1680s black slaves still comprised less than 7 percent of the population (see figure 3.1).

And while this was happening, free blacks such as Anthony Johnson fled the colony, looking elsewhere for freedom and the right to live their lives as they wished.

12. Bacon's Rebellion

> Everyone endeavors to get great tracts of land, and many turn land lopers, some take up 2000 acres, some 3000 acres, others ten thousand acres, nay many men have taken up thirty thousand acres of land, and never cultivated any part of it, only set up a hog house to leave the laps, thereby preventing others seating, so that too many rather than be tenants, seat upon remote barren land, whereby contentions arise between them and the Indians.[546]
>
> —William Sherwood's 1677 report on the causes and events surrounding Bacon's Rebellion.

For more than a decade the Long Assembly had held control of the Virginia legislature. During this time, all government appointments, including sheriff, justice, vestry, and minister, were under the control of the Governor and his Council, and many, especially sheriffs, were appointed to multiple offices for longer than the traditional one year term.[547] Gigantic tracts of land had been patented and taken over by these planters. Servitude had been lengthened, voting rights reduced, and slavery institutionalized. At the same time, the price of tobacco had plummeted, English law had restricted trade, and the bulk of good land had been acquired by the most powerful and connected Royalist legislators.

That this situation was going to explode into violence was first indicated by a small, abortive tax revolt that took place in Surry County. Though this protest was quickly squelched, it was a portent of things to come. In December, 1673 Mathew Swan, William Hancock, and John Sheppard met at the home of John Barnes. Unhappy and resentful of the taxes they were being forced to pay (which included excessive sheriff's and clerk's fees) and their belief that much of these taxes were being pocketed by government officials, these four men decided to gather as many men as possible and confront the authorities, demanding that the year's levies not be collected. They called a meeting at the parish church for Friday, December 12th, spreading the word among the Surry County community.[548]

Meeting in a place called Devil's Field that Friday, these men complained that taxes were "unjustly laid upon them, and they met with intent to remedy that oppression."[549] Councilor Thomas Swan was a specific target of resentment, supposedly receiving 5,000 pounds of tobacco for his own use out of public funds. One man, Roger Delk, stood before them all and said that "We shall burn all before one shall suffer."[550]

Because of the poor weather, not as many men were there as originally hoped. It was decided to meet again on Sunday, gathering at the Lawne's Creek church to demand that the local court not collect the year's taxes. If the justices refused, they would then try to stop the sheriff from collecting the taxes anywhere in Lawne's Creek parish.

In the interim, Lawrence Baker and Robert Spencer, the local authorities, discovered the plan and immediately arrested the ringleaders, including Mathew Swan, John Sheppard, John Barnes, William Hancock, and Roger Delk. The Surry County court sentenced them all to heavy fines, and had Swan and Delk bound over to Jamestown for more serious punishment, since these two men refused to retract their statements, Swan saying that "the court had unjustly proceeded in the levy . . . [and that] all or most of the county were of his mind," and Delk repeating his threat to "burn all" before the court and further complaining that "their taxes were so unjust and they would not pay it."[551] When the Council met in April it confirmed the decision of the local court and fined Mathew Swan an additional 2000 pounds of tobacco. The protests faded, and in September, Governor Berkeley canceled the fines, on the proviso that the protesters "acknowledge their fault in the said County Court, and pay the court charges."[552]

The Lawne's Creek protest hardly ranks as a rebellion of any serious note. Yet, coming as it did near the end of the Long Assembly's term, it is significant for what it indicated. Virginia society was not a happy place. Taxes were high, and resentments ran deep. Furthermore, none of the Lawn Creek protesters were office holders, even though some of them owned land. Roger Delk, as one example, owned 1000 acres, left to him by his father. Other property owners included John Greene (750 acres), William Little (200 acres), and William Hancock (700 acres).[553]

After more than a decade of unopposed rule, the ability of William Berkeley and the Long Assembly to maintain control over Virginia society was beginning to decay. Too many immigrant servants were earning their freedom and expecting as British citizens their share of the fruits of society. Yet, the government was closed to them. Worse, many had come to Virginia with the dream of owning their own landed estate, and because of the headright system and corruption in the Virginia Patent office[554] as well as advantageous deal-making by those in power, all the best land was closed to them.

In less than three years, Virginia would explode in rebellion, treason, violence, plunder and hatred.

It all began, as is usually the case, by a single act of minor injustice, escalating step by step until the situation became uncontrollable and bloody.

First, some Doeg and Susquahanock Indians claimed that Thomas Mathews of Stafford County "had abused and cheated them, in not paying them for such Indian truck as he had formerly bought of them."[555] Among the British this kind of dishonesty would have been dealt with peaceably in the local courts. Among the Indians, however, who could not read and were not familiar with the English legal system, restitution could only be obtained by direct and violent action. Crossing the Potomac in July 1675, they stole several of Mathews' hogs and fled by boat. Pursued by the English, eight Indians were "beaten or killed and the hogs retaken from them."[556]

Now the conflict began to escalate. An Indian war party quickly gathered and marched back to Mathews' farm, intent on revenge not only for Mathews' refusal to pay his debts but because of the murder on the Potomac. Arriving at the Mathews' plantation, the Indian party killed two of his servants, including a herdsman by the name of Robert Hen. As he was dying Hen managed to name the Doeg Indians as his attackers.[557]

The British colonists were enraged, and it was now their turn to send out a troop of men to find the killer to exact British vengeance. They chose George Mason and Giles Brent as their leaders, the same two men who in 1662 had been fined and suspended from office by the Restoration Assembly for "the several injuries and affronts done to Wahanganoche, king of Potowmeck Indians."[558] Now, in the summer of 1675, these two commanders led their men into the woods, probably convinced that all Indians were conspirators in Robert Hen's murder. Brent's men surrounded a cabin of Doeg Indians, while Mason laid siege to a Susquahanock cabin nearby. When the Doeg chief pleaded ignorance of the murder, Brent killed him. In the ensuing skirmish, the British killed ten Doeg Indians and another fourteen Susquahanock, despite George Mason's cries, "For the Lord's sake shoot no more, these are our friends the Susquahanocks!"[559]

Now it was the Indians turn to be enraged, and all along the frontier settlers found themselves under attack, many being killed or wounded. The report of the king's commission, issued the next year, estimated that "near 300 Christian persons [were] murdered by the Indians" during this period.[560]

Berkeley now appointed Royalist burgesses John Washington and Isaac Allerton to make "a full and thorough investigation" of these raids, and take appropriate action to punish the guilty parties.[561] Instead, in conjunction with 250 troops from Maryland, they gathered 1000 men and laid siege to another Indian village in Maryland in September, 1675. When five Indian chiefs came out to negotiate, they were accused of participating in the raids in both Virginia and Maryland, and despite their denial and the white flag under which they stood, the British commanders had them killed.[562] With this act began a full-scale war between the Indians and the Virginia colonists.

In George Bernard Shaw's play, *Caesar and Cleopatra*, Cleopatra has the leader of her opposition killed. When the murder is discovered a mob gathers outside Cleopatra's home, screaming for revenge. Though she begs Caesar to protect her, he laughs in contempt and says,

> Do you hear? These knockers at your gate are also believers in vengeance and in stabbing. You have slain their leader: it is right that they shall slay you. . . . And then in the name of that *right* (he emphasizes the word with great scorn) shall I not slay them for murdering their Queen, and be slain in my turn by their countrymen as the invader of their fatherland? Can Rome do less than slay these slayers, too, to show the world how Rome avenges her sons and her honor? And so, to the end of history, murder shall breed murder, always in the name of right and honor and peace, until the gods are tired of blood and create a race that can understand.[563]

Rather than trying to locate the specific individuals who stole the hogs, killed Roger Hen, or attacked the Indians on the Potomac, both sides made war on the other because of their tribal/ethnic affiliation. Issues of guilt or innocence and the responsibility of the individual were forgotten or never considered. The conflict became a racial war for both sides, fueled by vengeance, hatred and fear.

It is interesting to note that Royalist ideas of caste and birthright had much in common with such bigoted thinking, while it was Parliamentarian ideas that said you did not hate or condemn someone because of his birthright. And strangely enough, it was William Berkeley who stood for these Parliamentarian ideas in his defense of the Indians. Following the murder of the five Indian chiefs, Berkeley was horrified. "If they had killed my grandfather or grandmother, my father and mother and all my friends, yet if they come to treat of peace, they ought to have gone in peace."[564] He refused to give his permission for the British settlers to organize and go out to attack the Indians. Having successfully lived in peace with numerous Indians since 1644, Berkeley knew that some were friendly, others were not, and to wage indiscriminate war on all was wrong. Even when thirty-six settlers were killed on a single day in

January 1676, Berkeley refused to sanction an all out attack on the Indians, saying instead that

> There could not be anything done in order to revenge the blood of our poor neighbors and the wrong done to his most sacred majesty's loyal subjects, until an Assembly was to sit, which then was about the 11th of March.[565]

Berkeley was probably hoping that cooler heads would prevail at that March Assembly.

Unfortunately, the policies of Berkeley and the Royalists elites now came back to haunt them. Virginia had been a society of caste and forced labor for almost seventy years. Two full generations had passed since George Percy had "descried the land of Virginia," and ideas of violence, of oppression, of power, of racial hatred, had become commonplace in this society.[566]

No one had stood up when the headright system was established to take advantage of poor immigrants. No one had stood up when Puritans were banished. No one had stood up when the Quakers were imprisoned and oppressed. And no one had stood up when the first laws against the blacks were passed.

Now it was the Indians under attack, and though William Berkeley was actually right in his unwillingness to sanction an all out attack on all Indians, his moral standing was weak, considering his Royalist policies of the previous fifteen years. By accepting the oppression of black slaves and white servants, Berkeley could hardly claim a high moral right to protect Indians, especially when large numbers of settlers on the frontier were being attacked and killed.

Furthermore, Berkeley's refusal to sanction a military attack on the Indians was the final indignity felt by the thousands of Virginia citizens who had been over-taxed, over-regulated, and under-represented by their government for almost a decade and a half. Now their government was even willing to deny them the right to defend themselves. Berkeley's position became a lightning rod for all their resentments, grievances, and dissatisfactions. As Berkeley himself noted during the height of the rebellion, "How miserable that man is that governs a people where six parts of seven at least are poor, indebted, discontented, and armed."[567]

Among those who felt "poor, indebted, discontented, and armed" was one Nathaniel Bacon, Jr., the younger cousin of councilor Nathaniel Bacon, Sr. Having only arrived in the colony in 1674 as a young man in his twenties, Bacon had been unable to establish himself in a respectable life in England, and in fact had done some things that had made him the black sheep of his family. He had been removed from Cambridge University by his father for having "broken into some extravagancies."[568] He had participated in a plot to defraud a neighbor of his inheritance. His marriage to Elizabeth Duke, the daughter of Sir Edward Duke of Benhall, so angered her father that he disinherited her and never spoke to her again. Finally, in the summer of 1674 his father gave Bacon 1800 pounds and sent him packing to Virginia.[569]

With the help of his older cousin, Nathaniel Bacon, Sr., who had been in the colony for almost two decades and was firmly entrenched in the Royalist power structure, Bacon found the wheels greased for hisbenefit. He purchased land on the frontier, and on March 3, 1675, less than a year after his arrival, William Berkeley appointed him to the Council.[570]

Yet, Bacon did not seem interested in following the lead of his Royalist cousin. Instead, he seemed to identify more with the frontiersmen living on the edge of the wilderness. He attended only two of the five Council meetings held before March 1676 (see tables 41 and 42), while simultaneously requesting a license to trade with the Indians (which Berkeley rejected).[571] In September 1675, while Allerton and Washington were attacking the Susquahanocks, Bacon seized, without permission from Berkeley, "some friendly Appomattox Indians allegedly for stealing corn."[572] Berkeley rebuked him for this action, telling him his "rash heady actions" of attacking a friendly Indian tribe could only worsen the situation, angering them all and causing "a general combination of all the Indians against us."[573]

When the Long Assembly finally met in March 1676, it took what it considered prudent, defensive actions to protect the settlers on the frontier.[574] Forts would be built at the heads of the various rivers leading into the backcountry, with specific quantities of men and ammunition at each. The Assembly named the commanders of these forts. Also named were two men from each county for the purpose of impressing both supplies and men for these forts. Those named were almost all members of the Assembly. Most were powerful Royalists, familiar names such as councilors John Carter, Francis Willis, Philip Ludwell, Nathaniel Bacon, Sr., William Cole, and Joseph Bridger, or burgesses John Burnham, John West, John Page, Edward Ramsey, John Lear, and John Powell, or the sons of members, such as Edward Hill, Jr. and William Claiborne, Jr. These men were given unilateral power to take whatever "men and horse . . . provisions and other necessaries . . . sloops, boats, or other convenience of carriage"[575] they felt were required. These forts, combined with horse patrols, were expected to protect the frontier settlers and re-establish peace with the Indians.

Table 41-The Long Assembly, 1675

Current Governor:
Governor William Berkeley

Council-March 1675

Royalist:	Uncertain Parliamentarian:
*Bacon, Nathaniel, Sr.	*Ballard, Thomas
*Bridger, Joseph	+Beale, Thomas
*Chickley, Henry	+Swann, Thomas
+Corbin, Henry	
+Digges, Edward	
Bray, James	
Cole, William	
Ludwell, Phillip	

Council-June 1675

Royalist:	Uncertain Parliamentarian:
*Bacon, Nathaniel, Sr.	*Ballard, Thomas
*Bray, James	*Beale, Thomas
*Bridger, Joseph	*Swann, Thomas
*Chickley, Henry	
*Cole, William	
*Corbin, Henry	
*Ludwell, Phillip	

Council-October 1675

Royalist:	Parliamentarian:
*Bacon, Nathaniel, Sr.	Bacon, Nathaniel, Jr.
*Bray, James	
*Bridger, Joseph	Uncertain Parliamentarian:
*Cole, William	*Ballard, Thomas
*Corbin, Henry	*Beale, Thomas
*Ludwell, Phillip	*Swann, Thomas
Place, Rowland	Bowler, Thomas
Wormeley, Ralph, Jr.	

House-Did not meet in 1675

	Present Slave-holders	Including future slave-holders	Raised in Virginia
Council-March	8 of 11	8 of 11	1 of 11
Council-June	7 of 10	7 of 10	2 of 10
Council-October	8 of 13	9 of 13	2 of 13

Legend for all tables:
*-incumbent
+-served previously
Bold names have purchased slaves or have claimed headrights on blacks.
<u>Underlined names will eventually buy slaves or claim headrights on blacks.</u>
Italicized names either were born in Virginia or immigrated as minors.

Source for all tables: See Appendix B

    Unfortunately, the settlers on the frontier did not agree. Planters in Charles City County appealed to the governor for permission to send out a force to attack the Indians.[576] Berkeley refused, and in April 1676 a crowd of volunteers began to gather near the mouth of the Appomattox at Jordan's Point in defiance of Berkeley's orders, intent on war. Meanwhile, Bacon and several of his disenfranchised local planter friends got together to complain to themselves about the general situation, including the Assembly's dependence on forts (which they considered useless against the mobile Indian tribes that hid in the wilderness forests along the fall line) as well as Berkeley's refusal to consider an equally

Table 42-The Long Assembly, February to March, 1676

Current Governor:
Governor William Berkeley

Council-February 1676

Royalist:	Uncertain Parliamentarian:
*Bacon, Nathaniel, Sr.	*Beale, Thomas
*Bridger, Joseph	*Swann, Thomas
*Cole, William	

Council-March 1676

Royalist:	Parliamentarian:
*Bacon, Nathaniel, Sr.	+Bacon, Nathaniel, Jr.
*Bridger, Joseph	
*Cole, William	Uncertain Parliamentarian:
+Bray, James	*Beale, Thomas
+Chickley, Henry	*Swann, Thomas
+Ludwell, Phillip	+Ballard, Thomas
+Place, Rowland	+Bowler, Thomas
+Spencer, Nicholas	
+Wormeley, Ralph, Jr.	

House-March 1676-Held over without general elections

Royalist:	Unknown:	Uncertain Parliamentarian:
*Ball, William, Sr.	*Blackey, William	*Kendall, William
*Leare, John	*Blake, John	
*Page, John	*Eppes, Francis, Jr.	
*Presley, William, Jr.	*Ffarrer, William	
*Warner, Aug., Jr.-SPEAKER	*Griffith, Edward	
*Washington, John	*Holte, Robert	
*Whitaker, Walter	*Hone, Theophilus	
Burnham, John	*Jordan, George	
Hill, Edward, Jr.	*Powell, John	
	*Walker, Thomas	
Uncertain Royalist:	*Williamson, Roger	
*Baker, Lawrence	*Wyatt, Nicholas	
*Claiborne, William, Jr.	Bray, Robert	
*Filmer, Henry	Presley, Peter	
*Ramsey, Edward	Thruston, Malachi	
*Savage, John		
Dale, Edward		
Digges, William		
Littleton, Southy		
West, John, Jr.		

	Present Slave-holders	Including future slave-holders	Raised in Virginia	Incumbents	Including those who served previously
Council-February	3 of 5	3 of 5	1 of 5		
Council-March	10 of 14	10 of 14	2 of 14		
House	13 of 34	14 of 34	14 of 34	25 of 34	25 of 34

mobile force to strike back. They decided to visit the Jordan's Point encampment, "to take a quantity of rum with them to give the men to drink."577

As a Council member, Bacon's arrival at this camp caused an immediate stir, and a cry of "A Bacon! A Bacon! A Bacon!" arose in the camp. Soon the crowd was calling on Bacon to lead them, that "they would also go along with him to take revenge upon the Indians, and drink damnation to their souls to be true to him, and if he could not obtain a commission they would assist him as well and as much as if he had one."[578] As a Council member who seemed more in tune with the concerns of the local planters, Bacon quickly became the bolt of lightning into which all the pent-up energy of anger and rebellion flowed, now aimed directly at the lightning rod of William Berkeley and the Long Assembly. Just as the conflict between the Indians and the British had escalated, step by step, until it involved wholesale war, the conflict between William Berkeley and Nathaniel Bacon was about to escalate in the same way, eventually blossoming into a general revolt against the government.

Despite at least three written requests by Bacon, Berkeley refused to give him a commission to fight the Indians,[579] and in fact gathered 300 men in early May and marched to Henrico County "to call Mr. Bacon to accompt."[580] According to Berkeley and his Council, Bacon's following was a

> rabble crew, only the rascality and meanest of people, there being hardly two amongst them that we have heard of who have estates or are persons of reputation and indeed very few who can either write or read.[581]

The legitimate fear by frontier settlers of Indian raids was now less important to the Royalists than the defiance of their rule by this "rabble crew" of poor settlers.[582]

Before Berkeley arrived, however, Bacon had marched south with his own troop of approximately 250 men, searching for Indians to attack. In the three weeks he was gone, he indiscriminately attacked all Indians, including a hostile Susquahanock village as well as a friendly Occaneechee one.[583]

While Bacon was off killing Indians, Berkeley went home to Green Spring. As he waited for Bacon to return Berkeley appears to have reconsidered the political situation, issuing several proclamations on May 10th.[584] The first called Bacon a rebel and relieved him of his position on the Council, though it also offered complete pardons to everyone else if they returned to their homes. The second proclamation called for general elections, the first to be held in almost fourteen years. In this proclamation Berkeley seemed to be trying to appeal to the general population, taking responsibility

> for any act of injustice by me done or any reward, bribe, or present by me accepted or taken from any person whatsoever.... And supposing I whom am head of the Assembly may be of their greatest grievance I will most gladly join with them in a

> petition to his most Sacred Majesty to appoint a new Governor of Virginia and thereby to ease anddischarge me from the great care and trouble thereof in my old age.[585]

As honorable as this sounds, this proclamation was actually empty political rhetoric. Berkeley knew the king was highly unlikely to pay attention to any petition from the common citizenry. He also knew nothing he had done would be considered corrupt by the king, that his reputation with the Crown was that of an honest governor, and that the complaints were more specifically aimed at the abuses of his Royalist allies and friends in power. And since the voting process had been restricted to house-holders in 1670 and was administered by those very same Royalist sheriffs appointed by Berkeley, the governor was quite confident that the elections would leave the Assembly as strongly Royalist as before.

By this time Berkeley seemed to have also reconsidered his willingness to defend some Indians. In an intellectual transition that must have been easy for a Royalist who believed in the importance of birthright, Berkeley now accepted the general belief that all the Indians were in league against the British. As he said to Thomas Goodrich on May 15th, "I believe all the Indians our neighbors are engaged with the Susquahanocks and therefore I desire you to spare none that has the name of an Indian for they are now all our enemies."[586] To William Claiborne, Jr. on the same day he ordered him "to take all Indians for enemies that have left their plantations."[587]

Returning from the backcountry near the end of May, Bacon found that though he had been removed from the Council, his neighbors in Henrico County had elected him to the House of Burgesses (see table 43).[588] His war against the Indians had made him a hero, and though there are questions about the effectiveness of his campaign, the Indian raids on the frontier were beginning to subside. The focus of the conflict now shifted to the war between Berkeley and Bacon.

On June 5th the Assembly met in Jamestown. Bacon sailed down the James River in a sloop with forty armed men, anchored just outside town, and sent a message to the Governor, asking "if he might in safety come on shore." Berkeley responded by having the town's cannon fire on the sloop, forcing Bacon to move the sloop up river out of gunshot range. That evening Bacon secretly came on shore, attempting to meet with his supporters in town.[589] Berkeley sent Captain Thomas Gardner to arrest him, and Bacon tried to flee up the James River. Sailing up river in his larger warship, *Adam and Eve*, Gardner boarded the sloop and captured Bacon, bringing him back to Jamestown as a prisoner.[590]

Table 43-Bacon's Assembly, June 1676

Current Governor:
Governor William Berkeley

Council-No known record

House-June 1676

Royalist:	Neutral:	Parliamentarian:
*Hill, Edward, Jr.	Mathews, Thomas	Bacon, Nathaniel, Jr.
*Presley, William, Jr.		Blayton, Thomas
*Warner, Augustine, Jr.	Unknown:	Crewes, James
*Washington, John	*Holte, Robert	+Lawrence, Richard
+Allerton, Isaac (a)	*Presley, Peter	
+Lee, Robert, Jr.	*Wyatt, Nicholas	Uncertain Parliamentarian:
	+Haynes, Thomas	+Godwin, Thomas-SPEAKER
Uncertain Royalist:	Caufield, Robert	
*West, John, Jr.	Carter, John, Jr.	
Beverly, Robert	Church, Richard	
Bristow, Robert	Mason, Francis	
	Mason, George	
	Moseley, Arthur	
	Roberts, Thomas	
	Tiplady, John	

	Present Slave-holders	Including future slave-holders	Raised in Virginia	Incumbents	Including those who served previously
House	14 of 27	16 of 27	12 of 27	8 of 27	13 of 27

Legend for all tables: Source for all tables: See Appendix B
*-incumbent
+-served previously
Bold names have purchased slaves or have claimed headrights on blacks.
Underlined names will eventually buy slaves or claim headrights on blacks.
Italicized names either were born in Virginia or immigrated as minors.

By this time there were as many as two thousand Virginians crowded into Jamestown, eager to see what would happen. Most were probably Bacon supporters.[591] It is therefore not surprising that when Bacon agreed to kneel before the Governor, confess his sins, and beg the Governor's pardon, Berkeley immediately pardoned him, returned him to the Council, and even promised (but did not give) him the military commission Bacon had been asking for the last six weeks.[592]

Bacon did not stay for this June Assembly, asking instead to return to Henrico to visit his wife, who was reportedly ill.[593] While he was gone the Assembly, the first to be newly elected in so many years, addressed some of the obvious grievances felt by the general citizenry, but it did so in a manner resembling the actions of the Parliamentarian Assemblies under Richard Bennett. Some acts specifically reduced the power and control of the elite aristocracy of Virginia, while other acts confirmed the Royalist doctrine that rule can come from above, dictating how every citizen should live his or her life.

This Assembly is now called Bacon's Assembly by historians. While the known membership still appeared to favor the Royalists, the Assembly was no longer entirely dominated by these men. Several burgesses, such as James Crewes and Richard Lawrence, were quite hostile to Berkeley and the Royalists. James Crewes was a close friend of Bacon, had accompanied him on Bacon's first march against the Indians, and was eventually hung for being "a most notorious actor and assistor in the rebellion."[594] Richard Lawrence owned the tavern in Jamestown where Bacon had attempted to meet his supporters, helped lead the rebellion, and had had several disputes with Berkeley over the years. Sentenced to be hung for his part in the rebellion, he fled the colony to places unknown.[595]

This shift away from the Royalists had a direct effect on legislation. Almost half the acts passed by this Assembly restricted the power of the governor, the council, the local justices, and anyone else in a position of authority. The 1670 law restricting voting rights was repealed, allowing all free men the vote.[596] The sheriff's term was limited to one year, serving in more than one office was forbidden, and penalties set for taking bribes and interfering with elections.[597] The appointment of church vestries was taken from them, and instead, vestries were limited to twelve and were to be elected every three years.[598] Councilors were no longer exempt from taxes, given instead a salary.[599] County taxes were to be assessed by a committee of Justices plus elected representatives.[600] The appointment of tax collectors was taken from the Governor and Council and given to the county courts.[601] A general pardon was announced.[602] And finally, Edward Hill, Jr. and John Stith were expelled from all offices for abusing their power for personal gain.[603]

On the other hand, several laws passed by the June Assembly illustrated the continuing influence of Royalist ideas of top-down-rule. Demonstrations and protests were forbidden, and the Governor was given the power to use force to suppress them.[604] All trade with the Indians was forbidden.[605] The export of corn was banned.[606] Taverns were banned, except at two ferry crossings.[607]

Despite these restrictive measures, the June Assembly had made an attempt to reform the political abuses of the last two decades. Its members were property owners and freedmen from across the spectrum of Virginian society, many of whom had been denied access to power and wealth in the previous fourteen years.[608] The gist of their legislation was to open up the political process so they would not be excluded in the future.

Yet, the members of this new Assembly were chosen from a society that honored power and wealth and had no scruples about how that power and wealth could be obtained. Virginia-bred burgesses now made up almost a third of the House. Fourteen of the twenty-seven known burgesses of this Assembly were slave-owners or had claimed headrights on slaves, including Bacon and his closest allies. Because of this, there were limits on how much reform these men could tolerate, and so, they took no action to ease the servant or slave laws, or to change the headright system, all of which they now saw to their advantage.

Meanwhile, Nathaniel Bacon sailed back to Henrico County and began gathering a force of men. By Friday June 23rd he had returned to Jamestown with 500 frontiersmen, once again demanding the military commission promised by Berkeley.[609]

The town of Jamestown at this time had still not grown beyond a single row of about a dozen brick buildings adjacent to the James River. Bacon's small army must have filled the street, surrounded by the thousand or so citizens who had gathered there as well. In the small two story state house the Assemblymen, both councilors and burgesses, peeked out the windows as Bacon's troops circled the building, raising their fusils (an early type of musket) and shouting "We will have it! We will have it!", referring to Bacon's military commission. Others chanted, "No levies! No levies!", an obvious protest against the possibility of more taxes. One unnamed burgess waved his handkerchief in the air and called back pleadingly, "You shall have it, you shall have it!"

Meanwhile, Governor Berkeley charged outside and stood face to face with Bacon on the steps of the building, calling him a rebel and a traitor, utterly refusing him his commission. He then challenged Bacon to a duel, to settle "the difference singly between ourselves." When Bacon declined, the seventy-one year old Berkeley then bared his chest and shouted "Here! Shoot me, 'fore god, fair mark, shoot!"

Bacon replied, "No, may it please your honor, we will not hurt a hair of your head, or of any other man's. We are come for a commission to save our lives from the Indians, which you have so often promised, and now we will have it before we go." He then turned to his men. "Make ready and present."

As soon as the men snapped to, guns raised for firing, the burgesses in the windows cried out, "For God's sake hold your hands and forebear a little, and you shall have what you please."

Under pressure from the rest of the Assembly, Berkeley grudgingly backed down. William Sherwood related how

> the Governor declared he would rather loose his life than consent to the granting such unreasonable things as he [Bacon] demanded, but for the prevention of that ruin, which was then threatened upon their second request, order was given for such a commission as Mr. Bacon would have himself, and according to his own dictates.[610]

In this manner Bacon finally got his military commission from William Berkeley. In fact, he was able to force Berkeley to sign an additional thirty blank commissions so that Bacon could choose his own subordinate officers, who were given the same power to impress men and supplies as the Royalists had given themselves in the March Assembly. The Assembly also discontinued several of the forts established in March, put Bacon in charge of a 1,000 man army, and gave him the right to call for and choose his own volunteers.[611]

On Sunday, June 25, the Assembly dissolved. The next day Bacon took his commissions and his army, and the citizens of Virginia marched off again to make war on the Indians.

No sooner had Bacon disappeared into the wilderness than did William Berkeley declare Bacon a rebel once again, his military commission invalid as it was obtained by force. Berkeley then went to Gloucester County to try to Assemble his own army of 1,200 men "to follow and suppress that rebel Bacon."[612] As soon as Berkeley announced these intentions, however, his support dissolved.[613] Men refused to join an army in opposition to Bacon, who "was now advancing against the common enemy, who had in a barbarous manner murdered some hundreds of our dear brethren and country men, and would, if not prevented by God, and the endeavors of good men, do their utmost for to cut off the whole colony."[614]

Once again Berkeley had found himself deserted by his beloved colony. Bacon had defied Berkeley's advice, extorted his commission from Berkeley, and threatened Berkeley's government with violence and overthrow; yet, it was Bacon who the public flocked to support. Faced with Bacon's strong army and unable to raise a force of his own, Berkeley now fled across Chesapeake Bay to the Eastern shore.[615] For the rest of the summer of 1676, Virginia was divided by violent civil war, with Bacon and those shut out of government on one side and Berkeley and his Royalist allies on the other.

On July 30th Bacon issued a "Declaration to the People" and his "Manifesto," both of which condemned Berkeley and his friends for abusing their power, overtaxing the colony, and monopolizing the Indian trade. Since Berkeley had the sole power of licensing trade with the Indians, many had accused him of using this right to benefit his friends.[616]

On August 3rd and 4th Bacon held a convention at Middle Plantation, demanding and getting a series of further declarations accusing Berkeley of causing a civil war by his refusal to allow the settlers to fight the Indians.[617] During this gathering Bacon actually entertained the idea of a complete separation from England by declaring Virginia an independent state. Discouraged by those he broached the idea to, he never advocated the idea publicly.[618]

In August, Bacon led a campaign against the friendly Pamunkey Indians, killing some but achieving little else.[619] Meanwhile, the men he left behind to run the colony devoted their energy to the plunder and destruction of the mansions, plantations, and farms of the sixty or so Royalists who had deserted their homes and fled to the Eastern Shore with Berkeley. Berkeley's own estate, Green Spring, was practically gutted by Bacon's men. Henry Chickeley was kept imprisoned, and the estates of Royalists Augustine Warner, Sr., Edward Hill, Jr., Thomas Ludwell, Daniel Parke, Sr., and Lawrence Smith were looted.[620]

In late August Berkeley captured several large ships anchored in Chesapeake Bay, giving him control of the colony's major waterways. On September 8th he occupied Jamestown. In an attempt to recapture the town, Bacon and the rebels lay siege to it on September 15th, actually placing the wives of several Royalist councilors and burgesses on their makeshift fortifications for defensive purposes. In the ensuing battle, Berkeley's forces were easily defeated, running from the determined fire of Bacon's men. On September 19th Bacon occupied Jamestown and that very night he torched the town.[621]

Then suddenly, in October, Bacon died, not of a Royalist's bullet but by an illness they called the "Bloody Flux."[622] For the next two months the civil war became a series of small skirmishes as Berkeley used the ships under his control to drive the rebels back. On November 21st Captain Thomas Grantham arrived with the ship *Concord*, with its thirty guns. When Grantham moved against the main force of rebels (400 men garrisoned at the home of Royalist councilor John West), most quickly surrendered. The last hundred, however, included eighty black slaves and twenty servants, and these refused to give up unless Grantham promised them their freedom. Grantham did so, then disarmed them and promptly returned them to their masters.[623]

In January 1677, Berkeley convened a court martial which proceeded to convict and hang as many of Bacon's supporters as possible (see table 44).[624] Even after the king's commissioners arrived at the end of the month with 1,000 British troops and carrying a general pardon from the king for everyone but Bacon, Berkeley continued to sentence and hang his opposition, putting to death (by Berkeley's own count) at least thirteen men.[625] The commissioners reported to King Charles II that Berkeley apparently wanted "to hang upon this Rebellion more than ever suffered death for the horrid murder of that late glorious martyr of blessed memory [Charles I]."[626] Like the hero in his own play, Berkeley vowed to

> Prosecute till I have made
> All that were guilty of my loss of peace,
> Wash their impiety in their guilty blood.[627]

Table 44-The Court Martial Councils, 1677

Current Governor:
Governor William Berkeley

Council/Court Martial-January 1677

Royalist:	Uncertain Parliamentarian:
*Bacon, Nathaniel, Sr.	*Ballard, Thomas
*Ludwell, Phillip	
Beveley, Robert	
Hill, [Edward, Jr.]	
Kemp, Mathew	
Smith, Lawrence	
Warner, Augustine, Jr.	

Uncertain Royalist:
Armstead, Anthony
Claiborne, William, Jr.
Jennifer, Daniel
Littleton, Southy
Moryson, Charles
Page, [John]
Ramsey, Edward
West, John, Jr.

Council/Court Martial-March 1677

Royalist:	Uncertain Parliamentarian:
*Bacon, Nathaniel, Sr.	+Ballard, Thomas-SUSPENDED
*Ludwell, Phillip	
+Bray, James	
+Bridger, Joseph	
+Chickley, Henry	
+Cole, William	
+Moryson, Francis-Commissioner	
+Spencer, Nicholas	
+Wormeley, Ralph, Jr.	
Berry, John-Commissioner	
Jefferys, Herbert-Governor/Commissioner	

	Present Slave-holders	Including future slave-holders	Raised in Virginia
Council-January	9 of 16	10 of 16	6 of 16
Council-March	8 of 12	8 of 12	1 of 12

Legend for all tables:
*-incumbent
+-served previously
Bold names have purchased slaves or have claimed headrights on blacks.
Underlined names will eventually buy slaves or claim headrights on blacks.
Italicized names either were born in Virginia or immigrated as minors.

Source for all tables: See Appendix B

To Berkeley the rebellion had been as much a personal betrayal than a defiance of the king's authority.[628]. Nor did Berkeley act alone in these harsh measures. As the king's commissioners themselves noted,

> We observed that the royal party, that sat on the bench with us at the trial, to be so forward in impeachment, accusing, reviling, the prisoners at bar, with that inveteracy, as if they had been the worst of witnesses rather than justices of the commission; both accusing and condemning at the same time.[629]

This court martial included Royalists such as Nathaniel Bacon, Sr. Philip Ludwell, Edward Hill, Jr., Joseph Bridger, Ralph Wormeley, and Henry Chickeley, men whom had spent the last few decades trying to cement their status as a superior caste of rulers. Bacon's Rebellion had been a direct attack on their homes, their fortunes, and most importantly, their deepest held beliefs.

According to historian Thomas Wertenbaker, when rebel Anthony Arnold stood before William Berkeley, condemned to death for participating in the Rebellion, he proclaimed that

> Kings have no rights but what they got by conquest and the sword, and he who can by force of the sword deprive them thereof has as good and just title to them as the king himself. If the king should deny to do me right I would think no more of it to sheath my sword in his heart or bowels than of my mortal enemies.[630]

That these words so clearly echo those of Major Francis White's, written more than twenty-five years before as King Charles I was ascending the scaffold, helps explain why the older Royalists on the court martial felt such hatred and venom toward Bacon's allies.

What is important to remember about Bacon's Rebellion was not that it foreshadowed the American Revolution, but that it demonstrated the absurdity and idiocy of trying to administer a democratic government from above, without the consent and consideration of the citizenry. Eventually such rule can only lead to resentment, anger, corruption, violence, revolt, and finally, collapse.

The rebellion was a result of the unbending and unyielding nature of governmental philosophy held by William Berkeley and the members of the Long Assembly during Berkeley's second administration, and how that uncompromising rule oppressed the general citizenry of the colony. The ease in which Bacon was able to rally an army around him; the size of the crowds in Jamestown while the June Assembly met; the inability of Berkeley to raise his own army; the eagerness and pleasure that so many took in plundering the homes of the most powerful and wealthy Virginians; the presence of so many blacks and servants in Bacon's army: all these facts illustrate the anger and resentment felt by too many common Virginians against their legislators. And since those legislators had for so long denied them an opportunity to express this anger and resentment peaceably at the ballot box, they were eventually forced to use violence to express it.

Unfortunately for the blacks of Virginia, both free and enslaved, the rebellion represented a poor opportunity to stop the slow rise of slavery. Even when Bacon and his men controlled the Assembly in June, they did nothing to change the laws that encouraged the growth of slavery. They did not even act to reduce the term for indentured servants, which was a major reason for the decline in the immigration of white Englishmen to the colony at this time.

It is very difficult to repeal a law once passed. Such a law creates vested interests, individuals who benefit from its existence and therefore resist its repeal. Hence, those members of the June Assembly who owned slaves had financial reasons to leave the slave laws intact. For those who didn't own slaves, the law had no immediate or direct effect on them, and since there was no spiritual or ethical voice in the Virginian community to demand better from them, they could look the other way and say "I don't own slaves, I won't own slaves, so what difference does it make to me? If William Presley or Richard Lee or John Washington wants to trade in human beings, what can I do? Why make trouble for myself?"

Hence, even had the rebels not been defeated, even had the common folk maintained their control over the legislature of Virginia, it is unlikely that slavery would have been nipped in the bud by these new legislators. Virginian society still had no moral component. Issues of right or wrong still took second place to issues of power and financial gain.

Yet, had the rebels been able to maintain a greater share of power in the years following Bacon's Rebellion they might have opened the door of freedom just a crack. With the failure of the revolt, however, the last possible opposition to slavery and arrogant rule was finally stamped out. No one was left to oppose the steady and constant growth of slavery. What remained were deeply entrenched customs of servitude, oppressed labor, and birthright, flourishing in an atmosphere of ignorance and little spiritual guidance, functioning in a democracy in which the use of power earned respect.

Though slavery did not grow quickly over the next decade, its foundation was now deeply laid. By the 1690s, Virginia's native-born population was finally beginning to dominate the society, having grown up in a community with few religious institutions and no system for educating its children, ruled by an elite that believed in the concepts of caste, raised in broken homes with insecure relationships, and surrounded by a society that had already sanctioned the idea that human beings with black skins *belong* enslaved and to enslave these people was the best and fastest route to power and wealth. Such an education could only guarantee an increase in such a practice, and the result was an explosion of slave imports in the 1690s, accompanied by a series of new laws to more firmly institutionalize the custom.[631]

13. A Matter of Choice

> I tremble for my country when I reflect that God is just, that his justice cannot sleep forever.[632]
> —Thomas Jefferson in 1784 on the subject of slavery

On July 9, 1677, William Berkeley died in his brother's home in England. He had been recalled to London by King Charles II to explain his actions before, during, and after Bacon's Rebellion, and he had arrived in London only three weeks before. Determined "to clear his innocency" with his king, Berkeley never got the chance.[633] "He came here alive," said Secretary of State Henry Coventry, "but so unlike to live that it had been very inhumane to have troubled him with any interrogations. So he died without any accompt given of his government."[634] In the months that followed, Berkeley was strongly implicated by the king's commissioners as instigating Bacon's Rebellion by his governmental policies, and in later years this accusation as well as the Rebellion itself served to damage his reputation. Even now, more than three centuries later, no historian has ever honored his memory with a biography.

Nor was William Berkeley alone in this. Many of the leading men of Berkeley's generation were not remembered highly by their immediate descendants. As one example, only twelve months before his death, Edmund Scarburgh had been attacked by a servant, who supposedly said "he would work no more for Scarburgh's whores and bastards."[635] Five months after this Governor Berkeley suspended Scarburgh from all offices because of his ill-dealing with the Indians. When he died in 1671, twenty-one of his servants petitioned for their freedom, many claiming that they had been held in servitude beyond the expiration of their terms. The courts agreed, and most were granted their freedom.[636] And though the custom was quite common in Virginia, Scarburgh's eldest son, Charles, never named any of his children after his father, nor did the name Edmund appear among any of his descendants. It was as if Charles wished to forget the memory of his father.[637]

The men who ruled Virginia in the middle decades of the seventeenth century were incredibly courageous and daring individuals. They came to a wild and empty land and devoted their hearts and minds to creating in that wilderness a new society, modeled after the England they loved and admired. They came knowing that the journey involved incalculable risks, that at least one in nine would die in the first year from disease. All came intent on becoming as rich and as powerful as possible in as short a time as possible, and in doing so they would create a new British society in the process.

These leaders were not the poor, indentured servants who came to the colony because they had no other alternative. Most were like William Berkeley, born into the wealthy and powerful aristocracy of England, and though they were usually second sons, denied by English law a position in the nobility, the dangerous, risky and adventurous life on the frontier was not their only option.

Yet, they came. When Edmund Scarburgh's father immigrated to Virginia in the 1620's, he brought with him his wife and son with the obvious hope of building for them a new life and estate. When he himself died only a few years later, his son used that small estate to create for himself and for Virginia a thriving and multi-faceted enterprise, including foreign trade, ship-building, shoe-making, tanning, and salt-making in addition to tobacco cultivation and food production.[638] Not knowing the best way to make a new society out of nothing, he did the best he could, and succeeded by far in more ways than he failed.

It would be terribly unfair to these men if we did not recognize that most of them had sincere and well-meaning intentions for the colony. Berkeley especially dedicated his life to Virginia. He lobbied hard to encourage others to immigrate there, and when they arrived he tried to make Virginia as good place for them to live as possible. Even in disgrace after Bacon's Rebellion, none of the charges against Berkeley's administration were aimed at him personally. He was seen as having been honest and fair, though misguided concerning the corrupt legislators around him and his unwillingness to allow Virginians to fight the Indians.

Despite these good intentions, however, Berkeley and his generation led the people of Virginia down a terrible road of misery and oppression, and this is why their immediate descendants gave them so little honor or respect. When Bacon's Rebellion took place in 1676, it expressed better than any words the opinions of the Virginian citizenry, poor and wealthy alike, about their society and the government policies of their leadership. For almost two decades the Long Assembly and the Royalists that dominated it attempted to rule a democratic society with coercion and force, to use their governmental power to line their own pockets, and to pass laws that reduced the freedom of everyone else, both black and white. Bacon's Rebellion was the equivalent of a collective shout of "Enough!"

When Berkeley died in 1677, Virginia was not yet the slave society it would become in the next century. The slave population only comprised from 3 to 9 percent of the population, had only risen from 2 percent since Berkeley's arrival in the colony thirty-five years before, and would not begin growing significantly for another decade (see figure 3.1). Yet, the actions and decisions of men like Berkeley, Scarburgh, Bennett, Presley, and others to shape Virginia as they wished created a society that saw nothing wrong with enslaving people with black skin. For the next hundred years the importation of slaves would steadily increase, and by the time of the Revolutionary War, half the population of Virginia would be black slaves.

Since World War II, many historians have focused on what they saw as the inherent racism of British society as a principle cause for this rise in slave ownership. During the early 1960s, a blistering debate was waged between Carl Degler, who believed it was this inherent racism that caused slavery, and Oscar and Mary Handlin, who argued instead that slavery itself made Americans racist.[639] In subsequent years, the vigor of the debate has hardly diminished. Aldan Vaughan, Edmund Morgan, Winthrop Jordan, Leon Higginbotham, to name just a few, have all weighed in with their theories on the importance of racism and xenophobia to the origin of slavery.

What the historians who have advocated racism as a cause of slavery all seem to ignore, however, is the powerful distinction that lies between bigotry and oppression. While it might be quite easy to hate someone for being different than you, it is quite another thing to bring them into your home as a second-class citizen whom you consciously oppress.

Moreover, this assumption of a strong bigotry and xenophobia in British culture is not borne out by the facts. The treatment of the free blacks in early Virginia contradicts it. So does the willingness of every British colony, including Virginia, to accept immigrants from other European countries and allow them to become citizens. Even the Long Assembly routinely naturalized Dutch and German immigrants.[640]

And even if we totally accept this questionable modern interpretation that British society was inherently xenophobic, that most British citizens were bigoted against those who were outside their culture, this does not explain why the British colonists of Virginia were willing to buy slaves, while the British colonists of New England, Pennsylvania, and New York were not.

If you hate someone, you can merely avoid them. Many New England Puritans did not like Africans, considered slavery an acceptable condition for them, and were not afraid to express what we would call racist ideas about them.[641] Cotton Mather himself stated that "Indeed their [blacks'] stupidity is a discouragement. It may seem, unto as little purpose, to teach, as to wash an Ethiopian."[642]

Yet, despite the presence of such bigoted opinions, New Englanders simply chose not to purchase large numbers of slaves. In

Massachusetts the black population averaged 2 percent of the population (see figure 13.1), and by the Revolution, these blacks were free. Connecticut was similar, the black population always hovering around 3 percent (see figure 13.2). By the Revolution, 51 percent of these were free.

Only in Rhode Island did the black population ever exceed 15 percent, and was this high for only about thirty years (see figure 13.3). Even this number overstates the presence of slaves, since it is clear that a significant proportion of these blacks were free. By 1790, 78 percent of Rhode Island's blacks were free, and the percentage of slaves in the total population had shrunk to less than 1.4 percent.

In fact, the modern myth that slavery was widespread among New Englanders, that they gladly tolerated slavery in Virginia so they could sell them slaves, so exaggerates the facts that I wonder at the motives of those who advocate it. Not only did slavery never became a widespread and accepted custom in New England, the New Englanders who decided to sell slaves to Virginians in the 1680s felt compelled to do so in secret because of the unpopularity of the custom.[643] Moreover, New England legislators consistently gave the black slaves within their colony due process under the law. Slaves could own and transfer property, make contracts, and act as witnesses in court, even testifying against whites. Furthermore, unlike Virginia, a master was never given the right to

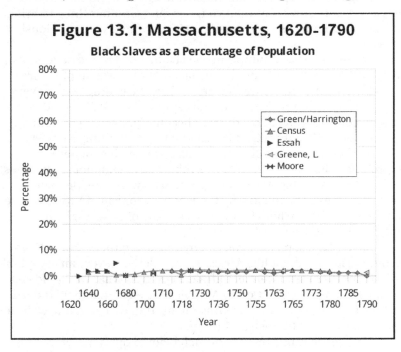

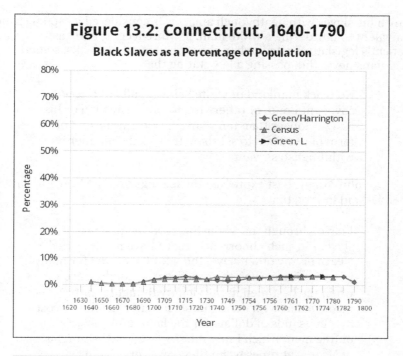

Figure 13.2: Connecticut, 1640-1790
Black Slaves as a Percentage of Population

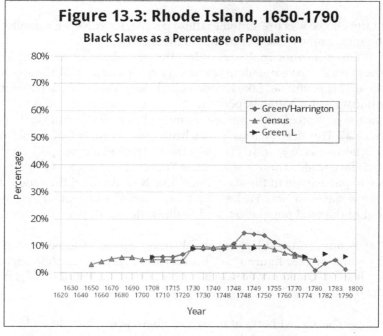

Figure 13.3: Rhode Island, 1650-1790
Black Slaves as a Percentage of Population

punish his slaves with death. Such an act was considered a capital crime, equivalent to murdering a white man.[644] Moreover, as early as 1652 Connecticut's legislature acted to ban the enslavement of blacks, something Virginians never did, passing a law stating that

> No black mankind or white being [shall be] forced by covenant bond, or otherwise, to serve any man or his assigns longer than ten years. . . . And at the end of term of ten years to set them free, as the manner is with English servants.[645]

John Adams best expressed the feelings of most New Englanders when he said in 1795 that

> I have, through my whole life, held the practice of slavery in such abhorrence, that I have never owned a Negro or any other slave, though I have lived for many years in times when the practice was not disgraceful, when the best men in my vicinity thought it not inconsistent with their character, and when it has cost me thousands of dollars for the labor and subsistence of free men, which I might have saved by the purchase of Negros at times when they were very cheap.[646]

When the choice was presented to these people, most New Englanders like Adams simply chose not to own someone else.

Other northern colonies reflect this as well. It is often claimed that New York's slave population was very comparable with the southern colonies. This is untrue. Like Rhode Island, New York's black population was almost always below 15 percent of the total population, and a significant portion of this number comprised free blacks, not slaves (see figure 13.4). This despite New York's history as a former Dutch colony during the years (1637 to 1654) when the Dutch had a monopoly on the African slave trade. Though some New Yorkers in later years were quite willing to participate in the slave trade, like New England they could sell only a few slaves in New York.[647] By 1790, blacks had declined to only 7.6 percent of the total population, and of these almost 18 percent were now free.

Pennsylvania's blacks probably never exceeded 10 percent of the total population (see figure 13.5). Furthermore, as early as 1688 several Pennsylvanian Quakers were already questioning the morality of slavery, stating that

> There is a liberty of conscience here which is right and reasonable, and there ought to be likewise liberty of

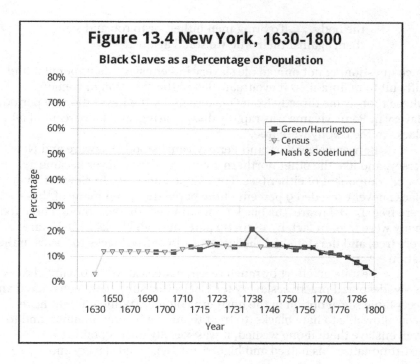

Figure 13.4 New York, 1630-1800
Black Slaves as a Percentage of Population

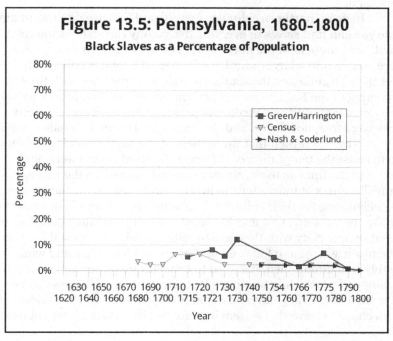

Figure 13.5: Pennsylvania, 1680-1800
Black Slaves as a Percentage of Population

the body. . . . To bring men hither, or to rob and sell them against their will, we stand against.[648]

Such questioning not only made slavery in Pennsylvania unpopular and difficult to maintain,[649] it eventually led to the abolition movement, whose roots came directly from Quakerism. By the Revolutionary period, slavery in Pennsylvania was rapidly disappearing, and 64 percent of all blacks in the colony were free.

Between New York and Pennsylvania stood Delaware and New Jersey, and like the other northern colonies, slavery never became a major component of either (see figures 13.6 and 13.7). In New Jersey blacks never exceeded 9 percent of the population, and many of these were free. In Delaware, the black population was higher, though here too many were free. In fact, by 1850, 89 percent of all blacks in Delaware were free, and the state had the largest ratio of free blacks to slaves in the nation.[650]

Despite an effort by much recent historical work to paint slavery as a widespread practice, slavery for blacks was simply not considered an acceptable social option in the northern colonies. While it might have been allowable to hate blacks, to treat them in a bigoted manner, and to even enslave them if one wished, it was socially unacceptable to institutionalize this hatred and bigotry into legalized slavery and oppression.

Instead, northern colonists merely said "No, we don't want any." While you can find slaves in every British colony throughout the colonial period, the numbers in the north never rise very high. In those areas where you do find a larger number of slaves, the total is still significantly lower than Virginia and the south, and this total declines with time and the coming of the Enlightenment and the American Revolution. Even more important, during the colonial period the majority of northern blacks were free, not slaves, and the law treated them as equals to whites.

Why was it then that the settlers of the northern colonies were able to resist the temptation of slavery while southerners were not?

In the final analysis, slavery's growth hinged on the willingness or unwillingness of individuals to make moral choices, for themselves, for their neighbors, for their children. The founders of Virginia in 1607 chose to make their colony, first and foremost, a money-making operation, and shaped their society with that goal in mind. Sandys' later establishment of the headright system followed in these initial footsteps, and while possibly well intentioned, it served to concentrate power and wealth in the hands of a few while treating a large percentage of the poor as no better than temporary slaves. Later administrations under Berkeley and others chose to leave this system in place, partly because it would have

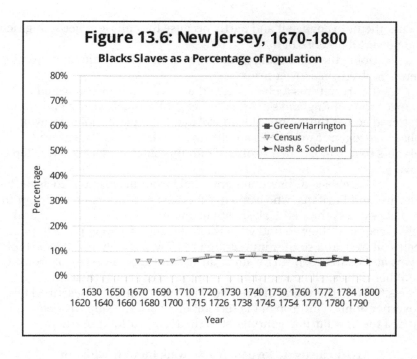

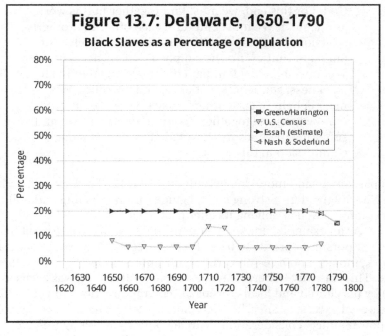

been difficult to change it and partly because they all were becoming rich and powerful because of it.

Doing the right thing would have been *hard*. Ethics and morality were simply given a lower priority.

The result was a place where the powerful, well-connected younger sons of England's aristocracy, eager for quick wealth and status, scattered across the countryside in isolated large plantations, growing tobacco in gigantic quantities and using forced, unpaid, indentured servants to maximize their profits. From this stage it was not a far step to slavery.

This choice differed fundamentally from the original goals of the Pilgrims and Puritans, who came to America not for greater profit but to actually express their ethical, spiritual, and moral beliefs. Those moral beliefs required them to find what they considered a good place to raise their children, and so they immigrated to New England as families, built towns and churches, lived on small farms, and grew a variety of crops to feed themselves, all in accordance with the strictures of their religious beliefs.[651] Note the vision of a perfect Puritan community described by Governor John Winthrop in his sermon on board the ship *Arabella* as it carried the first Puritan contingent to the New World in 1630:

> To do justly, to love mercy, to walk humbly with our God. For this end, we must knit together in this work as one man. We must entertain each other in brotherly affection. We must be willing to abridge ourselves of our superfluities, for the supply of others' necessities. We must uphold a familiar commerce together in all meekness, gentleness, patience, and liberality. We must delight in each other, make others' conditions our own, rejoice together, mourn together, labor, and suffer together, always having before our eyes our commission in this work.[652]

Such passionate and moral beliefs suffused New England society, and gave New Englanders a strong ethical guide for establishing and building their colony as well as solving its later problems.

As an example, unlike Virginians, the Puritans considered the too eager pursuit of profits to be sinful and greedy. They were not hostile to work, money, and free enterprise, but wanted moderation to be their guide. This was why Robert Keayne, a Puritan merchant, was accused in 1639 by his church and local citizens of "corrupt practice" and of "selling his wares at excessive rates." The community felt he had allowed greed to overpower ethical considerations. Already known for his aggressive business practice and sharp dealing, he was finally brought to court for having repeatedly demanded payment for a more expensive bag of eight-penny nails when the customer claimed he had only bought six-penny nails. Others also charged him with over-pricing his goods. He was fined by the General Court and threatened with excommunication by his church. These accusations, as well as others, practically destroyed Keayne, and he spent the last years of his life giving away his money in order to clear his name.[653]

It is hard to imagine such events happening in Virginia. Not only would no one have questioned Keayne's desire to overcharge, in Virginia a man who was questioned in this manner would have probably been unbothered by the accusation. When Puritan Richard Bennett became governor in 1652, the Assembly passed a law that limited all profits on the buying and selling of goods to under 50 percent. Not only was it repealed as soon as Edward Digges became Governor three years later, in the interim it had had little effect, and in fact was practically ignored.[654] Virginia was a place where each person was expected to get whatever he could, however he could, from whomever he could, regardless of the consequences or the ethics of those actions. Taking advantage of a 16-year-old immigrant boy to get eight years of free labor from him was considered perfectly acceptable in such a society.[655]

Since the Enlightenment, we have learned that democracy requires the free expression of religious, spiritual, and ethical ideas in order to put a check on this kind of abuse of power. As Jefferson said, "Where the press is free, and every man able to read, all is safe."[656]

In seventeenth century British society, religion served this function. Each religious order, from the Anglican to the Catholic to the Puritan to the Quaker, questioned the customs of society, raising uncomfortable ethical issues, and in the philosophical, spiritual, and intellectual debates that followed, drew society toward the pursuit of ethical and intellectual perfection. By permitting these different religious sects to co-exist peacefully while allowing them to freely and publicly air their moral disagreements, the individual citizens of British society were able to pick the best options from among them for creating a healthy, prosperous society for themselves and their children.

The warped nature of Virginia's society, its lack of women, towns, religion, and schools, was recognized as unacceptable by practically everyone who had either lived in or visited the colony. Berkeley himself perceived these problems, and sincerely tried, in his own Royalist way, to address them.

Hence, the arrival of both Puritans and Quakers to the colony should have been applauded by Berkeley and the Royalists, whether he or they agreed with their beliefs or not. Their ideas about what constituted a healthy society could only help to fix Virginia's social problems, if only because they would have forced the leadership in Virginia to consider a range of different approaches.

For example, to change Virginian society, Berkeley and his Royalist allies needed to encourage families to immigrate there, which is exactly what the Puritans and Quakers would have brought. Moreover, based on what the families of these religious groups looked for when they picked their destination, this usually meant a strong church, ownership of land, and clearly defined legal rights.[657] Therefore, to attract these families the Royalists would have had to give ministers and that "giddy multitude" some say in how the colony was run. They would have had to encourage the establishment of schools and alternative religious churches. They would have had to adopt land policies that would have allowed many poor settlers to quickly and easily obtain moderately-sized tracts of land. And they would have had to change the headright system to discourage the abuse of indentured servitude.

All of these actions would have encouraged the formation of small towns while simultaneously stabilizing the situation for the children in the colony. More importantly, however, they would have also limited the supply of forced labor, and distributed power downward to the general population, both of which would have reduced the central authority of Berkeley and his Royalists allies in the Assembly.

Instead, Governor Berkeley's first action upon arrival was to establish church government and to outlaw religious dissent. Under the false Royalist belief that church and religion must be imposed by the government in order to guarantee good social behavior, Berkeley and the Royalist majority discouraged, imprisoned, and expelled Puritan and Quaker ministers and their followers, first the Puritans in 1649 and then the Quakers in 1663. The result, however, was not a more effective Anglican church, but a spiritual and intellectual vacuum, especially since Berkeley and his Royalists allies would not allow their own Anglican church to develop independently. Combined with the legislature's reluctance to establish schools, Virginia became an uneducated community with no independent moral institutions for advocating the importance of ethics in daily life.

This situation was further compounded by Berkeley's decision upon his arrival to endorse the headright system, as demanded and maintained in the ensuing decades by the powerful leaders of the colony, thus encouraging the use of forced unpaid labor, exactly the opposite of what was needed to bring these religious families to Virginia.

Moreover, the Royalist obsession with birthright, privilege, and status meant that they put their faith in (what we today call) their group rights, rather than the uniqueness and abilities of each individual. If you could demonstrate membership in the right group, you would immediately earn the privileges of that group, regardless of your skills or talent. If you were instead a member of what Berkeley once called "the meanest quality and corruptest lives,"[658] you deserved nothing except the right to obey the orders of your superiors.

In a democracy, where each citizen is considered equal before the law, such allegiance to group rights, based on some subjective tribal standard, can only lead to conflict and injustice. All of the Founding Fathers at one time or another warned of this kind of tribalism, what they called factions, and how it was one of the chief causes destroying civil government. As James Madison said, "In a society under the forms of which the stronger faction can readily unite and oppress the weaker, anarchy may as truly be said to reign."[659] From our own perspective only a few decades after the end of a war-torn twentieth-century, it is not difficult to recognize the truth of this: consider for example the factional ethnic and racial hatreds that have killed millions in Lebanon, Bosnia, and Rwanda, to name just three examples from the 1990s.

In seventeenth-century Virginia, however, the Royalists considered their belief in group rights (i.e., the rights and privileges of the aristocracy) to be wholly beneficial to society. As an example, in the 1662 legal revisal Francis Moryson noted the need in criminal cases for judges to be chosen from

> men of the greatest abilities both for judgment and integrity . . . because that anything that concerns the life and limb requires the ablest jurors to inquire of it.

Moryson was not merely differentiating between the educated and uneducated, but the high and low born. As his friend Berkeley proudly noted only a year later in 1663,

> For men of as good families as any subjects in England have resided [in Virginia], as the Percys, the Berkeleys, the Wests, the Gages, the Throgmortons, Wyatts, Digges, Chickeleys, Molesworths, Morysons, Kemps, and hundred others, which I forbear to name, lest I misherald them in the catalog.[660]

For what they considered just and noble causes, they proceeded to not only set themselves above the rest of the poor white population, but to define blacks as inferior and incapable of handling freedom. The result was that for generations to come, Virginia would be sharply divided between a tiny caste of wealthy and privileged plantation families, owners of a vast army of black slaves, and the remaining population of poor and ignorant white farmers over whom they ruled.

The other British colonies did not do this. Instead, they relied upon Britain's first premise, that the individual is unique and must be allowed the right to rise and fall on his or her own merit. While the Virginian legislators chose to make group rights the law of the land, legitimizing a form of bigotry, the other colonies moved away from this premise, adopting instead the idea of individual freedom, determining a person's worth by what that person achieves.

Consequently, those who opposed these Royalist concepts of top-down-rule would not come to Virginia, would flee the colony for better pastures (either voluntarily or by force), or would stand by and do nothing. Quakers like Stephen Horsey were unwilling to raise their children in such an atmosphere and immigrated to Maryland. Others, like Obedience Robins and Edward Hill, Sr., did raise their children in Virginia, and were unable to teach them how to treat others with respect.

Richard Bennett best epitomizes the individual who chose to do nothing. A tolerant Puritan who in later years even had friendly ties with the Quakers, Bennett spent his entire life serving as either a burgess or councilman or governor. Yet, rather than use this power to encourage his religious beliefs, or revise the headright system, Bennett never seemed willing to challenge the will of the Royalist majority, making little public effort all his life to resist its desires. When he became governor from 1652 to 1655 he once again had the opportunity to change the colony's land policies or encourage the re-introduction of strong religious values, and instead chose to do nothing, exhibiting his consistent reluctance to lead. He remained a member of the Council until his death in 1675, and during the Long Assembly made no strong public protest that we know of against the slave and servant codes passed during that time. He stood silent while the Royalist view of birthright and caste took over the Assembly, cut off moral debate, barred dissenting religious views from the colony, crippled the power of their own Anglican ministers, and then moved to misuse their own power in an immoral way.

With no one questioning the morality of any of their decisions, this unhealthy majority continued to rule unchecked, deciding in 1662 that elections in a democracy were superfluous, and simply ended them. They were joined in this action by a generation of native-bred Virginians who didn't know the difference. These men of the Long Assembly then adopted increasingly racist and oppressive measures, distorting by far the original Royalist ideals of men like Berkeley and Edward Hill, Sr. By the 1670s, the majority in the Long Assembly saw nothing immoral with legalizing the enslavement of blacks for their own personal benefit as well as establishing a racial caste system based on a distorted, bigoted view of birthright. And when the general populace finally rebelled in 1676, they did so because of an uneducated, emotional, and selfish response to how their rulers were mistreating them, not because of some higher, moral principles taught them in school or church. In fact, they had no church or schools to go to, and most knew little of these things. Issues of ethics, morality, and right and wrong were buried in the passionate desire to destroy Indians and to plunder the homes of the hated local Justices.

By 1680 it would have taken a massive and dedicated effort to educate this unlettered populace and stave off the spread of slavery. Though difficult, it might still have been done, considering how alien a custom slavery was to British culture, how many Virginians were still British-born, and how few slaves were in the colony. King Charles II, as much a Royalist as his father, however, chose to team up with the dominant Royalist leadership in Virginia. Ethics was not a consideration. They all could make money on the sale of slaves, and they all now had ready-made the first generation of uneducated native-born Virginians, eager and willing to buy them.

And buy them they did. From the 1680s onward, slavery blossomed in Virginia, growing from approximately five percent of the population in 1680 to between 15 and 30 percent in 1700, and up to 50 percent by 1750, outstripping by far the northern colonies (see figure 13.8).[661] The descendants of those first British settlers no longer had any hesitancy about buying and owning slaves.

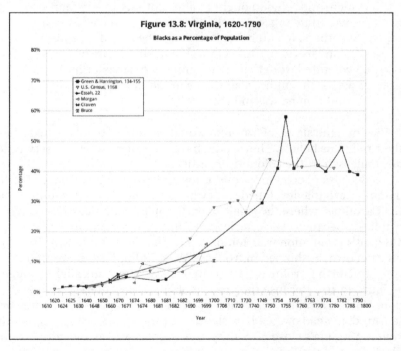

A Matter of Choice

Nor was this all. The influence of Virginia spread north and south. To the south, the restored king now moved to establish colonies in the Carolinas under the supervision of Royalist settlers from both the Caribbean and Virginia. The original charter for the Carolinas, issued after the restoration in 1660, named as the colony's proprietors such royal favorites as William Berkeley and his brother John as well as several Royalist slave-owners from Barbados and Jamaica, including Sir John Colleton, Sir Anthony Ashley Cooper, and the Duke of Albemarle.[662]

Given broad power to run the colony, these men quickly used their previous experiences in Virginia and the Caribbean to make the Carolinas into a single crop economy, built with slave labor for the sole purpose of exporting that single crop for their own personal profit. Similar to the Spanish and Portuguese sugar colonies of the Caribbean, the Carolinas became a variation of Virginia, a money-making operation for the sole benefit of its leading citizens. Berkeley was given power to name the governor and council, as well as the power to issue patents for land.[663] The colony's first constitution, written by John Locke in 1667, stated that

> Since charity obliges us to wish well to the souls of all men, and religion ought to alter nothing in any man's civil estate or right, it shall be lawful for slaves as well as others, to enter themselves and be as fully members as any freeman. But yet no slave shall hereby be exempted from that civil dominion his master hath over him, but be in all things in the same state and condition he was in before.[664]

Like Virginia, the slaves of Carolina would be held in bondage regardless of their religious beliefs. They would have no civil rights, their servitude firmly endorsed by the colony's constitution.

Because of economic conditions and geographic location, North and South Carolina developed somewhat differently. In the southern part of the Carolinas, where the single export crop became rice, the colony very quickly became a slave society.[665] Almost immediately the settlement's population was dominated by the slaves these Royalist plantation owners shipped in from the West Indies (see figure 13.9).

In North Carolina, the growth of slave labor took slightly longer (see figure 13.10). Because the proprietors took less interest in this region, North Carolina was settled by British immigrants, either free or indentured, instead of the black slaves the proprietors brought to South Carolina. Virginia's close proximity, however, as well as the initial control by Berkeley and the Royalist proprietors named by the king in 1660, influenced and shaped the colony's design. Land was given for the

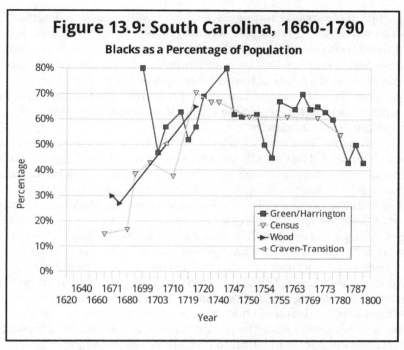

Figure 13.9: South Carolina, 1660-1790
Blacks as a Percentage of Population

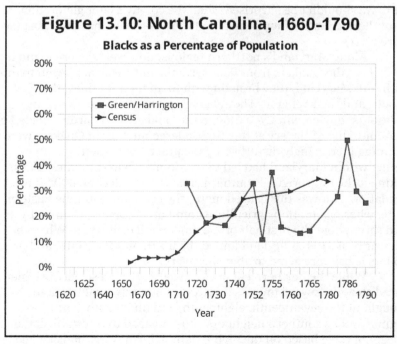

Figure 13.10: North Carolina, 1660-1790
Blacks as a Percentage of Population

importation of servants and slaves, and the government was designed from the top down.[666] Furthermore, too many of its settlers were familiar with the financial success of slave-trading Royalists like Edmund Scarburgh and worked to emulate that success. North Carolina became a tobacco colony built by the sweat of forced labor, spread out in dispersed plantations. In a little more than fifty years North Carolina's citizens, like Virginia's, had come to accept the idea of slavery and were buying slaves with as much enthusiasm.[667]

North of Virginia the influence of Berkeley and the Royalists was just as profound. The Catholic settlers, loyal to a church whose religious principles were dictated from above, had little problem with the Royalist concept of hierarchy, caste, and a frozen chain of being. Even the infusion of Puritan refugees from Virginia did not change this. More importantly, most of these settlers, Catholics and Puritans alike, were men who, like those in Virginia, centered their lives on money and getting rich quick. Southern Maryland soon became overwhelmed by the same social problems and tobacco culture that defined its southern neighbor, including skewed sex ratios and a poorly established educational system. Furthermore, the high mortality seen in Virginia was also seen in southern Maryland, and so children grew up in the same culture of parentless homes.[668] Since the terrain, climate and economy of tidewater Maryland were almost identical to its more populous neighbor to the south, we shouldn't be surprised when the colony's later generations in its southern counties found the temptation of slavery hard to resist (see figure 13.11).

Along Maryland's northern regions, however, slavery found itself blocked. As the Quakers from Pennsylvania and Delaware began to move south into Maryland, they brought with them their families, their schools, and their religion. They also brought with them a growing abolitionist movement. Even though the Quakers had hardly decided on the immorality of slavery at this time, a large number of Quakers were becoming increasingly disturbed by the practice. By the 1800s this conflict within Maryland had led to freedom for half the state's blacks, making it the only original southern state to see a decline in its slave population. Nor was this caused merely by economics: on the Eastern Shore, where plantation agriculture dominated, slavery declined by one third throughout the first half of the nineteenth century.[669] When the civil war broke out, Maryland joined the north, not the south, and in 1864 its legislature acted to abolish slavery.[670]

While the northern colonies as well as British society in general moved steadily toward universal suffrage and individual freedom throughout the seventeenth, eighteenth, and nineteenth centuries, Virginia and its southern neighbors chose instead to re-establish the concept of caste based on the divine right of birth. Virginians looked

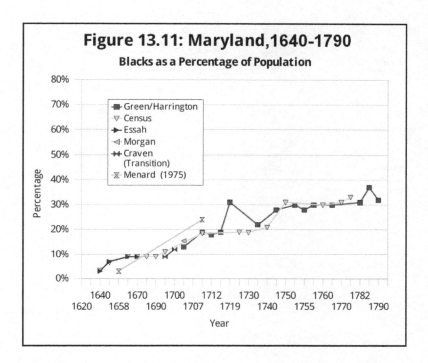

Figure 13.11: Maryland, 1640-1790
Blacks as a Percentage of Population

instead for easy solutions to the problems of low tobacco prices, labor shortages, racism, class conflict, and the violence and difficulty of settling a new land. The easy solution they chose was slavery.

A Matter of Choice

14. A Proclamation for Civilization in the Far Future

W. E. B. Du Bois, in studying the African slave trade, once asked: "How far in a State can a recognized moral wrong safely be compromised?" and answered his own question by saying that it is perilous for "any nation, through carelessness and moral cowardice, [to allow] any social evil to grow.... From this we may conclude that it behooves nations as well as men to do things at the very moment when they ought to be done."[671]

In the case of slavery and its origins in Virginia, Du Bois could easily have been referring to the headright system created by Edwin Sandys in 1619. Designed to encourage immigration to the colony, Sandys likely structured it as he did, with the land going to the person who paid the transport, because he assumed that those who could afford to pay the transport wouldn't be interested in immigrating, and that most of those who wanted to come to Virginia almost certainly couldn't afford it. By giving the person who paid the cost the benefit of land, he figured this would be the quickest way to encourage new settlers.

Yet, the immigration of the Pilgrims and Puritans to New England proved that this assumption was wrong. Had Sandys instead given the headright to the immigrant, he probably would have seen a quick wave of arrivals, though he might have not liked the fact that they were mostly religious refugees coming as families intent on building small private self-sustaining farms, and that their arrival would have done little to bring profits to the money-losing Virginia Company.

Once instituted, however, the headright system as Sandys designed it proved to be a societal disaster for Virginia, even if it did bring immigrants to the colony. It warped the immigration patterns, fostered the abuse of workers and the normalization of servitude, encouraged large dispersed plantations owned by a very few, and in general led Virginia down a bad social path.

Thus, Du Bois' advice here is most pertinent. These problems, caused by the headright system, became increasingly obvious throughout the 1620s and 1630s. Had the colony's early leaders been willing to face this issue promptly, and change the headright system – as it appears Governor Harvey and his superiors in England in the 1630s might have been trying to do – the entire history of Virginia, and the United States, could have been vastly different.[672]

They did not do this, however. To do so would have cost them money and power. Instead, for seven decades the colony's leaders tinkered around the edges of the problem, and while they tinkered the social fabric of Virginia drifted step by step in a direction that made black slavery and racial bigotry an accepted custom. And with each passing year the problem became increasingly harder and harder to fix.

Today we face the same circumstances. The human race is about to embark on the never-ending exploration of outer space. We, like the British in North America in the 1600s, will be establishing new human societies on the Moon, on Mars, on asteroids, and on innumerable other places in the vast emptiness of space, for now and as far into the future as any human can imagine.

Unfortunately, today – in the early decades of the twenty-first century – these first settlers in space will be saddled with as bad a legal framework as given to Virginia by Edwin Sandys in the early sixteenth century.

In 1967, during the height of the Cold War, the leaders of the United States, the Soviet Union, and a number of other nations negotiated the Outer Space Treaty, the first effort to provide a legal structure for the exploration of space by all nations.

The treaty had many positive outcomes. It required all signatories to publicly announce their launches and their planned activities in space. It demanded that they take responsibility for what they put in space, and make effort to prevent objects from falling out of control onto inhabited areas on the Earth. Signatories were also required to work together to aid and rescue all astronauts, whether in space or after their return to Earth. Nations were also tasked with preventing the contamination of either space or other planetary bodies.

The law also had many good intentions. It declared all astronauts as "envoys of mankind." It forbade anyone from bringing nuclear weapons into space, as well as establishing any military bases. Space was to be used "exclusively for peaceful purposes." The treaty also announced that

> Outer space, including the moon and other celestial bodies, shall be free for exploration and use by all States without discrimination of any kind, on a basis of equality and in accordance with international law, and there shall be free access to all areas of celestial bodies.

The treaty proclaimed space as the property of all humanity, and that "the exploration and use of outer space shall be carried out for the benefit and in the interests of all countries and shall be the province of all mankind."

Finally, and most important, the treaty forbade any nation from imposing its sovereignty on any territory in space.

> Outer space, including the moon and other celestial bodies, is not subject to national appropriation by claim of sovereignty, by means of use or occupation, or by any other means.

Space was to belong to all humanity, a vast utopia where borders would vanish and all humans would live together in peace and harmony.[673]

For the first half century of the space age this treaty sufficed. No one was yet capable of establishing colonies on the Moon, the asteroids, or the planets. The issue of property rights or national borders was not relevant. As written, the treaty provided a clean and clear framework for allowing the handful of space-faring nations (the United States, Russia, China, Europe) to avoid conflict and to work together, when necessary.

This is now changing. As the twenty-first century unfolds, nations across the globe are gearing up to establish the first interplanetary colonies. Private businesses are planning mining and manufacturing operations even as they fly tourists to space. Both will require firm legal rights to the territories settled and the investments made.

Yet, the Outer Space Treaty's ban on national sovereignty prevents this. The treaty's prohibition means that no framework exists for nations to establish safe and secure borders. No colony of any nation will have a secure right to its location, as according to the treaty all other nations have an equal right to the same territory.

Imagine if the U.S. establishes a lunar base near the Moon's south pole, a location where it is thought ice might be found in permanently shadowed craters, thus making that location very valuable real estate. What if the U.S. places its base in the best and most practical location for obtaining that ice?

What if, after the U.S. has established this base, the Chinese wish to also obtain access to this same real estate? According to the Outer Space Treaty, they will have that right, even if the resources there are somewhat limited and cannot be shared practically. As the treaty states, "there shall be free access to all areas of celestial bodies," and that possession cannot be determined by "use or occupation, or by any other means."

The U.S. would not take this lightly. Nor would China, if the roles were reversed. Without a mechanism for claiming ownership or sovereignty to its lunar base, both would have to use force to guarantee that ownership.[674]

The potential for military conflict is therefore enhanced, not weakened, by the Outer Space Treaty. Without a system to establish borders and sovereign rights, nations will have no choice but to fight over those borders and rights. The only rule for determining territorial sovereignty will be possession, and the use of military power to enforce it.

The law's prohibition on the establishment of sovereignty also means it forbids capitalist nations from establishing in space their laws on property rights and individual ownership. Future businesses establishing operations on other worlds will not be protected by any legal framework of any kind. Their property will not be protected by law. Nor will the individual rights of future settlers be protected in any manner. If you are an American citizen your rights under the Constitution and the Bill of Rights will instantly vanish when you leave the Earth.

You want to build a mining operation on the Moon? Go ahead, but there is no way you will be able to protect the ownership of your investment. It belongs not to you but, according to the Outer Space Treaty, is instead "the province of all mankind," and as such all others will be allowed "free access."

In the last decade, every country thinking of encouraging space exploration using private enterprise has recognized these issues. All have tried to write regulations that will allow for property rights in space, while still abiding by the provisions of the Outer Space Treaty.

All have so far failed. Luxembourg for example has been striving in the early 21st century to position itself as a leading investor for the private development of space. In its first attempt to rewrite its space law to overcome the treaty's restrictions on private ownership, that nation's legislature was forced to reject the language, noting that "private property claims are illegal or at least not legally binding in most of the international treaties and agreements relating to space and celestial bodies."[675]

Luxembourg's solution to this problem however was no solution. In their law's final draft, which their government finally accepted, they decided to ignore the problem. The revised law simply states that "Space resources are capable of being appropriated," which essentially admits that any private profit-oriented mission that launches under Luxembourg's authorization will have their blessing to take as much from any planetary body as they desire. No property rights are delineated, including the borders of any territory possessed, which is not surprising since the Outer Space Treaty forbids Luxembourg from doing so.[676]

Nor is Luxembourg alone. The United Kingdom, Japan, India, Europe, and the United States have all made failed attempts in recent years to pass laws that would recognize property rights in space within the restrictions of the Outer Space Treaty. Some, such as the United Kingdom and Japan, have, like Luxembourg, chosen to sidestep the issue, proposing laws that simply ignore the problem. India and Europe in turn have apparently decided that increased government control is a better solution, their proposed space laws concentrating the power and authority over future space ventures into the hands of government.

The United States government has meanwhile waffled. While its 2015 space law expressed full support for private property and called for the executive branch to begin a review of the Outer Space Treaty and its restrictions on ownership in space, the law itself did nothing to change the treaty or even to call for its revision.[677]

During the Trump administration there was an effort to bypass the treaty's limitations by requiring any nation that wished to partner with the U.S. in its Artemis lunar program to sign the Artemis Accords, a set of principles that were designed to respect property rights in space without violating the Outer Space Treaty.[678] By creating such bilateral agreements with other nations, the Trump administration hoped to do an end-around the treaty's restrictions.

Though eleven nations (as of June 2021) have signed on, it remains questionable whether this tactic will work. Others space powers, such as Russia and China, have publicly opposed the accords.[679]

At least one international law expert has even suggested that nations use "soft law" to overcome the treaty's limitations on sovereignty, property rights, and the establishment of borders.

> Soft law comprises rules or guidelines that have legal significance **but are not binding**. It sets standards of conduct for agreeing parties, much like those that protect the environment and endangered species. [emphasis added][680]

Faced with no lawful way for nations to establish their legal rights in space, this author proposes a system that will instead encourage outright contempt for the law.

Unfortunately, getting these countries to actually change the Outer Space Treaty carries no political weight. When the U.S. Senate held a series of hearings in 2017 to raise the issue of reworking the treaty, bureaucrats, lawyers, and industry experts from across the entire spectrum of the diplomatic and intellectual community rose up to defend the treaty and argue that no change was needed, or was possible. As Mike Gold, then Vice President of Space Systems Loral, noted at the second hearing,

> It would still be ill-advised for the U.S. to withdraw from the treaty or open it up to revisions. ... If the U.S. pulled out of the treaty it would create confusion and uncertainty, hindering new commercial developments as well as established private commercial space activities. Moreover, opening up the treaty to amendments would likely only result in more language being inserted into the Treaty that would run counter to U.S. interests.

Essentially, according to Gold, any effort to change the treaty `would carry with it too many unknowns, and was therefore too risky to try.[681] Others argued that it was simply premature to rework the treaty. Decades might still pass before human colonies were established on other worlds. Now was not the time to tinker with this document.

Congress has thus backed off from taking any action, and so the treaty stands. Like the early governors and planters in seventeenth century Virginia, the political leaders of the early 21st century would rather avoid dealing with the problems caused by the Outer Space Treaty than face reality and either amend the language of the treaty or replace it entirely. The international community has decided to kick the can down the road, making believe that the treaty, as written, can be made to work. All seem unaware that things can change very quickly (as illustrated by how fast SpaceX revolutionized the launch industry in just a few short years), and that by not dealing with the problem today they might find themselves in a very bad situation as soon as tomorrow.

And even if things don't change quickly? By not dealing with the gaps and weaknesses in the Outer Space Treaty now, when it is relatively easy to do, the leaders of the free world are still guaranteeing that future space colonists will not have the same rights and freedoms of those who live in that free world on Earth. They are also guaranteeing that the nations that achieve the first settlements in space will be forced into military conflict should they want to establish their sovereignty to those settlements. Like pirates, nations will be forced to grab as much as they can, as suggested by Luxembourg's space law, and will then use force to protect those holdings from any one else. Everyone will have to do this, because there will be no legal framework to establish their claims.

Thus, we are now repeating the same mistakes of Virginia, by not doing what Du Bois recommended, "to do things at the very moment when they ought to be done."

The Outer Space Treaty and the unwillingness of the nations of the world to rework it, however, is only a symptom of a much larger cultural problem.

The job of governments is to provide a clear and just legal framework for its citizens to follow, without telling those citizens what to do each step of the way. The U.S., as the leader of the free world, used to recognize this, based on its own colonial experience. After the Civil War, the nation used as its model the Puritan colonial approach. First the federal government created the territories that eventually became the western states, thus establishing a clear legal framework to encourage settlement. Then these territories established systems so that local towns and villages could incorporate and settlers could obtain private land to build their businesses and farms.

In conjunction with these local laws, the federal government in turn passed laws that encouraged these settlers to go west like the Puritans, as families who would build relatively small farms and businesses that were self-sufficient. The centerpiece of this legislation was the Homestead Act of 1862, where settlers who lived on a 160 acre plot for five years could claim ownership to it. With this framework in place, the federal government then got out of the way.

The result was that millions moved west to establish their own farms and businesses, owned mostly in small chunks as these individual and family settlers worked to create small towns and communities to raise their children and to recreate a viable society in the untamed wilderness. This was really how the west was won, by families and ordinary people, not by gunslingers and government programs.

This approach and the concepts of freedom behind it, however, were lost and forgotten during the 20th century. First, the settlement of the American west was largely completed, breaking the thread that connected our twenty-first century society with the settlement pattern learned in the past.

Second, the 20th century saw the coming of communism and its ideal of a centrally controlled society ruled from the top for the good of all. A host of nations, led by the Soviet Union, but followed in Eastern Europe, in Venezuela, in India, in China, and in many other places, all tried imposing a centralized socialist command society in order to avoid the unequal results that routinely come from freedom, capitalism, and private property. Even the leader of the capitalist free world, the United States, tried it, focusing much effort and money in the last half of the twentieth century and the first two decades of the twenty-first to using its federal government to regulate and control society for the betterment of all. Since the mid-twentieth century the U.S. has been reshaped with the passing of laws and the creation of federal agencies designed to regulate food, medicine, medical practice, environmental impact, workplace safety, housing, banking, and a host of other practices. In fact, in today's America almost every aspect of life now falls under the regulatory eye of the federal government.

Politically, the results of this worldwide move toward centralized government were routinely a disaster, demonstrated by the worst failures of such attempts. The Soviet Union went bankrupt, the people of Venezuela ended up starving, Eastern Europe revolted from communism, and China and India abandoned the attempt. Only when the last two nations adopted the free market and private enterprise did they finally begin to evolve from poor third world nations to rich first world powers with their own space age capabilities.

When the space race began in the 1960s, instead of simply laying out the simplest legal framework under which free citizens could follow their ambitions and dreams into space, the U.S. and every other space-faring nation moved, under these socialist concepts, to create "space programs," centralized government enterprises where the government itself dictated step-by-step what things would happen, when, and by whom.

The political nature of the Cold War undeniably forced this centralized policy on the United States. The country had to move quickly, and to catch up with the Soviet Union in space required a centralized government program. There wasn't time in the 1960s to allow for private enterprise to naturally evolve. Once that space race was won, however, the U.S. government did not release control to the private sector. If anything, it clamped down harder. First it ruled in the early 1960s that all U.S. communications satellites could only be built by a government-run corporation called Comsat. Then, in the 1970s, it required all satellites to be launched on the Space Shuttle. For almost three decades these actions effectively squelched both the American satellite industry as well as its rocket industry.

Not surprisingly, these centralized space programs of the latter half of the twentieth century also ended up producing little space exploration, and even less innovation. For the half century after the Apollo lunar landings no significant advancement in the manned exploration of the solar system took place. Instead, government spaceships went around and around the Earth, focused mostly on fueling government jobs on Earth rather then going where no human had gone before.

Only now, in the first decades of the twenty-first century, are some governments finally transitioning from this "space program" model to one that encourages private ownership and competing companies, each pushing their own ideas about space exploration. The result has been the first real innovation in space in decades.[682]

The problem today, in the twenty-first century, is that the free nations of the world, including the United States, no longer understand that in their effort to establish colonies in space it is essential for governments to lay out basic laws and then get out of the way. They no longer understand that they must let their free citizens do the exploring, the innovation, the inventing, and the settling.

Instead, they still think that it is their job to run things, even as some are simultaneously trying to encourage private enterprise in space. Rather than rework the Outer Space Treaty so that real private enterprise will flourish in space, nations still prefer to continue their space programs, making believe they can tinker at the edges of the treaty's restrictions even as they establish their government-controlled colonies in space.

What can be done to change this? Unfortunately, though many individuals both inside and outside government continue to push for the revision or rejection of the Outer Space Treaty, and though at this time such actions would be relatively easy because no established colonies in space yet exist, it is very unlikely that the treaty will be reworked, for many decades. Because the treaty outlaws national sovereignty in space, it acts to either focus power into the hands of the United Nations and the bureaucrats that administer its regulations and treaties, or leave nations and their power-hungry leaders entirely free to grab whatever they can in space, held back by no rules at all, including the rules designed to protect the rights of their citizens.

Meanwhile, as indicated by the quote above by the official from Space Systems Loral, the private commercial aerospace industry is terrified of any effort to abandon or rework the treaty because of the unknowns involved. Better the devil they know than the devil they don't.

In other words, vested interests already exist to defend the treaty, and they are acting to defend it as hard as they can. They have little interest in taking any risks or giving up any control, even if it might benefit the future space colonists on other worlds. And no one who might be in a position of power capable of pushing back against these interests seems willing to try.

It is therefore likely that the first few centuries of colonization throughout the solar system will not proceed peacefully or justly, as wished for by the good intentions of the Outer Space Treaty. Instead, that initial exploration will be a brutal legal nightmare for all involved, precisely *because* of the good intentions of the Outer Space Treaty.

Governments will scramble to grab as much as they can. And for private enterprise to succeed in space, the treaty's restrictions on property rights will force those operations, very expensive, time consuming and extremely risky, to focus on maximizing their profits so as to at least minimize the legal risks. Meanwhile, ordinary colonists will have few legal rights, because the rights citizens enjoy on Earth will not exist legally for them.

We will once again see some variation of Virginia playing out on the Moon, Mars, and on all the solar system's first colonies, big operations controlled by force and run by a handful of powerful individuals, either in the government or not, to the detriment and oppression of the majority of the colonists.

It will therefore take the settlers themselves, on colonies on the Moon, Mars, and the asteroids, to finally force a change in the treaty. It will take a revolt in space to throw off that yoke, a revolt demanding the same human rights for space-born civilians that many Earth-bound citizens in the west have taken for granted for centuries.

Sadly, this first revolt will more likely resemble Bacon's Rebellion in 1676 than it will the American Revolution of 1776. Established under the bad legal structure of the Outer Space Treaty and following a single-mindedness towards profit similar to that seen in Virginia, I do not expect the first interplanetary colonies to have the cultural depth and framework necessary to build a healthy, free, and prosperous society. Instead, they will rebel as the poor planters and slaves did in Virginia in 1676, in anger and hate for the oppression under which they have been subjected, with no deeper vision about a better way to do things.

I do expect, however, that the future space-faring societies that will follow these first raw colonies will eventually become more mature and sophisticated, just as the American colonies did in the hundred years following Bacon's Rebellion. Their experience will grow, as will their wealth and ability to function independently and robustly in space. They, like the British, will gain experience in their effort to establish new societies on new worlds.

Eventually, as the human race becomes increasingly proficient in the never-ending task of establishing new societies on alien worlds, certain patterns, as illustrated nicely by the British colonies in North America, will provide lessons on what should and should not be done.

First, they will at last recognize that some system for establishing borders and sovereign rights must be established. Consider for example the solution that Spain and Portugal found when both these Catholic nations were aggressively exploring and claiming territories in the Western Hemisphere. Soon after Columbus' first voyage they found their explorations in the Americas falling into conflict. So they went to the Pope, and after some negotiation and renegotiation in 1494 the Treaty of Tordesillas was signed, drawing a line 370 leagues west of the Cape Verde Islands, with Spain owning all new colonies to the west, and Portugal owning all new colonies to its east.

The solution wasn't elegant. It left out all other nations of the world, as well as the native populations in those territories. But it worked successfully to give the leaders and explorers of both Spain and Portugal a peaceful framework for their future colonies. As historian Daniel Boorstin wrote,

> Both Spain and Portugal showed remarkable goodwill in their efforts to obey the terms of the treaty, even though of that time the technology could not yet precisely locate the prescribed meridian.[683]

It is because of this treaty the eastern half of South America speaks Portuguese and the western half Spanish. During the settlement of South America, both countries respected the terms of the treaty and the borders it outlined, and focused their explorations in the areas allotted to them.

In space, I expect that future spacefaring societies will eventually do something similar. If they have wisdom, they will establish for all nations a kind of international homestead act, where each new colony would automatically establish some territorial rights to the nation that builds it. The act would include, like the American Homestead Act, a specific amount of territory that each colony would claim – not too big or too small. The size will be determined by the smallest territorial region capable of making the colony sustainable.

Within those claimed territories the laws of the claiming nation would then apply. Private citizens operating out of capitalist and free nations resembling the United States and the European states would know that their law and property rights would apply to them, should they build a business operation in space. Similarly, nations that favor centralized control, such as China and Russia, would have the right to establish their own legal frameworks on their own space colonies.

Such a system would essentially duplicate what the U.S. government did in successfully settling its west: Establish the basic legal framework and then get out of the way. It would also duplicate the basic goal of the Treaty of Tordesillas, delineating a system where borders are established within which nations can focus their exploration efforts.

What I have outlined is of course based on my own limited knowledge. The future is unknown, and will certainly be more complex than any historian can imagine. Future spacefarers will have a far better idea of what must be done, based on actual experience. Even so, their goal should be to create a clear system for allowing nations, both from Earth and in space, to claim territory, and to do so in small chunks that are also sufficiently large enough for the colony to thrive. Without such a system the establishment of any future space colony will be fraught with legal, social, and military problems.

What should the settlers themselves do?

The colonization of the solar system and the creation of new societies on new worlds is likely going to be the most fundamental problem dominating the rest of human history. If we go to Mars, to the asteroids, to the moons of Jupiter and Saturn, and then beyond to the stars, we shall, in each of these places, be building new societies as we spread life throughout the universe. In this effort we are going to repeatedly be faced with the same problems faced by the British settlers who established colonies along the eastern coast of North America in the 1600s.

We should pay attention to the lessons they learned.

First, we must recognize that there are many human factors, basic to human nature, that make for a healthy society and cannot be avoided. The lessons of the North American colonies illustrate those factors quite starkly, especially if we compare the very successful northern colonies with the failed southern colonies.

As already noted, it is essential that a new space colony have a good and clear legal system that interferes with the freedom of individuals and families as little as possible. This does not mean that there should be no laws, but that the founding laws should be clear, and allow for the new citizens to maximize not only their own personal ambitions but their own personal wealth.

With this in mind, the laws must allow for private property that will encourage settlers to work hard for their own gain. This was a lesson learned in both Virginia and in New England in the first years of colonization. Both colonies at first pooled the ownership of all land and property into a single corporation for which everyone worked, and with all profits equally shared. In both Virginia and New England this collectivist arrangement ended in unmitigated failure, leading to bankruptcy and starvation.

Only when each colony's leaders allowed the settlers to own their own private plots where they could earn money for themselves did both colonies begin to prosper.

At the same time, the laws that encourage private property must also encourage the arrival of as many small businesses as possible, owning small holdings and earning a smaller amount, rather than large corporate or government operations that control large chunks of property that optimize their profits to the utmost.

In comparing the British colonies, the need for small holdings becomes obvious. In Virginia the initial goal was to maximize profit, so the initial private holdings were large. The result was a distorted social structure with power concentrated in the hands of a very few, while everyone else was poor or enslaved.

In New England the initial goal was to build small self-sufficient farms, not to export a crop for profit. The result were small tight-knit communities of small farms owned by families raising children. Few were capable of exporting much from the colony, but the society was generally more sound and prosperous for most of its inhabitants.

Therefore, decentralize ownership as much as possible, from the start. While initially the difficulty of settlement on alien worlds might appear to favor larger owning entities, this is not written in stone. The Puritans proved that it is possible to begin a colony with power, ownership, and wealth decentralized. In space this will also be doable, if those building the colony have the courage to do so. Only later, when the colony is settled and more stable, should larger private holdings evolve, and only because of natural competition and free enterprise.

Keeping things small to begin with also recognizes the crucial fact that profit must not be the only goal of future space colonies. A focus on profit will force the new colonies to quickly find a single product that can be exported for profit, as was done with tobacco, rice, sugar, tea, and other single export colonies in the New World. The colony then becomes a company town, with power once again concentrated in the hands of a few, with everyone else either poor or enslaved. This in turn produces an unhealty social order.

Instead of profit, future colonists must focus on the idea of building a new human community. This is what both the Pilgrims and Puritans did.

> [W]e shall be as a City upon a Hill, the eyes of all people are upon us . . . we shall be made a story and a byword throughout the world.[684]

Profit was the last thing on their mind. In fact, these religious colonists generally became poorer because of their immigration. But they came wanting to build a new world, and to build it in such a way that it would become a society that future humans would want to emulate.

To build such a society requires some basic components. First, early colonists to new space colonies must go as families. Human societies do not fare well when the sex ratio is skewed, or the family unit is broken. Thus, even if it is not possible to go as families, at a minimum the settlers must go as a community of men and women, in equal ratios, so that families will naturally form and the children who will eventually be born in these new societies will have a stable family life in which to grow up.

Building a new human society also means the settlers must also go with the intent of raising healthy and well-adjusted children. Future space colonists must remember that they are **not** really exploring the unknown. What they are really doing is building new societies for their children and children's children. Such an effort carries great responsibility, and if we shirk that responsibility, our descendants will curse our memory.

The Pilgrims, Puritans, and Quakers all came with their children central and paramount to the immigration. They wanted to create a society where those children would be raised as upright, god-fearing citizens morally bound to do the right thing. As such, they shaped their communities with these ideas in mind. Since these ideas actually form the foundation of any vibrant human society, no matter whom the god, the result was that the northern British colonies quickly became successful human communities, despite their newness and the fact that they sat on the edge of the wilderness.

Future colonists must take the same approach. They might be exploring a new world and going where no one has gone before, but their primary purpose must always be the creation of a new society designed well for the future generations who will follow.

This focus on children also makes it essential that the families of space colonists be long-lived and stable. As shown in Virginia, broken homes do not encourage the proper raising of children, but instead cause their education and training in ethics and moral behavior to suffer. Long-lived marriages carry an additional benefit in that they also teach tolerance to the parents. No partnership is perfect. By committing to a permanent relationship, "for better or worse," married couples learn to accept the differences that exist between individuals. This in turn helps them tolerate the differences and disagreements that exist with others outside that partnership.

Finally, and most important, the initial settlers and their leadership need to go carrying with them a personal moral commitment to do the right thing. This point, in a sense a rephrasing of George Washington's exhortation in his final address, is probably the most difficult to achieve.

The Pilgrims, Puritans, and Quakers did it, however. They used as their moral blueprint the Bible, and most especially the Old Testament. This document outlines some very basic concepts about family, the raising of children, right and wrong, and personal individual responsibility. And believe it or not, it also celebrates individual choice and the liberty to choose one's beliefs, for yourself, teaching that, within a clearly defined moral code of right and wrong, each soul is free to choose, to grow, to rise or fall based on its own effort. England's first premise, decentralizing power and responsibility to the individual, was essentially an outgrowth of these Old Testament teachings, reinforced by the larger cultural teachings of Western Civilization that included the Greeks, the Romans, and British culture.

The future leaders of new space colonies therefore must recognize that they are not an annointed royal class destined to rule for generations. Instead, they are servants to the citizenry, empowered temporarily with the responsibility of creating that colony for the benefit of *all*. They, more than anyone else, must try to do the right thing, to build a society that has as its centerpiece a strong moral code, but that also, like England and the United States that followed, allows for the freedom of individuals to move up and down the ladder of success.

Note that I write these words about the Bible and the Old Testament not as a religious zealot but as a secular humanist. I do not believe in the spiritual god as portrayed in the Bible. Nonetheless, I recognize that the Old Testament, as interpreted within the foundations of Western culture, probably provides us with one of the best instruction manuals for building a good society as any that has ever been created. We will do very well indeed in outer space if we emulate what the Pilgrims, Puritans, and Quakers did in using it in North America.

In saying this I am also not claiming that there are no other blueprints that might be as beneficial. Hinduism, Buddhism, Shinto, and the Confucianism of China provide alternatives that have proven successful in creating relatively peaceful, prosperous, and thriving human societies. All of these religious texts, including the Bible, are to my mind intrinsic efforts by humanity to find the best and most just way for society to function.

At the same time, as I said in the introduction, some social orders are more humane and some moral concepts are more just. Not all religions or cultures bring with them a just or beneficial approach to human society. Some are power-based, and tend to form societies no different from that which developed in Virginia.[685] In the last two centuries we have seen the rise of a least one such culture (communism) and one such religion (Islam), bringing with them terror, violence, oppression, and poverty. We should recognize this reality, and honestly work to distinguish between the best and the worst, and choose the best.

As an American who has inherited the framework of Judeo-Christian-British culture, however, it is that culture that I know best. And as a historian who has looked at where this culture has both succeeded *and* failed, as illustrated by this very history you hold in your hand, I have come to the conclusion that this framework, when instituted correctly, works superbly as a foundation for future human colonies in space. Others might do the same with other philosophies, but I think the overall success of the United States, beginning in the 1600s as a handful of tiny colonies to become in the 1900s the world's most prosperous nation ever in the history of the human race, is proof enough for my conclusion.

So, to the future settlers of worlds beyond, I say: Build your future civilizations with families, aimed at raising children. Provide your citizens the right to private property while favoring at the beginning relatively small holdings. Encourage individual responsiblity, liberty of conscience, and freedom. And above all, enfuse your colonies with moral commitment, the insistence that truth, justice, and the protection of individual human rights are essential components of any new social order.

If you do these things, you will build well, and your descendents will thank you.

Or to paraphrase John Winthrop, you will become a city upon a hill, a byword throughout the universe.

One final thought: When in the coming centuries humans finally step out into the solar system to begin its settlement and colonization, every issue outlined in this history will be repeated again, compounded by the far more difficult technical challenges of building a viable colony on alien worlds.

What this history teaches above all, however, is that it is crucially important that no one take shortcuts, especially those in charge. Not fixing the Outer Space Treaty now, when it is still relatively easy to do, is cowardly and irresponsible, and is going to guarantee that the next few generations of spacefarers in colonies throughout the solar system will experience some of the same kind of injustices experienced by the poor immigrants who were made indentured servants in Virginia. This legal limbo in space might even create a situation where even worse evils, such as slavery, will be able to once again fester and grow.

There is great danger when the leadership in any human society irresponsibly creates new laws without care or deliberation, or allows bad laws to stand merely for personal expediency, because the consequences don't actually apply to the people who pass the laws or allow them to stand, but are faced instead by later generations. The harm that the headright system did to the few thousand poor immigrant men who came to Virginia in the first half of the 1600s dwindles in comparison to the harm this system did to blacks *and* whites born in the centuries that followed.

In any government, we make our laws and create our communities not for ourselves but for our children and our children's children, worlds without end into the future. If we act irresponsibly we might not do harm to ourselves, but we quite surely will bring terrible consequences to those who follow.

In the United States today we are seeing just this kind of negative consequence. The push in the twentieth century towards centralized government and a desire to rely on it for solutions has taught some very bad lessons to the generations born afterwards. Once Americans were greatly skeptical of government, and resisted depending on it at all costs. Once freedom and personal responsibility ruled the culture.

In post-World War II America however that skepticism of government has vanished, as has a strong commitment to freedom.

Instead, Americans today are quite willing to be ruled by their government in some of the most odious ways, and care not if that rule stamps out their freedom to pursue their happiness.

The result is a country that the Founding Fathers would not have recognized, and would likely have found horrifying.

Issues of morality and ethics are always present. A bad law is essentially an immoral statement, bringing with it oppression and social failure. To deny this fact does not free us from it, but instead indicates our willingness to choose badly, to allow future generations to suffer untold misery, because we today are lazy. Even in what seems to be the most trivial and mundane local city council measure, such governmental decisions are not only going to affect people's lives today but more importantly the lives of unborn children whom have no say in the matter, and to whom we have been given the profound and noble responsibility of bestowing a prosperous, safe, just, and free world to live in.

It is always a matter of choice. Modern government and the establishment of a healthy social order demands the serious, educated participation of each citizen. In Virginia in 1663 or in Philadelphia in 1776 or at Gettysburg in 1863 or in the United States now or in the future human settlements among the stars, we must face life and the choices it brings us, not merely for ourselves but to provide future generations the same blessings our ancestors so generously gave to us.

Each of us has a choice. We can, like Richard Bennett, stand by and let others choose for us. Or we can, like William Berkeley, impose our will by force so that the free market of ideas is unable to prosper and grow. Or we can, like the Puritans and the Quakers, express our moral beliefs with courage and good will, to eagerly try to convince the world that we are right, and accept the world's judgment without anger, violence, hatred or arrogance. We can do any of these things.

Each of us has a choice. It is always a matter of choice.

A Proclamation for Civilization in the Far Future

Appendix A: Additional Tables

The tables included throughout the text, as well as those shown here, are an attempt to illustrate the political make-up of Virginia's Grand Assembly during the period between William Berkeley's arrival in Virginia in 1642 and his death in 1677. The sources for all the tables can be found in Appendix B.

These tables are numbered sequentially in time. For narrative purposes, not all tables were included within the body of the text, and so those sessions, tables 25-27, 30-40 are included here instead.

Each new table represents a known assembly session as well as a change in the assembly, either through elections or appointments to the council. If no House elections have been held, this is stated on the table.

After 1670, the council is known to have met almost continually. Because attendance could change day to day, and even during a session as councilors came and went, I have simplified the listings by summarizing the total attendance for each month.

I have also tried to place council sessions in context with House sessions, and have therefore not listed the two bodies separately. When the council had numerous known sessions, as in the period from spring 1670 to fall 1673 (tables 32-39), I have placed the House and council sessions most closely linked by time on the same table, with the additional council meetings in their own table either before or after, depending on when those meetings took place.

The assemblymen, both councilors and burgesses, are divided into the several groups, with dissenters on the right side of each table with royalists on the left.

The four groups – **Royalist, Parliamentarian, Puritan, Quaker** – are made up of men who took actions or said something that demonstrated clearly their political or religious views on these four issues.

The four groups – **Uncertain Royalist, Uncertain Parliamentarian Uncertain Puritan Uncertain Quaker** – are made of men for whom our knowledge is less certain. These men usually took no direct action to indicate their views, but can be sufficiently linked to the above parties to indicate their probable views. Since the information is less certain, however, this must be indicated.

The group **Neutral** is reserved for those men who actually took actions at different times beneficial to both royalist and dissenting parties. Often we have a great deal of information about these individuals but none of it tells us their political position.

Appendix A: Additional Tables

Finally, the group **Unknown** is for councilors and burgesses for whom we know almost nothing.

Because so little information is available about many of these assemblymen, the above political labels are sometimes nothing more than my educated opinion, and are open to dispute. Though I recognize that more work needs to be done to determine the political positions of these men, I offer these tables as a good and reasonable start.

For all the tables in this appendix, the following legend applies:

*-incumbent
+-served previously
Bold names have purchased slaves or have claimed headrights on blacks.
<u>Underlined names will eventually buy slaves or claim headrights on blacks.</u>
Italicized names either were born in Virginia or immigrated as minors.

Table 25-1664-Long Assembly

Current Governor:
Governor William Berkeley

Council-March 1664

Royalist:	Neutral:	Puritan:
*Bacon, Nathaniel, Sr.	+*Stegg, Thomas, Jr.*	+**Bennett, Richard**
*Corbin, Henry		
*_Ludwell, Thomas_	Unknown:	Uncertain Puritan:
*Reade, George	+Bernard, William	+**Bland, Theoderick**
+Carter, John, Sr.	+Carter, Edward	
+**Lee, Richard, Sr.**		Uncertain Parliamentarian:
		+*Wood, Abraham*
Uncertain Royalist:		
*Smith, Robert		

House-September 1664-Held over without general elections

Royalist:	Unknown:	Uncertain Parliamentarian:
*Bridger, Joseph	*Barber, William	*Andrews, William, Jr.*
*Jenings, Peter	*Cockeram, William	***Ellyson, Robert**
*_Presley, William, Jr._	*_Ffarrer, William_	*Kendall, William
*Travers, Raleigh	*Griffith, Edward	
Thorowgood, Adam, Jr.	*Gwillen, George	
	*Hamelyn, Stephen**	
Uncertain Royalist:	*Hill, [Richard]	
Chiles, Walter, Jr.	*Knowles, John	
Claiborne, William, Jr.	*Lucas, Thomas, [Sr. or Jr.]*	
Ramsey, Edward	*Peyton, Valentine	
	*_Powell, John_	
	*Walker, Thomas	
	*Wallings, George	
	*Warren, Thomas	
	Weir, John	
	*Williamson, Roger	
	*Wynne, Robert-SPEAKER	
	*Yeo, Leonard	
	Browne, Devoreux	
	Yeo, Hugh	

	Present Slave-holders	Including future slave-holders	Raised in Virginia	Incumbents	Including those who served previously
Council	6 of 13	8 of 13	3 of 13		
House	10 of 31	16 of 31	10 of 31	28 of 31	28 of 31

Appendix A: Additional Tables

Table 26-1665-Long Assembly

Current Governor:
Governor William Berkeley

Council-March 1665

Royalist:	Unknown:	Uncertain Parliamentarian:
*Bacon, Nathaniel, Sr.	+Cary, Miles	+Swann, Thomas
*Carter, John, Sr.		
*Corbin, Henry		
*Reade, George		

Council-June 1665

Royalist:	Neutral:	Puritan:
*Bacon, Nathaniel, Sr.	+Stegg, Thomas, Jr.	+Bennett, Richard
*Carter, John, Sr.		
*Reade, George	Unknown:	Uncertain Puritan:
+Ludwell, Thomas	*Cary, Miles	+Bland, Theodorick
+Willis, Francis		
		Uncertain Parliamentarian:
		*Swann, Thomas
		+Wood, Abraham

Council-October 1665

Royalist:	Neutral:	Uncertain Puritan:
*Bacon, Nathaniel, Sr.	*Stegg, Thomas, Jr.	+Bland, Theodorick
*Ludwell, Thomas		
	Unknown:	
Uncertain Royalist:	*Cary, Miles	
+Smith, Robert		

House-October 1665-Held over without general elections

Royalist:	Unknown:	Uncertain Parliamentarian:
*Bridger, Joseph	*Barber, William	*Ellyson, Robert
*Jenings, Peter	*Browne, Devoreux	*Kendall, William
*Presley, William, Jr.	*Cockeram, William	Carver, William
*Thorowgood, Adam, Jr.	*Ffarrer, William	
*Travers, Raleigh	*Griffith, Edward	
+Fowke, Gerald	*Gwillen, George	
Meese, Henry	*Hill, [Richard]	
Washington, John	*Lucas, Thomas, [Sr. or Jr.]	
	*Powell, John	
Uncertain Royalist:	*Walker, Thomas	
*Chiles, Walter, Jr.	*Wallings, George	
*Claiborne, William, Jr.	*Warren, Thomas	
*Ramsey, Edward	*Weir, John	
Savage, John	*Williamson, Roger	
	*Wynne, Robert-SPEAKER	
	*Yeo, Hugh	
	*Yeo, Leonard	
	Southcoat, Thomas	

	Present Slave-holders	Including future slave-holders	Raised in Virginia	Incumbents	Including those who served previously
Council-March	5 of 6	5 of 6	1 of 6		
Council-June	7 of 11	7 of 11	4 of 11		
Council-October	4 of 6	5 of 6	2 of 6		
House	12 of 33	17 of 33	10 of 33	27 of 33	28 of 33

Table 27-March to July, 1666-Long Assembly

Current Governor:
Governor William Berkeley

Council-March 1666

Royalist:	Neutral:	Uncertain Parliamentarian:
*Bacon, Nathaniel, Sr.	*Stegg, Thomas, Jr.	+Swann, Thomas
*Ludwell, Thomas		Beale, Thomas
+Carter, John, Sr.	Unknown:	
+Reade, George	*Cary, Miles	
+Warner, Augustine, Sr.		
+Willis, Francis		
Uncertain Royalist:		
*Smith, Robert		

Council-July 1666

Royalist:	Neutral:	Puritan:
*Ludwell, Thomas	*Stegg, Thomas, Jr.	+Bennett, Richard
Uncertain Royalist:	Unknown:	Uncertain Puritan:
*Smith, Robert	*Cary, Miles	+Bland, Theoderick
		Uncertain Parliamentarian:
		*Swann, Thomas

House-June 1666-Held over without general elections

Royalist:	Unknown:	Uncertain Parliamentarian:
*Bridger, Joseph	*Barber, William	*Carver, William
*Jenings, Peter	*Browne, Devoreux	*Kendall, William
*Meese, Henry	*Ffarrer, William	
*Presley, William, Jr.	*Griffith, Edward	
*Thorowgood, Adam, Jr.	*Hill, [Richard]	
*Travers, Raleigh	*Lucas, Thomas, [Sr. or Jr.]	
*Washington, John	*Powell, John	
Leare, John	*Southcoat, Thomas	
Spencer, Nicholas	*Walker, Thomas	
	*Warren, Thomas	
Uncertain Royalist:	*Weir, John	
*Chiles, Walter, Jr.	*Williamson, Roger	
*Claiborne, William, Jr.	*Wynne, Robert-SPEAKER	
*Ramsey, Edward	*Yeo, Hugh	
*Savage, John	*Yeo, Leonard	
+Filmer, Henry	+Blackey, William	
Baker, Lawrence	+Blake, John	
Parke, Daniel, Sr.	+Holte, Robert	

	Present Slave-holders	Including future slave-holders	Raised in Virginia	Incumbents	Including those who served previously
Council-March	6 of 11	7 of 11	4 of 11		
Council-July	4 of 7	5 of 7	3 of 7		
House	18 of 36	20 of 36	11 of 36	28 of 36	32 of 36

Appendix A: Additional Tables

Table 30-The Long Assembly, 1668

Current Governor:
Governor William Berkeley

Council-April 1668
Royalist:
*Corbin, Henry
**Ludwell, Thomas*
*Reade, George
+Carter, John, Sr.

Uncertain Royalist:
*Smith, Robert

House-September 1668-Held over without general elections

Royalist:	Unknown:	Uncertain Parliamentarian:
**Allerton, Isaac (a)*	*Barber, William	*Carver, William
*Bridger, Joseph	*Blackey, William	*Kendall, William
*Jenings, Peter	*Blake, John	
*Leare, John	**Ffarrer, William*	
*Meese, Henry	*Griffith, Edward	
**Presley, William, Jr.*	*Hill, [Richard]	
**Scarburgh, Edmund*	*Holte, Robert	
**Thorowgood, Adam, Jr.*	*Hone, Theophilus	
*Travers, Raleigh	**Lucas, Thomas, [Sr. or Jr.]*	
*Washington, John	**Powell, John*	
+Spencer, Nicholas	*Southcoat, Thomas	
	*Walker, Thomas	
Uncertain Royalist:	*Warren, Thomas	
*Baker, Lawrence	*Weir, John	
**Claiborne, William, Jr.*	*Williamson, Roger	
*Filmer, Henry	*Wynne, Robert-SPEAKER	
**Parke, Daniel, Sr.*	**Yeo, Hugh*	
**Ramsey, Edward*	*Yeo, Leonard	
**Savage, John*		

	Present Slave-holders	Including future slave-holders	Raised in Virginia	Incumbents	Including those who served previously
Council	5 of 5	5 of 5	1 of 5		
House	19 of 37	20 of 37	12 of 37	36 of 37	37 of 37

Table 31-The Long Assembly, 1669

Current Governor:
Governor William Berkeley

Council-April/May 1669

Royalist:	Neutral:	Puritan:
*Ludwell, Thomas	+Stegg, Thomas, Jr.	+Bennett, Richard
+Bacon, Nathaniel, Sr.		
+Warner, Augustine, Sr.		Uncertain Puritan:
		+Bland, Theodorick
Uncertain Royalist:		
*Smith, Robert		Uncertain Parliamentarian:
		+Beale, Thomas
		+Swann, Thomas

Council-October 1669

Royalist:	Neutral:	Uncertain Parliamentarian:
*Bacon, Nathaniel, Sr.	*Stegg, Thomas, Jr.	*Beale, Thomas
+Corbin, Henry		*Swann, Thomas

House-October 1669-Held over without general elections

Royalist:	Unknown:	Uncertain Parliamentarian:
*Allerton, Isaac (a)	*Blackey, William	*Carver, William
*Bridger, Joseph	*Blake, John	*Kendall, William
*Jenings, Peter	*Ffarrer, William	
*Leare, John	*Griffith, Edward	
*Meese, Henry	*Hill, [Richard]	
*Scarburgh, Edmund	*Holte, Robert	
*Spencer, Nicholas	*Hone, Theophilus	
*Thorowgood, Adam, Jr.	*Lucas, Thomas, [Sr. or Jr.]	
*Travers, Raleigh	*Powell, John	
*Washington, John	*Southcoat, Thomas	
Whitaker, Walter	*Walker, Thomas	
Wormeley, Ralph, Jr.	*Warren, Thomas	
	*Weir, John	
Uncertain Royalist:	*Williamson, Roger	
*Baker, Lawrence	*Wynne, Robert-SPEAKER	
*Claiborne, William, Jr.	*Yeo, Hugh	
*Filmer, Henry	*Yeo, Leonard	
*Parke, Daniel, Sr.	+Baldry, Robert	
*Ramsey, Edward	Wild, Daniel	
*Savage, John		

	Present Slave-holders	Including future slave-holders	Raised in Virginia	Incumbents	Including those who served previously
Council-April/May	6 of 9	6 of 9	4 of 9		
Council-October	2 of 5	2 of 5	2 of 5		
House	19 of 39	20 of 39	12 of 39	34 of 39	35 of 39

Appendix A: Additional Tables

Table 32-Spring 1670-Long Assembly

Current Governor:
Governor William Berkeley

Council-April 1670

Royalist:	Puritan:
*Bacon, Nathaniel, Sr.	+Bennett, Richard
*Corbin, Henry	
+Digges, Edward	Uncertain Puritan:
+*Ludwell, Thomas*	+Bland, Theodorick
+*Warner, Augustine, Sr.*	
+Willis, Francis	Uncertain Parliamentarian:
Chickley, Henry	*Beale, Thomas
	Swann, Thomas
Uncertain Royalist:	
+Smith, Robert	

Council-June 1670

Royalist:	Puritan:
*Bacon, Nathaniel, Sr.	*Bennett, Richard
*Digges, Edward	
Ludwell, Thomas	Uncertain Puritan:
*Willis, Francis	+Bland, Theodorick
Uncertain Royalist:	Uncertain Parliamentarian:
+*Parke, Daniel, Sr.*	*Swann, Thomas*

	Present Slave-holders	Including future slave-holders	Raised in Virginia
Council-April	9 of 12	9 of 12	3 of 12
Council-June	6 of 8	6 of 8	3 of 8

230

Table 33-The Long Assembly, Fall 1670

Current Governor:
Governor William Berkeley

Council-October 1670

Royalist:	Puritan:
*Bacon, Nathaniel, Sr.	*Bennett, Richard
*Digges, Edward	
*Ludwell, Thomas	Uncertain Puritan:
*Willis, Francis	*Bland, Theodorick
+Chickley, Henry	
Jenings, Peter	Uncertain Parliamentarian:
	*Swann, Thomas
Uncertain Royalist:	+Ballard, Thomas
*Parke, Daniel, Sr.	+Beale, Thomas
+Smith, Robert	

House-October 1670-Held over without general elections

Royalist:	Unknown:	Uncertain Parliamentarian:
*Allerton, Isaac (a)	*Baldry, Robert	*Kendall, William
*Bridger, Joseph	*Blackey, William	
*Jenings, Peter	*Blake, John	
*Leare, John	*Ffarrer, William	
*Scarburgh, Edmund	*Griffith, Edward	
*Thorowgood, Adam, Jr.	*Hill, [Nicholas]	
*Travers, Raleigh	*Holte, Robert	
*Washington, John	*Hone, Theophilus	
*Whitaker, Walter	*Lucas, Thomas, [Sr. or Jr.]	
*Wormeley, Ralph, Jr.	*Powell, John	
+Presley, William, Jr.	*Southcoat, Thomas	
Ball, William, Sr.	*Walker, Thomas	
Lee, Richard, Jr.	*Weir, John	
	*Wild, Daniel	
Uncertain Royalist:	*Williamson, Roger	
*Baker, Lawrence	*Wynne, Robert-SPEAKER	
*Claiborne, William, Jr.	*Yeo, Hugh	
*Filmer, Henry	*Yeo, Leonard	
*Ramsey, Edward	Eppes, Francis, Jr.	
*Savage, John	Moseley, William	

	Present Slave-holders	Including future slave-holders	Raised in Virginia	Incumbents	Including those who served previously
Council	9 of 13	9 of 13	3 of 13		
House	20 of 39	21 of 39	14 of 39	34 of 39	35 of 39

Appendix A: Additional Tables

Table 34-Spring 1671-Long Assembly

Current Governor:
Governor William Berkeley

Council-March 1671

Royalist:
*****Bacon, Nathaniel, Sr.**
*****Digges, Edward**
*Jenings, Peter
*****Ludwell, Thomas***

Uncertain Royalist:
*****Parke, Daniel, Sr.***

Puritan:
*****Bennett, Richard**

Uncertain Puritan:
*****Bland, Theodorick**

Uncertain Parliamentarian:
*****Ballard, Thomas**
**Swann, Thomas*

Council-April 1671

Royalist:
*****Bacon, Nathaniel, Sr.**
*****Digges, Edward**
*Jenings, Peter
*****Ludwell, Thomas***
+Chickley, Henry
+Corbin, Henry
+Reade, George
+*Warner, Augustine, Sr.*
+Willis, Francis

Uncertain Royalist:
*****Parke, Daniel, Sr.***

Puritan:
*****Bennett, Richard**

Uncertain Parliamentarian:
*****Ballard, Thomas**
**Swann, Thomas*

Council-May 1671

Royalist:
*****Bacon, Nathaniel, Sr.**
*****Chickley, Henry**
*****Digges, Edward**
*****Ludwell, Thomas***

Uncertain Royalist:
*****Parke, Daniel, Sr.***
+Smith, Robert

Uncertain Parliamentarian:
*****Ballard, Thomas**
+Beale, Thomas

Uncertain Puritan:
*****Bland, Theodorick**

	Present Slave-holders	Including future slave-holders	Raised in Virginia
Council-March	7 of 9	7 of 9	3 of 9
Council-April	10 of 13	10 of 13	4 of 13
Council-May	8 of 9	8 of 9	2 of 9

Table 35-The Long Assembly, Fall 1971

Current Governor:
Governor William Berkeley

Council-September/October 1671

Royalist:		Puritan:
*Bacon, Nathaniel, Sr.		+Bennett, Richard
*Chickley, Henry		
*Digges, Edward		Uncertain Puritan:
*Ludwell, Thomas		*Bland, Theodorick
+Corbin, Henry		
+Jenings, Peter		Uncertain Parliamentarian:
+Reade, George		*Ballard, Thomas
		*Beale, Thomas
Uncertain Royalist:		+Swann, Thomas
*Parke, Daniel, Sr.		+Wood, Abraham
*Smith, Robert		

Council-November 1671

Royalist:	Unknown:	Uncertain Puritan:
*Bacon, Nathaniel, Sr.	Pate, John	*Bland, Theodorick
*Digges, Edward		
*Ludwell, Thomas		Uncertain Parliamentarian:
		*Beale, Thomas
Uncertain Royalist:		*Swann, Thomas
*Parke, Daniel, Sr.		

House-September 1671-Held over without general elections

Royalist:	Unknown:	Parliamentarian:
*Allerton, Isaac (a)	*Baldry, Robert	Lawrence, Richard
*Ball, William, Sr.	*Blackey, William	
*Leare, John	*Blake, John	Uncertain Parliamentarian:
*Lee, Richard, Jr.	*Eppes, Francis, Jr.	*Kendall, William
*Presley, William, Jr.	*Ffarrer, William	
*Washington, John	*Griffith, Edward	
*Whitaker, Walter	*Hill, [Nicholas]	
*Wormeley, Ralph, Jr.	*Holte, Robert	
	*Hone, Theophilus	
Uncertain Royalist:	*Lucas, Thomas, [Sr. or Jr.]	
*Baker, Lawrence	*Powell, John	
*Claiborne, William, Jr.	*Southcoat, Thomas	
*Filmer, Henry	*Walker, Thomas	
*Ramsey, Edward	*Wild, Daniel	
*Savage, John	*Williamson, Roger	
+Mason, Lemuell	*Wynne, Robert-SPEAKER	
	+Browne, William	
	Haynes, Thomas	

	Present Slave-holders	Including future slave-holders	Raised in Virginia	Incumbents	Including those who served previously
Council-Sept/Oct	11 of 15	11 of 15	4 of 15		
Council-Nov	6 of 8	6 of 8	3 of 8		
House	14 of 34	16 of 34	12 of 34	30 of 34	32 of 34

Appendix A: Additional Tables

Table 36-Spring 1672-Long Assembly

Current Governor:
Governor William Berkeley

Council-March/April 1672

Royalist:	Unknown:	Puritan:
*Bacon, Nathaniel, Sr.	*Pate, John	+Bennett, Richard
*Digges, Edward		
*Ludwell, Thomas		Uncertain Parliamentarian:
+Chickley, Henry		*Beale, Thomas
+Corbin, Henry		*Swann, Thomas
Spencer, Nicholas		+Ballard, Thomas

Uncertain Royalist:
*Parke, Daniel, Sr.
+Smith, Robert

Council-May 1672

Royalist:		Uncertain Parliamentarian:
*Bacon, Nathaniel, Sr.		*Ballard, Thomas
*Corbin, Henry		
*Digges, Edward		
*Ludwell, Thomas		

Uncertain Royalist:
*Parke, Daniel, Sr.

	Present Slave-holders	Including future slave-holders	Raised in Virginia
Council-Mar/April	11 of 13	11 of 13	3 of 13
Council-May	6 of 6	6 of 6	2 of 6

Table 37-Fall 1672-Long Assembly

Current Governor:
Governor William Berkeley

Council-September/October 1672

Royalist:	Unknown:	Puritan:
*Bacon, Nathaniel, Sr.	+Pate, John	+Bennett, Richard
*Corbin, Henry		
*Digges, Edward		Uncertain Parliamentarian:
*Ludwell, Thomas		*Ballard, Thomas
+Chickley, Henry		+Beale, Thomas
		+Swann, Thomas
Uncertain Royalist:		+Wood, Abraham
*Parke, Daniel, Sr.		

Council-November 1672

Royalist:		Uncertain Parliamentarian:
*Bacon, Nathaniel, Sr.		*Ballard, Thomas
*Digges, Edward		*Beale, Thomas
*Ludwell, Thomas		
Uncertain Royalist:		
*Parke, Daniel, Sr.		

House-September 1672-Held over without general elections

Royalist:	Unknown:	Parliamentarian:
*Allerton, Isaac (a)	*Baldry, Robert	*Lawrence, Richard
*Ball, William, Sr.	*Blackey, William	
Leare, John	*Blake, John	Uncertain Parliamentarian:
*Lee, Richard, Jr.	*Browne, William	*Kendall, William
*Presley, William, Jr.	*Eppes, Francis, Jr.	
Washington, John	*Ffarrer, William	
*Whitaker, Walter	*Griffith, Edward	
*Wormeley, Ralph, Jr.	*Haynes, Thomas	
+Page, John	*Hill, [Nicholas]	
Warner, Augustine, Jr.	*Holte, Robert	
	*Hone, Theophilus	
Uncertain Royalist:	*Lucas, Thomas, [Sr. or Jr.]	
*Baker, Lawrence	*Powell, John	
*Claiborne, William, Jr.	*Walker, Thomas	
*Filmer, Henry	*Williamson, Roger	
*Mason, Lemuell	*Wynne, Robert-SPEAKER	
*Ramsey, Edward	+Moseley, William	
*Savage, John	Wyatt, Nicholas	

	Present Slave-holders	Including future slave-holders	Raised in Virginia	Incumbents	Including those who served previously
Council-Sept/Oct	9 of 12	9 of 12	4 of 12		
Council-Nov	5 of 6	5 of 6	2 of 6		
House	15 of 36	17 of 36	13 of 36	32 of 36	34 of 36

Appendix A: Additional Tables

Table 38-Spring/Summer 1673-Long Assembly

Current Governor:
Governor William Berkeley

Council-March 1673

Royalist:	Puritan:
*Bacon, Nathaniel, Sr.	+Bennett, Richard
*Digges, Edward	
*Ludwell, Thomas	Uncertain Parliamentarian:
+Spencer, Nicholas	*Ballard, Thomas
	*Beale, Thomas
	+Swann, Thomas

Council-April 1673

Royalist:	Uncertain Parliamentarian:
*Bacon, Nathaniel, Sr.	*Ballard, Thomas
*Digges, Edward	*Swann, Thomas
*Ludwell, Thomas	
Uncertain Royalist:	
+Parke, Daniel, Sr.	

Council-May 1673

Royalist:	Puritan:
*Bacon, Nathaniel, Sr.	+Bennett, Richard
*Digges, Edward	
*Ludwell, Thomas	Uncertain Parliamentarian:
	*Ballard, Thomas
Uncertain Royalist:	*Swann, Thomas
*Parke, Daniel, Sr.	+Beale, Thomas

Council-June 1673

Royalist:	Uncertain Parliamentarian:
*Bacon, Nathaniel, Sr.	*Ballard, Thomas
*Ludwell, Thomas	*Beale, Thomas
	*Swann, Thomas

Council-July 1673

Royalist:	Uncertain Parliamentarian:
*Bacon, Nathaniel, Sr.	*Ballard, Thomas
*Ludwell, Thomas	
+Digges, Edward	
Uncertain Royalist:	
+Parke, Daniel, Sr.	

	Present Slave-holders	Including future slave-holders	Raised in Virginia
Council-March	6 of 8	6 of 8	2 of 8
Council-April	5 of 6	5 of 6	3 of 6
Council-May	6 of 8	6 of 8	3 of 8
Council-June	3 of 5	3 of 5	2 of 5
Council-July	5 of 5	5 of 5	2 of 5

Table 39-Fall 1673-Long Assembly

Current Governor:
Governor William Berkeley

Council-October/November 1673

Royalist:	Puritan:
*Bacon, Nathaniel, Sr.	+Bennett, Richard
*Digges, Edward	
*Ludwell, Thomas	Uncertain Parliamentarian:
+Bridger, Joseph	*Ballard, Thomas
+Chickley, Henry	+Beale, Thomas
+Corbin, Henry	+Swann, Thomas
Uncertain Royalist:	
+Parke, Daniel, Sr.	

House-October 1673-Held over without general elections

Royalist:	Unknown:	Parliamentarian:
*Allerton, Isaac (a)	*Baldry, Robert	*Lawrence, Richard
*Ball, William, Sr.	*Blackey, William	
*Leare, John	*Blake, John	Uncertain Parliamentarian:
*Page, John	*Browne, William	*Kendall, William
*Presley, William, Jr.	*Eppes, Francis, Jr.	
*Warner, Augustine, Jr.	*Ffarrer, William	
*Washington, John	*Griffith, Edward	
*Whitaker, Walter	*Haynes, Thomas	
*Wormeley, Ralph, Jr.	*Hill, [Nicholas]	
	*Holte, Robert	
Uncertain Royalist:	*Hone, Theophilus	
*Baker, Lawrence	*Lucas, Thomas, Jr.	
*Claiborne, William, Jr.	*Moseley, William	
*Filmer, Henry	*Powell, John	
*Mason, Lemuell	*Walker, Thomas	
*Ramsey, Edward	*Williamson, Roger	
*Savage, John	*Wyatt, Nicholas	
	*Wynne, Robert-SPEAKER	
	Lee, John	

	Present Slave-holders	Including future slave-holders	Raised in Virginia	Incumbents	Including those who served previously
Council-Oct/Nov	9 of 11	9 of 11	3 of 11		
House	14 of 36	17 of 36	14 of 36	35 of 36	35 of 36

Appendix A: Additional Tables

Table 40-1674-Long Assembly

Current Governor:
Governor William Berkeley

Council-September/October 1674

Royalist:	Uncertain Parliamentarian:
*Bacon, Nathaniel, Sr.	*Ballard, Thomas
*Bridger, Joseph	*Swann, Thomas
*Chickley, Henry	+Wood, Abraham
*Corbin, Henry	
*Digges, Edward	
*Ludwell, Thomas	

Uncertain Royalist:
*Parke, Daniel, Sr.

Council-November 1674

Royalist:	Uncertain Parliamentarian:
*Bacon, Nathaniel, Sr.	*Ballard, Thomas
*Bridger, Joseph	
*Chickley, Henry	
*Ludwell, Thomas	

Uncertain Royalist:
*Parke, Daniel, Sr.

House-September 1674-Held over without general elections

Royalist:	Unknown:	Parliamentarian:
*Allerton, Isaac (a)	*Baldry, Robert	*Lawrence, Richard
*Ball, William, Sr.	*Blackey, William	
*Leare, John	*Blake, John	Quaker:
*Page, John	*Eppes, Francis, Jr.	+Porter, John, Sr.
*Presley, William, Jr.	*Ffarrer, William	
*Warner, Augustine, Jr.	*Griffith, Edward	Uncertain Parliamentarian:
*Washington, John	*Haynes, Thomas	*Kendall, William
*Whitaker, Walter	*Hill, [Nicholas]	
*Wormeley, Ralph, Jr.	*Holte, Robert	
	*Hone, Theophilus	
Uncertain Royalist:	*Lucas, Thomas, Jr.	
*Baker, Lawrence	*Moseley, William	
*Claiborne, William, Jr.	*Powell, John	
*Filmer, Henry	*Walker, Thomas	
*Ramsey, Edward	*Williamson, Roger	
*Savage, John	*Wyatt, Nicholas	
	*Wynne, Robert-SPEAKER	
	+Jordan, George	
	Appleton, John (a)	

	Present Slave-holders	Including future slave-holders	Raised in Virginia	Incumbents	Including those who served previously
Council-Sept/Oct	8 of 10	8 of 10	4 of 10		
Council-Nov	6 of 6	6 of 6	2 of 6		
House	15 of 36	16 of 36	14 of 36	33 of 36	35 of 36

Appendix B: Source Information for Tables

The following list of every burgess and councilor known to have served in the Virginia General Assembly from 1642 to 1677, as well as all governors from that period, was mostly used to determine slave ownership or headright claims as well as the factional affiliation of each burgess, councilor, or governor as indicated in the tables. The names were culled from several sources, including the following:

Hening, William Waller, ed. *The Statutes at Large; Being a collection of all the Laws of Virginia from the 1st Session of the Legislature in the Year 1619, 13 vols.* Richmond 1809-1823.

Leonard, Cynthia Miller, ed. *The General Assembly of Virginia, July 30, 1619-January 11, 1978, A Bicentennial Register of Members.* Richmond: Virginia State LIbrary, 1978.

Billings, Warren M., "Virginia's Deplored Condition, 1660-1676: the Coming of Bacon's Rebellion." Ph.D. diss., Northern Illinois University, 1968.

McIlwaine, H. R., ed. *Journals of the House of Burgesses of Virginia 1619-1776. 13 vols.* Richmond, 1925-1930.

McIlwaine, H. R., ed. *Minutes of the Council and General Court of Colonial Virginia.* Richmond, 1926.

"Acts, Orders, and Resolutions of the General Assembly of Virginia, at sessions of March 1643-1646." *Virginia Magazine of History and Biography* 23 (July 1915) 225-255.

For some of the more well known individuals, such as Governor Berkeley, a full list of sources would take several pages, and the reader should instead refer to the endnotes within the text.

Because of the vagaries of spelling in 17th century documents, many of these men's names were spelled in several different ways. It is sometimes an educated guess whether someone like William Whitiker is the same man as William Whittacre. In most cases, it is possible to use the county of the individual as an indication that such different spellings really refer to the same man, and in such cases I have done so. Sometimes, however, as in the case of John Shepherd and John

Appendix B: Source Information for Tables

Sheppard, it is almost impossible to separate the different individuals, and here I have included my sources under both names, hoping that future researchers can untangle the information.

Books are listed by their author or editor. If an author has more than one book to his name, I use the shortened endnote form as used in the text (author's last name, plus a shortened title if the author is cited for more than one book). Volume numbers are shown immediately after author name. Full listings can be found in the bibliography. The following journal references, however, have been abbreviated as follows:

NE - *New England History and General Registry.* (preceded by volume number and followed by page number).

V - *Virginia Magazine of History and Biography* (preceded by volume number and followed by page number).

W(1)(2) - *William and Mary Quarterly,* first or second series (preceded by volume number and followed by page number).

Abrahaall, Robert: Nugent-1: 212,242,281-2,306,346-7,366,408,485,490,502; Nugent-2: 162,169,188,199,225,266; Tyler-Encyclopedia-1: 169; Billings-Deplored: 279
Allerton, Isaac: Nugent-1: 367,401,403,409,534,536; Nugent-2: 88,246:7; Tyler-Encyclopedia-1: 140; Billings-Deplored: 280; Quitt: 48-9,52-3; Sprinkle: 145; Sibley: 253-6; 22W(1)80
Andrews, William, Jr.: Nugent-1: 261,300,371,401,410,534,539; Tyler-Encyclopedia-1: 172; Billings-Deplored: 277,279; Whitelaw: 141; Haight: 432; Jester: 73-4,257
Andros, Edmund: Tyler-Encyclopedia-1:56
Appleton, John: Nugent-1: 334,408,435,438; Tyler-Encyclopedia-1: 172; Billings-Deplored: 284; 4W(1)80; 15W(1)180
Armstead, Anthony: Nugent-2: 177,230,397; Tyler-Encyclopedia-1: 172; Billings-Deplored: 273; McIlwaine-Minutes: 454-5
Ashton, Peter: Nugent-1: 302,357,377,398,406,409,434,531; Nugent-2: 42,46; Tyler-Encyclopedia-1: 174; Billings-Deplored: 280; 25V165; Hening-1: 422
Aston, Walter: Nugent-1: 31,93,148,154-5,165;24V67-68; 24V67-68; Tyler-Encyclopedia-1: 174
Bacon, Nathaniel, Jr.: Billings-Deplored: 269; Sprinkle: 72
Bacon, Nathaniel, Sr.: Nugent-1: 275,381-2,478,547-8; Nugent-2: 2,265; Tyler-Encyclopedia-1: 54; Kukla-Politics: 202; Hening-1: 422,512; Andrews: 28; McIlwaine-Minutes: 219; Bruce-Economics-2: 88n; Billings-Deplored: 270; 11W(1)30-1; 23W(1)177
Bagnall, James-Norfolk: Nugent-1: 145,168; Nugent-2: 175,263-4; Billings-Deplored: 281; 7W(1)253
Baker, Lawrence: Nugent-1: 156,193,280,350,422,561; Nugent-2: 43-4; Tyler-Encyclopedia-1: 176; Sprinkle: 183-5; Billings-Deplored: 283; 2W(1)122-124
Baldridge, Thomas: Nugent-1: 207,211; Tyler-Encyclopedia-1: 176; 1V75-81
Baldry, Robert: Nugent-1: 259; Tyler-Encyclopedia-1: 176; Billings-Deplored: 285; Karraker: 188
Baldwin, Robert: Nugent-1: 40,56-7,179,258
Ball, Henry: No information found.
Ball, William, [Sr. or Jr.]: Nugent-1: 244,347-8; Nugent-2: 20,163,251-2,262-3,282-3; Tyler-Encyclopedia-1: 178; Tyler-Encyclopedia-5: 1017; Billings-Deplored: 275-276; Quitt: 29-30; Hayden: 46-56

Ballard, Thomas: Nugent-1: 309,354,379-81; Nugent-2: 5; Tyler-Encyclopedia-1: 130; McIlwaine-Minutes: 223,454-5; Billings-Deplored: 269; 23W(1)276
Barber, William: Nugent-1: 70,121,178,214,318,442-3,480; Billings-Deplored: 285; Tyler-Encyclopedia-1: 179-80; 23W(1)208
Barker, William: Nugent-1: 20,35,70,78-9,100,103,108,110,121,301,338,531; Tyler-Encyclopedia-1: 180
Barrett, William: Nugent-1: 175,177,306,324,344,382-3; Nugent-2: 376-7; Tyler-Encyclopedia-1: 181
Batte, William: Nugent-1: 115,146,152,175,179,197,435-6; Nugent-2: 35,72; Tyler-Encyclopedia-1: 182
Baugh, John: Nugent-1: 64,87-9,157,192,352,516; Tyler-Encyclopedia-1: 182-3
Bayly, Arthur: Nugent-1: 78-9,86,97,121; Tyler-Encyclopedia-1: 183
Baynhan, Alexander: Nugent-1: 501; Tyler-Encyclopedia-1: 183
Beale, Thomas: Nugent-1: 351-375; Nugent-2: 7,45; McIlwaine-Minutes; Tyler-Encyclopedia-1: 127-8; Sprinkle: 51; Billings-Deplored: 67,270
Beazley, [Job] or [Robert]: Nugent-1: 309,318; Tyler-Encyclopedia-1: 184
Bennett, Phillip: Nugent-1: 150,162,177; Tyler-Encyclopedia-1: 184; Carson: 81,83; 10W(1)35-38
Bennett, Richard: Nugent-1: 23,45,66,88,104,109,127,136,139,141,196,247,385, 476; Hall: 235,254-5,302-3; Catterall-4: 16; Whitelaw: 29-31,247,408,630; Tyler-Encyclopedia-1: 47; Carson: 81; Bruce-Economic-1: 122; Browne-Provincial: 321-322; 25V393; 48NE114-115; 13W(2)1-9
Berkeley, William: Nugent-1: 27,54,115,119,128,154-7,160,229,390-3,401-5,410, 414,415,419; Tyler-Encyclopedia-1: 46; Wood:14-20; Carson: 52-3,79,81, 83,84, 89-96; Kukla-Politics: 127-9,139; Currier-Briggs-3: 685; McIlwaine-Journals-1: 75-76; 20V?, 21V369-370; 6W(1)214, 8W(1)165, 10W(1)358, 22W(1)239
Berkeley, William, Jr.: Billings-Bio: 38
Bernard, Thomas: Nugent-1: 67,93,127,159; Tyler-Encyclopedia-1: 185
Bernard, William: Nugent-1: 131,127,219-20,390,455; Tyler-Encyclopedia-1: 117-8; Jester: 93-5; Billings-Deplored: 269; Hening-I:508
Berry, John: Nugent-1: 530; Tyler-Encyclopedia-1: 185
Beverly, Robert: Nugent-1: 555; Nugent-2: 56-8,73,138-40,157,163,185,190,201, 335,395-6; Tyler-Encyclopedia-1: 136-8; Billings-Deplored: 278; McIlwaine-Minutes: 454-5; Bruce-Economic-2: 88n,92
Bishopp, John, Sr.: Nugent-1: 31,98,105,148,216,218,281,353,367; Tyler-Encyclopedia-1: 186
Blackey, William: Nugent-1: 169,257-8,310,342,355,387,422; Nugent-2: 70; Billings-Deplored: 279
Blake, John: Nugent-1: 253,280,358,443-4,454,492; Nugent-2: 96-7,120,142; Billings-Deplored: 279; Tyler-Encyclopedia-1: 188
Bland, Theoderick: Nugent-2: 38; Tyler-Encyclopedia-1: 129,188; Jester: 98-9; 87V132; Billings-Old: 158-9; Billings-Deplored: 216
Blayton, Thomas: Tyler-Encyclopedia-1: 188-9; Sprinkle: 84; 3V248; Billings-Deplored: 272
Bond, John: Nugent-1: 101,192,433,459; Tyler-Encyclopedia-1: 190; Billings-Deplored: 274; Billings-Deplored: Hening-2: 39
Booth, Robert: Nugent-1: 95,284,276,278; Tyler-Encyclopedia-1: 190-1
Borne, Robert: Nugent-1: 167,324,347,381
Boucher, Dan: Nugent-1: 303,322; Tyler-Encyclopedia-1: 191; 7W(1)236
Bowler, Thomas: Nugent-1: 322,389-90; Nugent-2: 154,167; Tyler-Encyclopedia-1: 133; Billings-Deplored: 270; Sprinkle: 51
Bracewell, Robert: Nugent-1: 424; Nugent-2: 79; Hening-1: 378
Branch, John: Nugent-1: 38,160,339; Tyler-Encyclopedia-1: 192

Appendix B: Source Information for Tables

Bray, James-Norfolk: Nugent-2: 93,94; Tyler-Encyclopedia-1: 131; Billings-Deplored: 269; McIlwaine-Minutes: 457-9
Bray, Robert: Nugent-2: 126,214,331; Billings-Deplored: 277; Tyler-Encyclopedia-1: 193
Breman, Thomas: Nugent-1: 233,303
Brewer, John: Nugent-1: 26,507; Tyler-Encyclopedia-1: 194
Brewster, Richard: Nugent-1: 80-1,142; Tyler-Encyclopedia-1: 194; 27V295
Bridger, Joseph: Nugent-1: 433,559; Nugent-2: 53,79,129,186; Tyler-Encyclopedia-1: 130; Billings-Deplored: 269; Morton: 166-167; 7W(1)212,213,242-4, 2W(2)186-187; 2V380, 5V65
Bristow, Robert: Nugent-1: 340,505,536,553; Nugent-2: 133-4; Tyler-Encyclopedia-1: 195; Sprinkle: 126; Billings-Deplored: 273; 13V59-62
Broadhurst, Walter: Nugent-1: 189,199,420; Tyler-Encyclopedia-1: 196; 4W(1)76-77; 17V441
Brocas, William: Nugent-1: 83,88,112,160,190,193,245,265-6,273,281,312-313; Tyler-Encyclopedia-1: 105; Rutman: 48; 12V387-389; 20W(1)138, 22W(1)13
Browne, Devoreux: Nugent-1: 290-1,455; Nugent-2: 97,104,106-7; Tyler-Encyclopedia-1: 197; Whitelaw: 109,600,634,965,968-9,1288; Jester: 296; Billings-Deplored: 271; Ames-Studies: 25,138
Browne, Henry: Nugent-1: 61,149-51,154,160; Tyler-Encyclopedia-1: 104; 16W(1)227
Browne, William: Nugent-1: 29,471; Nugent-2: 61-2,222; Billings-Deplored: 283; Tyler-Encyclopedia-1: 197
Burnham, John: Nugent-1: 148; Tyler-Encyclopedia-1: 199; Billings-Deplored: 278; Rutman: 81,107,208, 267
Burnham, Rowland: Nugent-1: 97,114,144,185,215,219-20; Tyler-Encyclopedia-1: 199; Quitt: 27-8; Rutman: 48; 2W(1)270
Burroughs, [Christopher or Charles]: Nugent-1: 38,104,180,221; Tyler-Encyclopedia-1: 149; Kukla-Politics: 112; McIlwaine-Journals: 73; 39V1-20
Bushrod, Thomas: Tyler-Encyclopedia-1: 200-1; Richter: 212; Catterall: 58; Billings-Deplored: 285; 23V48-49; 1W(1)14,93-4,177,195; 11W(1)30-33
Butler, William: Nugent-1: 36,87,146,182,207,209,211-2,337,347-8,447; Nugent-2: 201; Tyler-Encyclopedia-1: 201
Calithrop, Christopher: Nugent-1: 12,26,39,44; Tyler-Encyclopedia-1: 202; Jester: 113-6; Billings-Deplored: 285; Richter: 119,135-8,196,236-7
Cant, David: Nugent-1: 366,385,496,506; Nugent-2: 41,43; Billings-Deplored: 273; Tyler-Encyclopedia-1: 203
Carter, Edward: Nugent-1: 387,445,453,480,536; Billings-Deplored: 270; Tyler-Encyclopedia-1: 125
Carter, John [Lancaster County]: Nugent-1: 256-7,414,504; Nugent-2: 37; Tyler-Encyclopedia-1: 122; Billings-Deplored: 269; Bruce-Economics-2: 87-88,124
Carter, John, Sr. [Nansemund County]: Nugent-1: 46,132,150,156,225,233,326, 378-9,518,536; Nugent-2: 37; Tyler-Encyclopedia-1: 122; Tyler-Encyclopedia-4: 499; Hening-1: 501,512; Bruce-Social: 73; Billings-Deplored: 269; Rutman: 54
Cary, Miles: Nugent-1: 286,294,326,353-4,374,533; Nugent-2: 247,368; Tyler-Encyclopedia-1: 130; Tyler-Encyclopedia-4: 94; Jester: 326-7; Billings-Deplored: 266,270
Catchmaie, George: Nugent-1: 501; Billings-Deplored: 279; Tyler-Encyclopedia-1: 206
Carver, William: Nugent-1: 477-8; Nugent-2: 169; Tyler-Encyclopedia-1: 205; Finestone: 23; Billings-Deplored: 277; Sprinkle: 51
Caufield, Robert: Nugent-2: 231,237; Billings-Deplored: 283; Tyler-Encyclopedia-1: 206
Caufield, William: Nugent-1: 337,342,421; Billings-Deplored: 283; Tyler-Encyclopedia-1: 206-7
Chandler, John: Nugent-1: 15,44,273,341; Nugent-2: ?; Tyler-Encyclopedia-1: 207
Charlton, Stephen: Nugent-1: 79,82,129,200-1; Tyler-Encyclopedia-1: 207; Whitelaw: 28-9,409,424-5,432,1396; Currier-Briggs-3: 678; Haight: 433; Ames-Studies: 96-7,104-5; Bruce-Institutions: 78; Sprinkle: 51; 77V259-276

Chessman, John: Nugent-1: 7,37,53,69,127-9,232,262-3,419,469; Tyler-Encyclopedia-1: 115; Billings-Deplored: 271; Richter: 168n,196; Bruce-Economics: 89; 1W(1)97-9; 11W(1)29-30; 14V86-7

Chew, John: Nugent-1: 2-3,44,62,100-1,223; Tyler-Encyclopedia-1: 208; Kukla-Politics: 114-5; Currier-Briggs-3: 685; Jester: 127-8; 1V87-88

Chickley, Henry: Nugent-1: 253,291,334,384; Tyler-Encyclopedia-1: 50-1; Carson: 38; McIlwaine-Minutes: 207; Quitt: 35-6; Currier-Briggs-8: 674; Billings-Deplored: 269; Sprinkle: 138-9

Chiles, Mathew: No information found.

Chiles, Walter, Jr.: Nugent-2: 112; Tyler-Encyclopedia-1: 208; Kukla-Politics: 169-179; McIlwaine-Journals: 127; Billings-Deplored: 275; 19V211-3

Chiles, Walter, Sr.: Nugent-1: 87,103-4,140,187; Tyler-Encyclopedia-1: 114-5; Billings-Deplored: 275; 19V104-6

Church, Richard: Nugent-2: 40,281; Billings-Deplored: 277; Tyler-Encyclopedia-1: 210

Claiborne, William, Sr.: Nugent-1: 6,70,149,165,223-4,244-5,247,290,292,358-9, 376,406; Hening-1: 508; Hening-2: 10; Hale; Whitelaw: 26,30-1,168-9,630,1386; Tyler-Encyclopedia-1: 96-7; Tyler-Encyclopedia-4: 148; Hale: 244,261-6; Carson: 81; Haight: 433; Jester: 131-3; McIlwaine-Journals: 85; 56V328-343,431-460; Billings-Deplored: 279; Browne-Provincial: 4:22-3, 281,390,458-9

Claiborne, William, Jr.: Nugent-2: 236; Tyler-Encyclopedia-1: 211-2; Jester: 133; McIlwaine-Minutes: 454-5; Hale: 209; 5W(1)109; 1V60, 56V445

Cocke, Richard: Nugent-1: 54,58,120-1,137,266,441,513; Tyler-Encyclopedia-1: 214; 3V406; 9W(2)49-59

Cocke, William: Nugent-1: 156,331,359-60,458-9; Tyler-Encyclopedia-1: 215; 52V214-16; 9W(2)49-59

Cockeram, William: Nugent-1: 114,336,341,562; Billings-Deplored: 283; Tyler-Encyclopedia-1: 215

Cole, William, Jr.: Nugent-1: 6,27-8,31,82,150,281,442,549; Nugent-2: 270,291, 365; Tyler-Encyclopedia-1: 133; Jester: 141-4; Andrews: 23; Billings-Deplored: 270; Sprinkle: 124; McIlwaine-Minutes: 457-9

Collclough, George: Nugent-1: 210,359-60; Tyler-Encyclopedia-1: 215; Billings-Deplored: 280; Billings-Old: 165-9

Corbin, Henry: Nugent-1: 358,369,388,392,409,432; Nugent-2: 21,32-3,64,73; Tyler-Encyclopedia-1: 128; Rutman: 50-51,93; Billings-Deplored: 266,269; 28V281-283, 29V375

Corker, John: Nugent-1: 81,124,139,374; Billings-Deplored: 285; Tyler-Encyclopedia-1: 218

Corker, William: Nugent-1: 341; Nugent-2: 95-6; Billings-Deplored: 285; Tyler-Encyclopedia-1: 218

Crew, Randall: Nugent-1: 102,122,177; Tyler-Encyclopedia-1: 219; Bruce-Economics-2: 75

Crewes, James-Norfolk: Nugent-1: 466; McIlwaine-Minutes: 455; Billings-Deplored: 274; Sprinkle: 84

Crowshaw, Joseph: 62,92,117,152,166,222,227,249,250,361; Tyler-Encyclopedia-1: 219; Jester: 145-6; Billings-Deplored: 285; 30V274-279, 37V139-141; 1W(1)193-5

Culpepper, Lord Thomas: Tyler-Encyclopedia-51-2

Curtis, John: Nugent-1: 43,47,64,78,334,339,371,385,407,465-6,481,499,567; Tyler-Encyclopedia-1: 220; Quitt: 62-65; Billings-Deplored: 276; Rutman: 56

Dacker, William: no information found.

Dale, Edward: Nugent-1: 137; Tyler-Encyclopedia-1: 220; Billings-Deplored: 276; Rutman: 132-133

Davis, Thomas: Nugent-1: 6,17,57,76,156,158,162-3,23,230,252,451,490,552; Tyler-Encyclopedia-1: 221; Jester: ?; Hening-2: 15

Davis, William: Nugent-1: 16,25,27-8,33,69,82,108-9,112,160,194-5,213-4,235-6, 240,243-4,263,308,315,336,351,373,446; Tyler-Encyclopedia-1: 221

Death, Richard: Nugent-1: 80,125,169; Tyler-Encyclopedia-1: 222; 7W(1)219

Appendix B: Source Information for Tables

Denson, William: Nugent-1: 85,277,356; Tyler-Encyclopedia-1: 222
Dew, Thomas: Nugent-1: 95,105,118,151; Nugent-2: 83,221; Tyler-Encyclopedia-1: 120-1; Whitelaw: 30-1; Billings-Deplored: 269; Hening-1: 508
Digges, Edward: Nugent-1: 208-9,214,236,243,555-6; Tyler-Encyclopedia-1: 47-8; Hale: 300; Jester: 154-6; Billings-Deplored: 67,271; Hening-1: 426; McIlwaine-Minutes: 207; 20W(1)8-13, 22W(1)74; 14V305, 19V350,357-8
Digges, William: Tyler-Encyclopedia-1: 223; Billings-Deplored: 285; Andrews: 434; Sprinkle: 124
Dipnall, Thomas: Nugent-1: 250
Douglas, Edward: Nugent-1: 181,226,344,414; Tyler-Encyclopedia-1: 225; Whitelaw: 97-8,641; Haight: 434
Dunston, John: Nugent-1: 40,109
Edlowe, Mathew: Nugent-1: 59-60,491; Tyler-Encyclopedia-1: 228; Billings-Deplored: 275; 15W(1)282-3
Edwards, William: Nugent-1: 22,29,47,133,143,149,176,244-5,250,253,353,355, 433,442,468,524-5,554; Nugent-2: 99,186,216,283,322,342,353-4,371,373,401; Tyler-Encyclopedia-1: 228
Elliot, Anthony: Nugent-1: 154,157,208,223,309,381; Tyler-Encyclopedia-1: 124; Kukla-Politics: 202; Billings-Deplored: 268; Hening-1: 318,356
Ellyson, Robert: Nugent-1: 240-1,338,348; Tyler-Encyclopedia-1: 229; 72V42-49; Billings-Deplored: 275; Hening-1: 422
English, John: Nugent-1: 67,244-5,372,400,512; Tyler-Encyclopedia-1: 230; 7W(1)240
Eps, Francis, Jr.: Nugent-1: 505; Nugent-2: 18,31,84,52,134, 206-207; Billings-Deplored: 272; 3V393-4; Tyler-Encyclopedia-1: 230
Eps, Francis, Sr.: Nugent-1: 31,84,281; Tyler-Encyclopedia-1: 115; Jester: 160-4; Bruce-Economic-2: 75
Eyres, Robert: Nugent-1: 129,178; Tyler-Encyclopedia-1: 231
Fallowes, Thomas: No information found.
Fantleroy, Moore: Nugent-1: 154,194-5,319; Tyler-Encyclopedia-1: 233; Jester: 266; Hening-1: 422,427; Billings-Deplored: 281; 20W(1)132-5; 1V75-81
Fawdown, George: Nugent-1: 23,111; Tyler-Encyclopedia-1: 233
Ffarrer, William: Nugent-1: 41,60,516; 8V97-8,20608, 50V?; Billings-Deplored: 275; Tyler-Encyclopedia-1: 93,232-3
Filmer, Henry: Nugent-1: 149,230,561-2; Nugent-2: 9,134; Tyler-Encyclopedia-1: 234; Billings-Deplored: 284; 21V153-154
Fleet, Henry: Nugent-1: 13,177,194,259,296,311,316,332,353; Tyler-Encyclopedia-1: 236-7; Hale: 253,291-2; Hening-1: 318-9; McIlwaine-Journals: 85; Jester: 172-5,382-3; Billings-Deplored: 276; Andrews: 41,55-6,72
Fletcher, George: Nugent-1: 251,266; Tyler-Encyclopedia-1: 236; Hening-1: 374
Flint, Thomas: Nugent-1: 8-10,13,70-1,81; Tyler-Encyclopedia-1: 237
Fludd, John: Nugent-1: 32,86,104,176,199,228,268; Tyler-Encyclopedia-1: 237; Jester: 175-7; Hening-1: 328; Hening-2: 138
Ford, Richard: Nugent-1: 118,219,275,377; Tyler-Encyclopedia-1: 238; Billings-Deplored: 275; Hening-1: 342-3
Foster, Richard: Nugent-1: 94,249,306,377,430,473; Billings-Deplored: 277
Fowke, Gerald: Nugent-1: 297-8,377-8,388,432,469-70,473,489,533; Nugent-2: 46; Tyler-Encyclopedia-1: 238-9; Hayden: 155; Hening-2: 149-51; Billings-Deplored: 284; 15W(1)179
Fowke, Thomas: Nugent-1: 297-8; Billings-Deplored: 284; Tyler-Encyclopedia-1: 238-9
Fowler, Francis: Nugent-1: 33,72,113,154,483; Tyler-Encyclopedia-1: 239
Frances, Rawleigh: No information found. Probably Rawleigh Travers.
Francis, Thomas: Nugent-1: 104,318,543; Tyler-Encyclopedia-1: 240
Franklin, Ferdinand: Tyler-Encyclopedia-1: 240

Freeman, Bridges: Nugent-1: 18,36,63,124,224,299,522; Tyler-Encyclopedia-1: 113; Kukla-Politics: 114-5; Hening-1: 342-3
George, John: Nugent-1: 32,93,98-9,127,249,459; Nugent-2: 20,40; Tyler-Encyclopedia-1: 241; Billings-Deplored: 274; 7W(1)240-1
Gill, Stephen: Nugent-1: 52,114,122,132,142,179,180,213,354; Tyler-Encyclopedia-1: 242; Kukla-Politics: 114-5; 24W(1)40
Godwin, Thomas: Nugent-1: 367,371,462; Nugent-2: 31; Billings-Deplored: 279; Tyler-Encyclopedia-1: 242-243
Gooch, William: Nugent-1: 199,311,322-3,384,408; Billings-Deplored: 271; Tyler-Encyclopedia-1: 121
Goodwin, James: Nugent-1: 367,371,462; (Under Godwin): Nugent-1: 197-8,319,327; Nugent-2: 31; Billings-Deplored: 285; Tyler-Encyclopedia-1: 242-244
Gough, Mathew: Nugent-1: 111; Tyler-Encyclopedia-1: 245
Gough, Nathaniel: Tyler-Encyclopedia-1: 245
Gray, Francis: Nugent-1: 28,231-2,288,337,355,483,501; Tyler-Encyclopedia-1: 246; 43V144-149; Billings-Deplored: 272; Jester: 195
Griffith, Edward: Nugent-1: 429; Tyler-Encyclopedia-1: 248; Billings-Deplored: 284
Gwillen, George: Nugent-1: 174,201; Nugent-2: 171
Gwin, Hugh: Nugent-1: 132,141,182,248,263,301,362,417; Tyler-Encyclopedia-1: 249; Catterall: 77
Hackett, Thomas: Nugent-1: 131,208,278-9,284,353,495
Ham, Jeremy: Nugent-1: 275; Tyler-Encyclopedia-1: 250; Hening-1: 501-2
Hamelyn, Stephen: Nugent-1: 102,140,203,550; Billings-Deplored: 272; Tyler-Encyclopedia-1: 250
Hammond, John: Nugent-1: 98,315; Tyler-Encyclopedia-1: 251; McIlwaine-Journals: 84; Hall: 277-308
Hamond, Manwaring: Nugent-1: 187,299,383,408; Nugent-2: 25; Tyler-Encyclopedia-1: 126; Carson: 38-9; Billings-Deplored: 270; Hening-1: 545
Haney, John: Nugent-1: 207,234,349,462,505; Tyler-Encyclopedia-1: 255; Nugent-2: 43
Hardy, George: Nugent-1: 67,140,146,176-7; Tyler-Encyclopedia-1: 251; 7W(1)234
Harlow, John: Nugent-1: 46,130,139,217,273; Tyler-Encyclopedia-1: 252
Harmer, Ambrose: Nugent-1: 125; Tyler-Encyclopedia-1: 110; Kukla-Politics: 146-7
Harris, Thomas: Nugent-1: 9-10,33,37,40,60,68,101,152,166,177,203,278,281,319, 332,334,348,365,371-3,386,436,492,527,543; Tyler-Encyclopedia-1: 252; Jester: 202-3; Bruce-Economics-2: 75
Harris, William: Nugent-1: 6,11-2,51,53-4,77,150,171,305,307,372,383,388,398-9,404,492-3,554,559; Nugent-2: 41,84,141; Tyler-Encyclopedia-1: 252; Jester: 202-5; Hening-1: 422,426; Billings-Old: 307
Harrison, Benjamin: Nugent-1: 15-16,25,56,152,159,163,186,224; Nugent-2: 122,245,275-276,309; Tyler-Encyclopedia-1: 253; 51V82, 53
Harwood, Thomas: Nugent-1: 14-5,25,30,36,75,83-4,150,259,274; Tyler-Encyclopedia-1: 118-9; Tyler-Encyclopedia-4: 447; Kukla-Politics: 147-8; Currier-Briggs-2: 397; Jester: 205-8; 2V183-184
Hatcher, William: Nugent-1: 40,59,89,347; Nugent-2: 154; Tyler-Encyclopedia-1: 254; 16W(2)457-9; Sprinkle: 51,54
Hay, Wiliam: Nugent-1: 370,393,465-6,503,514; Tyler-Encyclopedia-1: 255; Currier-Briggs-2: 398; Billings-Deplored: 286; Richter: 105,112,177,188
Haynes, Thomas: Nugent-1: 368; Nugent-2: 56,193; Billings-Deplored: 276
Haywood, John: Nugent-1: 25,83,114,199,233,334,370,414,440,454, 515,548; Tyler-Encyclopedia-1: 257; Currier-Briggs-2: 398; Richter: 176
Heyrick, Henry: No information found.
Higginson, Humphrey: Nugent-1: 80,136,159,160,298; Tyler-Encyclopedia-1: 112; Billings-Old: 165-9; 5W(1)186

Appendix B: Source Information for Tables

Hill, Edward, Jr.: Nugent-2: 40,222,268,271,344; Sprinkle: 124,129; Whitelaw: 968; McIlwaine-Minutes: 454-5; Hening-2: 364-5; Billings-Deplored: 75-6,273; 3V250
Hill, Edward, Sr.: Nugent-1: 93,324,382,405,457,460; Billings-Deplored: 268; Tyler-Encyclopedia-1: 119; Kukla-Politics: 145; Hening-1: 321,422-3; Browne-Proceedings: 238-9, 266; Browne-Provincial: 314-6,321-2,340-1,389,408, 444,512-3; Hale: 261-66
Hill, John: Nugent-1: 21,153,172,181; Tyler-Encyclopedia-1: 257-8; Hallman: 50; Billings-Deplored: 64-5; Hening-2: 198; Jones, R: 274; 1V75-81, 40V240
Hill, [Nicholas or Richard]: As Nicholas: Nugent-1: 38,77,175,444; Nugent-2: 145; Tyler-Encyclopedia-1: 258; Billings-Deplored: 271,274; 7W(1)238,314; As Richard: Nugent-1: 92,261, 413,449,482-3,507-8,516; Nugent-2: 26,99,108,317
Hobbs, Francis: Nugent-1: 196; Billings-Deplored: 274; Tyler-Encyclopedia-1: 258
Hockaday, William: Nugent-1: 131,133,167,181,210,257,276,303,311,342,376,511, 563; Tyler-Encyclopedia-1: 258; McIlwaine-Minutes: 223
Hoddin, John: Nugent-1: 148,272
Hoe, Rice: Nugent-1: 37,86-7,110,119,148; Tyler-Encyclopedia-1: 260; Jester: 211-3; Billings-Old: 169; 48V270-274; Hening-1: 262
Holland, John: Nugent-1: 189,207,252,444; Nugent-2: 238-9,257,389; Tyler-Encyclopedia-1: 259
Holmewood, John: Nugent-1: 203,365; Tyler-Encyclopedia-1: 259; Billings-Deplored: 273; Catterall-4: 16-17
Holte, Robert: Nugent-1: 103,123,151,215,231,320,487,524; Kukla-Politics: 115; Billings-Deplored: 275; Hening-1: 422
Hone, Theophilus: Nugent-1: 275,322,540; Nugent-2: 169-70; Tyler-Encyclopedia-1: 260; Billings-Deplored: 275; Sprinkle: 146
Horsey, Stephen: Nugent-1: 170,224-5,259,297,325; Whitelaw: 28-31,1414,1416; Kukla-Politics: 174; Jones-Quakers: 335-7; 19V105
Horsmenden, Warham: Nugent-1: 314,388; Tyler-Encyclopedia-1: 122-3; Kukla-Politics: 202; Jester: 287-9; Billings-Deplored: 268; Hening-1: 501-2,505,512
Hoskins, Anthony: Nugent-1: 233,264,294; Whitelaw: 32,122,714,728-9; Haight: 435; Ames-Studies: 200n
Hoskins, Bartholomew: Nugent-1: 7,178-9,182,287,298; Tyler-Encyclopedia-1: 261; 1V75-81,311 39V122
Howard, Francis, Lord Effingham: Tyler-Encyclopedia-1: 53-54
Hull, Peter: Nugent-1: 119,166,176,183,261,509
Hutchinson, Robert: Nugent-1: 97,126,128,138,145,151; Tyler-Encyclopedia-1: 262; Tyler-Encyclopedia-2: 586-587
Iversonne, Abraham: Nugent-1: 50,208,217,328,403,548; Billings-Deplored: 274
Jeffreys, Herbert: Tyler-Encyclopedia-1: 49-50
Jenings, Peter: Nugent-1: 372,392,411; Tyler-Encyclopedia-1: 131,265; Hening-2: 150; McIlwaine-Minutes: 221; Billings-Deplored: 268; 19V360-361
Jennifer, Daniel: Nugent-2: 135,162,177-8,215,259,303,307,358; Billings-Deplored: 271; McIlwaine-Minutes: 454-5
Johnson, Joseph: Nugent-1: 28,56,137,183,429,456; Tyler-Encyclopedia-1: 265
Johnson, Thomas: Nugent-1: 22,29,35,94,100,163-4,169,277,329,384,394,404,413, 464,534-5,568; Tyler-Encyclopedia-1: 266; Whitelaw: 28-31,528-9,628; Haight: 436
Jones, Anthony: Nugent-1: 27,52,329,381,509,530,550; Tyler-Encyclopedia-1: 266:7; 7W(1)219
Jones, William: Nugent-1: 123, 157, 455, 464,482-3,506; Nugent-2: 64; Tyler-Encyclopedia-1: 269; Whitelaw: 219-20,256,347,428; Billings-Deplored: 279; Haight: 436
Jordan, George: Nugent-1: 61,133,326,334-5,386,457; Nugent-2: 171; Tyler-Encyclopedia-1: 269; Quitt: 273-4; Hening-1: 342-3; Billings-Deplored: 283; 7W(1)231-2n; 16W(1)288-9
Kemp, Mathew: Nugent-2: 161,202-3,229,397; Tyler-Encyclopedia-1: 138; McIlwaine-Minutes: 454-5; Billings-Deplored: 278; 1V60

Kemp, Richard: Nugent-1: 27,28,31,45,67; Tyler-Encyclopedia-1: 47,106; Quitt: 90-93; Currier-Briggs-1: 214; Rutman: 48; 10W(1)167; 2V174, 8V302-6, 11V170, 20V74
Kendall, William: Nugent-1: 236,284,321,410,419,434-5,533; Nugent-2: 3-4; Tyler-Encyclopedia-1: 271; Whitelaw: 56-7,151-2,188-9,199,677,1252,1319; Haight: 437; Ames-Studies: 98-100,106; Billings-Deplored: 279; Sprinkle: 51; 19V10-11
Knight, Peter: Nugent-1: 83,85,104,184,248,258,277,282,295,340,343,346,389, 394,460; Tyler-Encyclopedia-1: 273; Billings-Deplored: 274,281; Hale: 263-6,293
Knowles, John: Nugent-1: 450-1,537,565; Leonard: 39
Lambert, Thomas: Nugent-1: 22,84,173; Billings-Deplored: 277; Tyler-Encyclopedia-1: 273
Langley, Ralph: Nugent-1: 510; Hening-1: 422; Billings-Deplored: 286; Karraker: 77-8; 19W(1)197-9
Lawrence, Richard: Nugent-1: 367,441,467,472,478,527,530-1; Nugent-2: 22; Sprinkle: 51; Rutman: 84-5; Andrews: 30-1,85n,95-7; 16W(3)244-249
Leare, John: Nugent-1: 333; Nugent-2: 209,246,373; Tyler-Encyclopedia-1: 139-40; Sprinkle: 124; Billings-Deplored: 279; Bruce-Economics-2: 125; 17V228-231
Lee, Henry: Nugent-1: 43,68,83,189,196,218,237,311,456,528; Tyler-Encyclopedia-1: 276; 24W(1)48
Lee, [Hugh]: Nugent-1: 205,242,351,364; Nugent-2: 134,146,207,378; 1V75-81; 49V23
Lee, John: Nugent-1: 189,209-10,258,261,288-90,346,378,501; Nugent-2: 51,172, 389; Billings-Deplored: 284
Lee, Richard, Jr.: Nugent-2: 61,152; Tyler-Encyclopedia-1: 134-5; Quitt: 40-42; Billings-Deplored: 284; 21W(2)29-32
Lee, Richard, Sr.: Nugent-1: 51-2,92,96,122,126,131,149,154-6,162,178,213,219, 235,241,258,276,306,329,330,343,346,390,404,435-6,467-8,522; Tyler-Encyclopedia-1: 116-7; Quitt: 36-9; Kukla-Politics: 114-5; Rutman: 48; Billings-Deplored: 270; Carson: 36; 62V3-49; 10W(1)171-2
Littleton, Nathaniel: Nugent-1: 155,331; Whitelaw: 30-1,77-8,281,629,675; Tyler-Encyclopedia-1: 110; Haight: 437; Hening-1: 356; 85V278-9, 90V488; Ames-Studies: 99,105
Littleton, Southy: Nugent-1: 327,452; Nugent-2: 94,129,139,157-9,182,195-6; Haight: 437; McIlwaine-Minutes-454-5; Billings-Deplored: 271; 18V22
Llewellin, Daniel: Nugent-1: 78,81-2,95,130,138,203,240; Tyler-Encyclopedia-1: 279; Jester: 274-5; 4V6; 13V53
Lloyd, Cornelius: Nugent-1: 27,50,52-3,100,157,166,170,172; Tyler-Encyclopedia-1: 279; Jester: 223; Hening-1: 318; Billings-Old: 161; McIlwaine-Journals: 71; 1V75-81, 9V312
Lloyd, Edward: Nugent-1: 42,233,472; Tyler-Encyclopedia-1: 279; Jester: 223-225; Hall: 218-9,228,325
Lobb, George: Nugent-1: 99-100,387; Tyler-Encyclopedia-1: 279
Loveing, Thomas: Nugent-1: 89,118,137,224,294; Billings-Deplored: 275; Tyler-Encyclopedia-1: 280
Lucas, Thomas, Jr.: Nugent-1: 262,362,497,518,521; Tyler-Encyclopedia-1: 280
Lucas, Thomas, Sr.: Nugent-1: 161,17-2,240,261-2,345,381; Tyler-Encyclopedia-1: 280; Leonard: 39
Luddington, William: Nugent-2: 176-7
Ludlow, George: Nugent-1: 96,161,201,214,229,239,243,246,262,295-6,456; Tyler-Encyclopedia-1: 113; Kukla-Politics: 114-5; 54V255-257, 21V350-54; 1W(1)190
Ludwell, Philip: Nugent-1: 429,548; Nugent-2: 33,130,386; Tyler-Encyclopedia-1: 145-6; Quitt: 78-80; 4W(2)156, 19W(1)199-214; 1V174-186; Billings-Deplored: 269; McIlwaine-Minutes: 454-5
Ludwell, Thomas: Nugent-1: 165,178,429; Nugent-2: 57,84-5,92; Tyler-Encyclopedia-1: 126-7; 19W(1)199-214; 21V36-42; Billings-Deplored: 269; Hening-2: 39
Lyggon, Thomas: Nugent-1: 440,516; Nugent-2: 49,92,116,124; Billings-Deplored: 274; 16W(2)303,308-10

Appendix B: Source Information for Tables

Mansell, David: Nugent-1: 29,106,117,158,297; Tyler-Encyclopedia-1: 283
Major, Edward: Nugent-1: 55,159,167,170,559; Tyler-Encyclopedia-1: 282; Kukla-Politics: 163; 9W(2)49-59
Martin, John: Nugent-1: 8,162,220,299,528,539; Nugent-2: 83-4; James-Norfolk-3: 141; James-Norfolk-4: 79; Billings-Deplored: 277; 14V177
Mason, Francis: Nugent-1: 34,134,151,166,209; Tyler-Encyclopedia-1: 285-6; Billings-Deplored: 283; Hallman: 49
Mason, George: Nugent-1: 332-3; Tyler-Encyclopedia-1: 286; Hening-2: 149-51; Sprinkle: 84; Billings-Deplored: 282
Mason, James: Nugent-1: 98,173,177,234,281; Billings-Deplored: 283; Tyler-Encyclopedia-1: 286
Mason, Lemuell: Nugent-2: 169,195,324,381; Tyler-Encyclopedia-1: 286; Billings-Deplored: 278; Jester: 242; 2V381-5
Mathew, Thomas: Nugent-1: 109,128-9,259,452,484; Nugent-2: 192-3; Tyler-Encyclopedia-1: 286; Billings-Deplored: 281; Andrews: 11-41
Mathews, Samuel, Jr.: Nugent-1: 291,342,344,348-9,373; Tyler-Encyclopedia-1: 48-9, 119; Kukla-Politics: 158-9; Jester: ?; Hening-I: 426; 14W(2)105-113
Mathews, Samuel, Sr.: Nugent-1: 4,133,134,144,160,; Tyler-Encyclopedia-1: 48-9; Carson: 81; Quitt: 32-33; Force: 8:15; Kukla-Politics: 132-7; Hall: 59-61,259; 14W(2)105-113; 1V91
Meares, Thomas: Nugent-1: 55,179,182,191-2,370; Tyler-Encyclopedia-1: 289; Jones-Quakers: 277,330; Hallman: 49; Catterall: 13
Meese, William: Nugent-1: 514,534,559-60; Billings-Deplored: 282; Tyler-Encyclopedia-1: 136
Melling, William: Nugent-1: 43,53,68,287,329,435,539; Nugent-2: 63; Tyler-Encyclopedia-1: 289; Whitelaw: 158,255,633,696,1034,1670; Billings-Deplored: 280; Haight: 438
Michell, William: Nugent-1: 28,39,81,260,324,353,356,437,462,487; Tyler-Encyclopedia-1: 292; Hening-1: 501-2; Ames-Studies: 96
Minifie, George: Nugent-1: 24,120,123-4,128,378; Tyler-Encyclopedia-1: 104-5; Jester: 248-9; ?W(1)85; 1V86-7, 9V267, 14V421-2
Molesworth, Guy: Nugent-1: 220; Tyler-Encyclopedia-1: 292-293; Billings-Deplored: 268
Montague, Peter: Nugent-1: 67,164,173,176,383,385; Tyler-Encyclopedia-1: 293; Jester: 250-3; Rutman: 95-6; 2W(1)271
Moone, John: Nugent-1: 17,32,77,101,140,513-4; Tyler-Encyclopedia-1: 293; Jester: 253-4; Hallman: 142; Currier-Briggs-2: 408; 7W(1)222
Morgan, Francis: Nugent-1: 72,95-6,132,171,253,259-60,272, 276,293,391,424; Tyler-Encyclopedia-1: 294
Morley, William: Nugent-1: 328,368; Tyler-Encyclopedia-1: 294
Morrison, Richard: Nugent-1: 173,223; Tyler-Encyclopedia-1: 111-2
Moryson, Charles: Nugent-2: 63; Tyler-Encyclopedia-1: 294; McIlwaine-Minutes: 454-5
Moryson, Francis: Nugent-1: 92,195,240,305,391; Tyler-Encyclopedia-1: 49; Carson: 38; Hening-1 429; Hening-2: 10,13,34,148; Billings-Deplored: 269; 83V93
Moseley, Arthur: Nugent-1: 275,474; Nugent-2: 87; Billings-Deplored: 278; Tyler-Encyclopedia-1: 294
Mosley, William: Nugent-1: 138,191,246,275,431-2,468,500; Nugent-2: 5,82,101, 157,252,257,355; Tyler-Encyclopedia-1: 294; James-Norfolk-3: 141; James-Norfolk-4: 79, 109; Billings-Deplored: 278; 5V328, 35V218-220, 62V346-347
Mottrum, John, Sr.: Nugent-1: 122,132,198,248,252,260,299,331,337,359-60,395, 488,490; Tyler-Encyclopedia-1: 295; Hale: 265; Billings-Old: 165-9; Bruce-Economics-2: 113-4; 17W(1)53-57
Nicholson, Francis: Tyler-Encyclopedia-1: 54-56
Paine, Florentine: Nugent-1: 131,219; Tyler-Encyclopedia-1: 302

Conscious Choice

Page, John: Nugent-1: 230-1,252,279,340; Nugent-2: 30,261; Tyler-Encyclopedia-1: 136; Jester: 231; Sprinkle: 159; McIlwaine-Minutes: 454-5; Billings-Deplored: 286; Richter: 218
Parke, Daniel, Sr.: Nugent-1: 323,399,492,558; Tyler-Encyclopedia-1: 132; McIlwaine-Minutes: 221; Bruce-Economics-2: 123; Billings-Deplored: 271; 10W(1)167,172-3; 5V65, 14V174-5
Pate, John: Nugent-1: 216,260,286,478,509,550; Nugent-2: 58,157; Tyler-Encyclopedia-1: 131; Hening-2: 15; Billings-Deplored: 268; 19V256
Pate, Richard: Nugent-1: 49,204; Tyler-Encyclopedia-1: 302; 19V255-257
Perry, Henry: Nugent-1: 78,120,128,378; Nugent-2: 26; Tyler-Encyclopedia-1: 119; Jester: 267-9; Billings-Deplored: 268; Bruce-Economic-2: 75
Pettus, Thomas: Nugent-1: 159,162,225,273,389; Tyler-Encyclopedia-1: 111; Billings-Deplored: 269; 3V153-4
Peyton, Valentine: Nugent-1: 470,489-90; Tyler-Encyclopedia-1: 304-5; Billings-Deplored: 284; Leonard: 40
Pierce, William: Nugent-1: 9,14,16,27,29,57-8,62,123,131,143,147,149,152,176, 185,202,225,259,412,444-5,478-9,506,512,546-7; Tyler-Encyclopedia-1: 101-2; Jester: 261-3; Billings-Old: 160; Catterall: 77
Pitt, Robert: Nugent-1: 85,143,145,171,290,433,456,465,488,561; Tyler-Encyclopedia-1: 143; Billings-Deplored: 271,274; 7W(1)253
Place, Rowland: Nugent-2: 59-60,170-1; Billings-Deplored: 268; Tyler-Encyclopedia-1: 133-4
Poole, Henry: Nugent-1: 62,136
Porter, John, Sr.: Nugent-1: 142,239,274,328,446,450,507; Nugent-2: 136,266; Quitt: 51-2; Jones-Quakers: 274-5,334-5; Hening-2: 198; Billings-Deplored: 278; 1W(1)39 **Pott, John**: Nugent-1: 3,10,15,142,225
Powell, John: Nugent-1: 49,89,140,155,172,250,264,302,344,388,526,559; Nugent-2: 300,310; Billings-Deplored: 273; Tyler-Encyclopedia-1: 307
Poythers, Francis: Nugent-1: 60,129,175; Tyler-Encyclopedia-1: 307; Browne-Provincial: 408; Billings-Deplored: 273; 14W(2)77-9
Presley, Peter: Nugent-1: 185,522,532; Nugent-2: 4,101; Tyler-Encyclopedia-1: 308; Quitt; 23W(1)185
Presley, William, Jr.: Nugent-1: 198,351; Nugent-2: 4; Tyler-Encyclopedia-1: 308; Andrews: 25n; Billings-Old: 165-9; 23W(1)184-5; 1V75-81
Presley, William, Sr.: Nugent-1: 185,298; Tyler-Encyclopedia-1: 308; 23W(1)184-5
Price, Arthur: Nugent-1: 131,181,214,241,312
Prince, Edward: Nugent-1: 30,117,127,138,166; Tyler-Encyclopedia-1: 309
Pritchard, Thomas: Nugent-1: 198,278,380-381
Pyland, James-Norfolk: Nugent-1: 140,480; Tyler-Encyclopedia-1: 309; Hening-1: 374-5; 7W(1)213
Ramsey, Edward: Tyler-Encyclopedia-1: 310; Billings-Deplored: 275; McIlwaine-Minutes: 454-455; 14V420
Ramsey, Thomas: Nugent-1: 29,74; Tyler-Encyclopedia-1: 310; 14V420
Ransom, Peter: Nugent-1: 156,238,261,278,303,499-500; Tyler-Encyclopedia-1: 312
Reade, George: Nugent-1: 96,127-7,144,177,180,201,237,389; Tyler-Encyclopedia-1: 123-4; Hening-1: 421; Billings-Deplored: 271; 14W(1)117-9
Revell, Randall: Nugent-1: 233,272-3; Tyler-Encyclopedia-1: 313; Whitelaw: 146,154,625,630,1386,1388,1417; Jones-Quakers: 335; Haight: 440
Reynolds, [Charles] or [Christopher]: Nugent-1: 47,363; Tyler-Encyclopedia-1: 312; 7W(1)221
Ridley, Peter: Nugent-1: 114,283; Tyler-Encyclopedia-1: 313; Hening-1: 342-3
Robbins, John: Nugent-1: 7,16,19,102,106-7,136,144,159,301,309,385; Tyler-Encyclopedia-1: 314; Bruce-Economics-2: 75
Roberts, Thomas: Nugent-1: 99,127,428; Billings-Deplored: 286

Appendix B: Source Information for Tables

Robins, Obedience: Nugent-1: 152,224-5,401,407; Tyler-Encyclopedia-1: 121; Whitelaw: 26,28,147,168,177,188,629,639; Haight: 440; Billings-Deplored: 270; Jester: 256-7
Rogers, John: Nugent-1: 118,128,199,250,260,292,357,419,423,458-9,562; Billings-Deplored: 281; Tyler-Encyclopedia-1: 316
Roper, William: Nugent-1: 46; Tyler-Encyclopedia-1: 317; Haight: 440-1
Sadler, Roland: No information found.
Savage, John: Nugent-1: 30,463,468,481-2,515,524-5; Tyler-Encyclopedia-1: 319; Whitelaw: 217-9,228,281-3,574; Billings-Deplored: 280; Jester: 292
Scarburgh, Edmund: Nugent-1: 35,119,101,134,183,225,286,328,418:9,425,452-3, 536; Tyler-Encyclopedia-1: 320; Whitelaw: 30-1,33,37,427-430,624-5,639-41,655-7,770,1040,1149-53,1286,1386-7,1389,1414-8; Kukla-Politics: 146; Haight: 442; Hening-1: 356; 46V170-172; See text for additional references.
Seward, John: Nugent-1: 24,69,85,106,126,164,171,439; Tyler-Encyclopedia-1: 322; 7W(1)255
Shepeard, Robert: Nugent-1: 28,94,204,427,446; Tyler-Encyclopedia-1: 323; Hening-1: 342-3; Bruce-Economics-2: 95
Shepherd, John: Nugent-1: 96,118,130,142-3,156,161,257-8,456,558; Tyler-Encyclopedia-1: 323; Jester: 313
Sheppard, John: Nugent-1: 96,118,130,142-3,156,161,257-8,456,558; Tyler-Encyclopedia-1: 323; Jester: 313; McIlwaine-Minutes: 219
Sidney, John: Nugent-1: 169; Tyler-Encyclopedia-1: 323; Hening-1: 501-2; Billings-Deplored: 278; 1V75-81, 39V14
Smith, Arthur: Nugent-1: 76,82-3,125,526; Tyler-Encyclopedia-1: 325, Jester: 303-4; 7W(1)248; 2V390-1
Smith, John [Francis Dade]: Nugent-1: 17,134,172,194,263,315,337,357; Tyler-Encyclopedia-1: 326; 4W(1)46-52; 23W(1)292-3
Smith, Lawrence: Nugent-2: 12,39,61,93,123,148-9,160-1,373; Tyler-Encyclopedia-1: 115, 326; 81V354; McIlwaine-Minutes: 454-5
Smith, Nicholas: Nugent-1: 172,181,290,329,375,395,478,491,559; Nugent-2: 386; Tyler-Encyclopedia-1: 327; Billings-Deplored: 275; 7W(1)249
Smith, Robert: Nugent-1: 25,78,204,217,228,318,322,334,354-5,384,388,392,478, 517; Nugent-2: 32,55; Billings-Deplored: 269; Tyler-Encyclopedia-1: 128-9
Smith, Toby: Nugent-1: 57,125-6,155,162,174,217,239,273,304,348,519; Tyler-Encyclopedia-1: 327; 20W(1)138
Soane, Henry: Nugent-1: 222,240-1,277,280,336-8,499,506,548; Nugent-2: 346; Billings-Deplored: 275; Tyler-Encyclopedia-1: 327
Southcoat, Thomas: Tyler-Encyclopedia-1: 328; Leonard: 37
Sparrow, Charles: Nugent-1: 181,198; Billings-Deplored: 273; Tyler-Encyclopedia-1: 328
Spencer, Nicholas: Nugent-1: 28,81,327-8,341,395,547; Tyler-Encyclopedia-1: 53; Quitt: 76-77; 266; Morgan-American Slavery: 201-2; Billings-Deplored: 270; 45NE67-8; 1V60; 3W(2)134-136
Speke, Thomas: Nugent-1: 207,219,252-3,356-7; Tyler-Encyclopedia-1: 328; 1V75-81
Stegg, Thomas, Jr.: Nugent-1: 230,425,478,485,537; Nugent-2: 57,69; Tyler-Encyclopedia-1: 129; Kukla-Politics: 84,110; Billings-Deplored: 266; 19V362
Stegg, Thomas, Sr.: Nugent-1: 52,59,118-9; Tyler-Encyclopedia-1: 114; Kukla-Politics: 84,110; 20V76-8
Stephens, George: No information found.
Stoughton, Samuel: Nugent-1: 162-3
Streeter, Edward: Nugent-1: 358
Stringer, John: Nugent-1: 213,418,434,524-5,549; Nugent-2: 19; Tyler-Encyclopedia-1: 333; Whitelaw: 106,109,229,237,371,421,1235-6,1389,1395,1414; Haight: 443; Billings-Deplored: 280; Ames-Studies: 96

Conscious Choice

Swann, Thomas: Nugent-1: 90,103,148,326,386,465,520; Nugent-2: 55; Tyler-Encyclopedia-1: 125; Jester: 323-4; Hening-1: 406, 501-2; Sprinkle: 52; Billings-Deplored: 270; 27V154-156; 13W(3)362-363
Taberer, Thomas: Nugent-1: 277-8,504; Nugent-2: 217; Tyler-Encyclopedia-1: 335; Billings-Deplored: 275; 7W(1)212,221,248-9
Tayler, Phillip: Nugent-1: 74,150,259-60,313; Whitelaw: 170,188,206,256,281-2; Morgan-American Slavery: 155; Haight: 444
Taylor, Thomas: Nugent-1: 24,37,54,58,65,67,109,116,122,128-9,138,141,149198,216,276,439,460,485,523,538; Tyler-Encyclopedia-1: 337; Jester: 326=7
Taylor, William: Nugent-1: 97,125,142,213,276,331,419,435; Nugent-2: 68,136; Tyler-Encyclopedia-1: 117
Thomas, William: Nugent-1: 33,175,227,237,268-9,325,352,375,395,398,409,436, 485,502,532; Billings-Deplored: 281; Billings-Old: 169
Thornbury, Thomas: Nugent-1: 20,212; Tyler-Encyclopedia-1: 340
Thorowgood, Adam, Jr.: Nugent-1: 21-3,36,70-1,79-80,110,129,136,143,445; Tyler-Encyclopedia-1: 105-6,341; Quitt: 273; Billings-Deplored: 278; Jester: 329-31; 2V416-417
Thruston, Malachi: Nugent-2: 110,266,276,279,341; Billings-Deplored: 278; Tyler-Encyclopedia-1: 342
Tiplady, John: Nugent-1: 116,142; Nugent-2: 288; Billings-Deplored: 286
Townsend, Richard: Nugent-1: 120,160,248,259,271,291,536; Tyler-Encyclopedia-1: 106; Jester: 334-5; 9V173-4; 1W(1)82, 10W(2)67
Travers, Rawleigh: Nugent-1: 241,333,374,430,436,469,535-6; Nugent-2: 50; Tyler-Encyclopedia-1: 344; Billings-Deplored: 276; 1V75-81
Travis, Edward: Nugent-1: 56-7,83,108,224,231,270-1,450,503-4; Tyler-Encyclopedia-1: 344
Trussell, John: Nugent-1: 32,79,185,205,382; Tyler-Encyclopedia-1: 345; Jester: 337-9; Bruce-Institutional-2: 426n; 1V75-81
Underwood, William, Sr.: Nugent-1: 47,64-5,142-3,191,384,429; Nugent-2: 199; Billings-Deplored: 282; Tyler-Encyclopedia-1: 347
Upton, John: Nugent-1: 25,69,71-2,76,98-9,143,147-8,440,535; Tyler-Encyclopedia-1: 347; 7W(1)220-1
Wade, Armiger: Nugent-2: 369; Tyler-Encyclopedia-1: 348; Billings-Deplored: 286; Richter: 119,138,247
Walker, John: Nugent-1: 69,166-7,198,222-3,244-5,271,301,308,322,323-4; 350,362,368,381,385,449,481,520; Tyler-Encyclopedia-1: 104; Hening-1: 422,427, 512; Catterall: 79; 17W(1(1)84
Walker, Peter: Nugent-1: 159,390; Whitelaw: 108,138,1376; Haight: 445
Walker, Thomas: Nugent-1: 181-2,278,505-6,556; Nugent-2: 306-7,317; Billings-Deplored: 274; Tyler-Encyclopedia-1: 350
Wallings, George: No information found.
Warne, Thomas: Nugent-1: 99-100,115,152,387
Warner, Augustine, Jr.: Nugent-2: 42,110-111; Tyler-Encyclopedia-1: 135; Andrews: 72n; Hening-1: 340; Billings-Deplored: 274; 81V347
Warner, Augustine, Sr.: Nugent-1: 22,32,92,142,227,264,301,365,385-6; Tyler-Encyclopedia-1: 124; Billings-Deplored: 268; Currier-Briggs-3: 685; 81V319-367; ? W(1)98-99
Warren, John: Nugent-1: 186,199,324,403-4; Leonard: 39
Warren, Thomas: Nugent-1: 34,138-9,168-9,176,393,459; Nugent-2: 30,276; Billings-Deplored: Billings-Deplored: 284; Tyler-Encyclopedia-1: 352; 8W(1)161
Washington, John: Nugent-1: 429,446,448-9; Nugent-2: 48,91,178; Tyler-Encyclopedia-1: 352-3; Tyler-Encyclopedia-5: 692; Quitt: 44-48; Billings-Deplored: 285; 4W(1)80; 13W(1)145-8; 15W(1)176

Appendix B: Source Information for Tables

Waters, William: Nugent-1: 164,166,178,184,260,382,417-8,448,517; Nugent-2: 108; Tyler-Encyclopedia-1: 354; Whitelaw: 34,143,147,172,427,947; Haight: 444-5; Jester: 346-8; Billings-Deplored: 280; Ames-Studies: 96
Watson, Abraham: Nugent-1: 204,343,364
Weale, John: No information found.
Webb, Giles: Nugent-1: 165,274,282,450; Nugent-2: 31,378; Billings-Deplored: 279; Tyler-Encyclopedia-1: 354
Webb, Stephen: Nugent-1: 23-4,48,103,124,139,219,358; Tyler-Encyclopedia-1: 355; 3V57-58
Webb, Wingfield: Nugent-1: 204
Webster, Richard: Nugent-1: 116,130; Tyler-Encyclopedia-1: 355; Hening-1: 501-2
Weir, John: Nugent-1: 42-3,291,294,368,424,431,465,484-5,488,557,560-1; Nugent-2: 2,5,19,22,82; Tyler-Encyclopedia-1: 355; Billings-Deplored: 282; Leonard: 39
Wells, Richard: Nugent-1: 60,71,208,228,292,315-6,378,453,460,492,528; Hall: 228
West, Francis: Nugent-1: 18,70
West, John (York-Gloucester counties): Nugent-1: 14,21,54,160,185,213,232,258, 268,295,450,528,567; Tyler-Encyclopedia-1: 45-6; Carson: 81; Quitt: 214-5; Jester: 349-51; Catterall: 80; Sprinkle: 52; 1V60,423-4
West, John (Accomack county): Nugent-1: 407,413,454; Nugent-2: 104,106,115-6, 176,183,204,241-2,346; Tyler-Encyclopedia-1: 356; McIlwaine-Minutes: 454-5
Westropp, John: Nugent-1: 201,250; Tyler-Encyclopedia-1: 357; 15V55-6
Wetherall, Robert: Nugent-1: 160,173,294,415
Whitaker, Walter: Nugent-2: 360; Tyler-Encyclopedia-1: 358; Rutman: 81,88; Billings-Deplored: 279; 5V66
Whittaker, William: Nugent-1: 317,471; Tyler-Encyclopedia-1: 125-6; Jester: 209; Hening-1: 422; 14W(3)309-343
Whittbye, William: Nugent-1: 158,209,229,258,263,300,420; Tyler-Encyclopedia-1: 358; Kukla-Politics: 169-79; Currier-Briggs-2: 432; McIlwaine-Minutes: 215
Wilcox, John: Nugent-1: 8,50,271,286,444; Tyler-Encyclopedia-1: 360; Hening-1: 421
Wild, Daniel: Nugent-1: 291; 474,486; Nugent-2: 34; Billings-Deplored: 286
Wilford, Thomas: Nugent-1: 365; Tyler-Encyclopedia-1: 359; 1V75-81
Williams, Major: No information found. Probably William Worleich as per Leonard.
Williamson, James-Norfolk: Nugent-1: 190-1,298,353,384; Tyler-Encyclopedia-1: 360; Andrews: 281
Williamson, [Roger or Robert]: Nugent-1: 79,427,558; Nugent-2: 292,304; Billings-Deplored: Leonard: 38; Billings-Deplored: 275; Tyler-Encyclopedia-1: 360
Willis, Francis: Nugent-1: 188,403-4,546-7,565; Tyler-Encyclopedia-1: 124-125; Kukla-Politics: 202; Hening-1: 421,427,512; Billings-Deplored: 269; McIlwaine-Journals: 107
Willowby, Thomas: Nugent-1: 10,34,54,112,119,302; Tyler-Encyclopedia-1: 109-10; Hening-1: 321; Jester: 359
Windham, Edward: Nugent-1: 22-3; Tyler-Encyclopedia-1: 361
Wood, Abraham: Nugent-1: 88,95,110,137,255,301-2,388,411; Nugent-2: 211; Tyler-Encyclopedia-1: 122; Hening-1: 421,427; Jester: 362-3; Billings-Deplored: 268; 15W(1)234-241
Woodhouse, Henry: Nugent-1: 57,181,287; Tyler-Encyclopedia-1: 362-3; Jester: 365-6; 13V202-3; 1W(1)229
Woodlife, John: Nugent-1: 60,68,93,130; Tyler-Encyclopedia-1: 363; Hecht: 309
Worleich, William: Nugent-1: 34,39,148,186,195; Tyler-Encyclopedia-1: 364; Billings-Deplored: 273; Richter: 138
Worleigh, George: Nugent-1: 152,258; Tyler-Encyclopedia-1: 363
Wormeley, Christopher: Nugent-1: 91,99,118,164,182; Tyler-Encyclopedia-1: 106-7; Sprinkle: 147; Rutman: 48; Bruce-Economic-2: 75; 1V60, 36V98-101
Wormeley, Ralph, Jr.: Nugent-2: 169,208-9,313,332,373; Bruce-Economics-2: 88n; Wright: 188-189; Sprinkle: 124; 36V283-291

Wormeley, Ralph, Sr.: Nugent-1: 132,164,181-2,200,206,220; Tyler-Encyclopedia-1: 110; Quitt: 35-6; Kukla-Politics: 114-115; Rutman: 48; 36V98-101
Wyatt, Anthony: Nugent-1: 430,534; Nugent-2: 62; Tyler-Encyclopedia-1: 364; Hening-1: Billings-Deplored: 273; 422; 3V160
Wyatt, Francis: Nugent-1: 25,123,126,148; 31V237-244
Wyatt, Nicholas: Nugent-2: 299,302; Billings-Deplored: 273; 3V160
Wynne, Robert: Tyler-Encyclopedia-1: 365; 14V173, 43V148-149
Yardly, Argoll: Nugent-1: 87,96,126,289-90,549; Whitelaw: 28,30,219,288,290; Tyler-Encyclopedia-1: 111; Morgan-American Slavery: 156; Haight: 446; 70V410-419; Billings-Deplored: 280; 1W(1)190
Yardly, Francis: Nugent-1: 179,296; Tyler-Encyclopedia-1: 365
Yeardley, George: Nugent-1: 2,96,126
Yeo, Hugh: Nugent-1: 259,297,321-2,418,513; Nugent-2: 158; Tyler-Encyclopedia-1: 366; Whitelaw: 428,478,720; Billings-Deplored: 272; 31V319-20
Yeo, Leonard: Nugent-1: 62,224,257-8; Billings-Deplored: 273; Tyler-Encyclopedia-1: 366
Zouch, John: Nugent-1: 90,137,554; Fischer-Albion: 216; 12V87-89, 21V200-1, 51V25-35

Appendix B: Source Information for Tables

Bibliography

Primary Sources

Abrams, M. H. and others, eds., *The Norton Anthology of English Literature*, revised, volume 1. New York: W.W. Norton & Co., 1968.

"Acts, Orders, and Resolutions of the General Assembly of Virginia, at sessions of March 1643-1646." *Virginia Magazine of History and Biography* 23 (July 1915): 225-255.

Andrews, Charles M., ed. *Narratives of the Insurrections, 1675-1690*. New York: Scribners, 1915.

Berkeley, William. *A Discourse & View of Virginia*. 1663; reprint, Norwalk, Conn.: W. H. Smith, 1914.

_____. *A Lost Lady*. 1638; reprint, London: Malone Society, 1987.

"Berkeley's Instructions." *Virginia Magazine of History and Biography* 3 (July 1895): 15-20.

Billings, Warren M. *The Old Dominion in the Seventeenth Century, A Documentary History of Virginia, 1606-1689*. Chapel Hill: University of North Carolina Press, 1975.

_____, ed. "Some Acts not in Hening's Statutes, the Acts of Assembly, April 1652, November 1652, July 1653." *Virginia Magazine of History and Biography* 83 (January 1975): 22-76.

_____, ed. "A Quaker in Seventeenth-Century Virginia: Four Remonstrances by George Wilson." *William and Mary Quarterly*, 3rd Ser., 33 (January 1976): 127-140.

Bond, Ronald B. *Certain Sermons or Homilies (1547) and A Homily against Disobedience and Wilful Rebellion (1570), a critical edition*. Toronto: University of Toronto Press, 1987.

Braithwaite, Richard. *The English Gentleman*. London: John Dawson, 1641; reprint, Microfilming Corporation of America, 1976.

Breen, T. H. ed., "George Donne's 'Virginia Reviewed:' A 1638 Plan to Reform Colonial Society." *William and Mary Quarterly*, 3rd Ser., 30 (July 1973): 449-466.

Breen, T.H., James H. Lewis, and Keith Schlesinger, eds. "Motive for Murder: A Servant's Life in Virginia, 1678." *William and Mary Quarterly*, 3rd Ser., 40 (January 1983): 106-120.

Browne, William Hand, ed. *Archives of Maryland, Proceedings and Acts of the General Assembly of Maryland, January 1637/8-September 1664*. Baltimore: Maryland Historical Society, 1883.

_____. *Archives of Maryland, Judicial and Testamentary Business of the Provincial Court, 1637-1650*. Baltimore: Maryland Historical Society, 1887.

"Causes of Discontent in Virginia, 1676." *Virginia Magazine of History and Biography* 1 (1893): 289-292; 3 (July 1895): 35-42.

"Census of Tithables in Surry County in the Year 1668." *William and Mary Quarterly*, 1st Ser., 8 (January 1900): 160-5.

Clayton, Thomas, ed. *The Works of Sir John Suckling, the Non-dramatic Works*. Oxford: Clarendon Press, 1971.

Cooper, Thomas, ed. *The Statutes at Large of South Carolina, Vol. 1*. Columbia, South Carolina: A. S. Johnston, 1836.

Currier-Briggs, Noel, ed. *English Adventurers and Virginian Settlers*. London: Phillimore & Co., 1969.

Bibliography

D'Avenant, Sir William. *The Dramatic Works of Sir William D'Avenant*. London: H. Sotheran & Co., 1872.

"Defense of Edward Hill." *Virginia Magazine of History and Biography* 3 (1896): 239-252, 341-349, 4 (July 1896): 1-15.

Dunn, Mary Maples, and Richard S. Dunn, ed. *The Papers of William Penn, Volume 1, 1644-1679*. Pennsylvania: University of Pennsylvania Press, 1981.

Evans, Lawrence B., ed. *Writings of George Washington*. New York: Putnam's, 1908.

Fausz, J. Frederick, and Jon Kukla, ed. "A letter of Advice to the Governor of Virginia, 1624." *William and Mary Quarterly*, 3rd Ser., 34 (January 1977): 104-129.

Finestone, Harry, ed. *Bacon's Rebellion, the Contemporary News Sheets*. Charlottesville, Virginia: University of Virginia Press, 1956.

Force, Peter, ed. *Tracts and Other Papers, relating principally to the Origin, Settlement, and Progress of the Colonies in North America, from the Discovery of the Country to the Year 1776, 5 vols*. New York: Peter Smith, 1947.

Gray, Robert. *A Good Speed To Virginia*. New York: Scholars' Facsimiles & Reprints, 1937.

Hall, Clayton Coleman, ed. *Narratives of Early Maryland, 1633-1684*. New York: Scribners, 1910.

Haller, William, ed. *Tracts on Liberty in the Puritan Revolution, 1638-1647, 3 vols*. New York: Columbia University Press, 1934.

Hening, William Waller, ed. *The Statutes at Large; Being a collection of all the Laws of Virginia from the 1st Session of the Legislature in the Year 1619, 13 vols*. Richmond 1809-1823.

Hotten, John, ed. *The Original Lists of Persons of Quality, religious exiles, political rebels, serving men sold for a term of years, apprentices, children stolen, maidens pressed, and others who went from Great Britain to the American plantations, 1600-1700; with their ages, the localities where they formerly lived in the mother country, the names of the ships in which they embarked, and other interesting particulars, from mss. preserved in the State Paper Department of Her Majesty's Public Record Office, England*. Baltimore: Genealogical Pub. Co., 1983.

Huehns, G. ed. *Clarendon, Selections from* The History of the Rebellion and Civil Wars *and* The Life by Himself. London: Oxford University Press, 1955.

"Isle of Wight County Records." *William and Mary Quarterly*, 1st Ser., 7 (April 1898): 205-315.

James, Edward, ed. "The Church in Lower Norfolk County." Chapters in *Lower Norfolk County Virginia Antiquary*, volumes 1-5. New York: Peter Smith, 1951.

_____. ed., "Henry Woodhouse," *William and Mary Quarterly*, 1st Ser., 1 (1892-3): 227-32.

Johnson, Edward. "A Chapter from a Puritan Writer." *William and Mary Quarterly*, 1st Ser., 10 (January 1901): 35-8.

Kukla, Jon. "Some Acts Not in Hening's Statutes, Acts of Assembly October 1660." *Virginia Magazine of History and Biography* 83 (January 1975): 77-97.

Leonard, Cynthia Miller, ed. *The General Assembly of Virginia, July 30, 1619-January 11, 1978, A Bicentennial Register of Members*. Richmond: Virginia State Library, 1978.

Ludwell, Philip. "Philip Ludwell's Account of Bacon's Rebellion." *Virginia Magazine of History and Biography* 2 (October 1893): 174-186.

Ludwell, Thomas. "A Description of the Government of Virginia." *Virginia Magazine of History and Biography* 21 (January 1913): 36-42.

Mathews, Samuel, Sr. "The Mutiny in Virginia, 1635: Letter from Capt. Sam'l Mathews concerning the eviction of Harvey, Governor of Va." *Virginia Magazine of History and Biography* 1 (April 1894): 416-424.

McIlwaine, H. R., ed. *Journals of the House of Burgesses of Virginia 1619-1776. 13 vols*. Richmond, 1925-1930.

_____. *Minutes of the Council and General Court of Colonial Virginia*. Richmond, 1926.

Morgan, Edmund, ed., *Puritan Political Ideas, 1558-1794*. Indianapolis: Bobbs-Merrill, 1965.

"Northampton County Records in 17th Century." *Virginia Magazine of History and Biography* 4 (1896): 401-410; *Virginia Magazine of History and Biography* 5 (1897): 33-41.

Nugent, Nell Marion. *Cavaliers and Pioneers, Abstracts of Virginia Land Patents and Grants, volumes 1 and 2.* Volume 1: Baltimore: Genealogical Publishing Co., 1979; Volume 2: Richmond: Virginia State Library, 1977.

"Papers from the Records of Surry County." *William and Mary Quarterly* 1st Ser., 3 (October 1894): 121-6.

Peacham, Henry. *A Compleat Gentleman*. London: Clarendon Press, 1906.

"Proclamations of Nathaniel Bacon." *Virginia Magazine of Biography and History* 1 (1893): 55-63.

Rolfe, John. *A True Relation of the State of Virginia*. New Haven: Yale, 1951.

Salley, Alexander S., Jr., ed. *Narratives of Early Carolina, 1650-1708*. New York: Charles Scribner's Sons, 1911.

Saunders, William L., ed. *The Colonial Records of North Carolina, Volume 1, 1662 to 1712*. Raleigh: P. M. Hale, 1886.

Scarborough, Edmund. "Colonel Scarborough's Report, Being an Account of His Efforts to Suppress the Quakers in What is Now Part of Maryland, then Claimed by Virginia." *Virginia Historical Magazine* 19 (1911): 173-180.

Sherwood, William. "Virginia's Deploured Condition." *Collections of the Massachusetts Historical Society*, 4th Ser., 9 (1871): 162-187.

Smith, John. *The Complete Works of Captain John Smith, 1580-1631*, ed. Philip L. Barbour. Virginia: Institute of Early American History and Culture, 1986.

Smith, Sir Thomas. *De Pepublica Anglorum, a Discourse on the Commonwealth of England*. 1583; reprint, Cambridge: Cambridge University Press, 1906.

Stock, Leo Francis. *Proceedings and Debates of the British Parliaments respecting North America, Volume I, 1542-1688*. Washington, D.C.: Carnegie Institution, 1924.

Stith, William. *The History of the First Discovery and Settlement of Virginia*. South Carolina: The Reprint Co., 1965.

Tyler, Lyon G., ed. *Narratives of Early Virginia, 1606-1625*. New York: Scribner's, 1907.

"Speech of Sir William Berkeley and Declaration of the Assembly, March 1651." *Virginia Magazine of Biography and History* 1 (1893): 75-81.

Winthrop, John, "Governor John Winthrop of Massachusetts Bay Gives a Model of Christian Charity, 1630." In *Major Problems in American Colonial History*. ed. Karen Ordahl Kupperman, 145-157. Lexington, Massachusetts: D. C. Heath and Company, 1993.

Washburn, Wilcomb E., ed. "Sir William Berkeley's 'A History of Our Miseries'." *William and Mary Quarterly*, 3rd Ser., 14 (July 1957): 402-413.

Wright, Louis B., ed. *A Voyage to Virginia in 1609; two narratives: Strachey's 'True reportory' and Jourdain's Discovery of the Bermudas*. Charlottesville: Univ. Press of Virginia, 1964.

Wright, Louis, and Elaine W. Fowler, eds. *West and By North*. New York: Delacorte Press, 1971.

Secondary Sources, Published books

Allen, J.W. *English Political Thought, 1603-1660. volume 1*. London: Methuen & Co., 1938.

Ames, Susie M. *Studies of the Virginia Eastern Shore in the 17th Century*. Richmond, Va.: Dietz Press, 1940.

Bibliography

Anderson, Fred. *A People's Army, Massachusetts Soldiers and Society in the Seven Years' War*. New York: W.W. Norton & Co., 1984.

Aptheker, Herbert. *American Negro Slave Revolts*. New York: Columbia Universiy Press, 1943; reprint, New York: International Publishers, 1963.

_____. *Negro Slave Revolts in the U.S., 1526-1860*. New York: International Publishers, 1939.

Aylmer, G.E. *The King's Servants: The Civil Service of Charles I, 1625-1642*. New York: Columbia University Press, 1961.

Bailyn, Bernard. *The Ideological Origins of the American Revolution*. Cambridge, Massachusetts: Harvard University Press, 1967.

Ballagh, James. *History of Slavery in Virginia*. Baltimore: John Hopkins Univ. Press, 1902; reprinted in *Early Studies of Slavery by States*. Northbrook, Illinois: Metro Books, 1972.

Beckett, J. V. *The Aristocracy in England, 1660-1914*. Oxford: Basil Blackwell, 1986.

Bentley, Gerald Eades. *The Jacobean and Caroline Stage, Dramatic Companies and Players, Vol 1-5*. Oxford: Clarendon Press, 1941.

Bernard, Jessie. *The Future of Marriage*. New Haven: Yale University Press, 1982.

Beverly, Robert. *The History & Present State of Virginia*. 1705; reprint, ed. Louis B. Wright, Chapel Hill: Univ. of North Carolina, 1947.

Billings, Warren M., John E. Selby, and Thad. W. Tate, eds. *Colonial Virginia: a History*. White Plains, N.Y.: KTO Press, 1986.

Boskin, Joseph. *Into Slavery, Racial Decision in the Virginia Colony*. Philadelphia: Lippincott, 1976.

Breen, T. H. *Puritans and Adventurers, Change and Persistence in Early America*. New York: Oxford University Press, 1980.

_____. *Tobacco Culture, the Mentality of the Great Tidewater Planters on the Eve of Revolution*. Princeton: Princeton University Press, 1985.

Breen, T.H., and Stephen Innes. *Myne Owne Ground*. New York: Oxford Univ. Press, 1980.

Brewer, John and John Styles. *An Ungovernable People, the English and their law in the seventeenth and eighteenth centuries*. New Brunswick, New Jersey: Rutgers University Press, 1980.

Brown, Alexander. *The Genesis of the United States, 2 vols*. Cambridge: Riverside Press, 1890.

Bruce, Phillip. *Economic History of Virginia in the 17th Century, 2 vols*. Gloucester, Mass.: Peter Smith, 1935.

_____. *Institutional History of Virginia in the 17th Century, 2 vols*. Gloucester, Mass.: Peter Smith, 1964.

_____. *Social Life of Virginia in the 17th Century*. Richmond: Whittet & Shepperson, 1907.

Burk, John. *The History of Virginia from its First Settlement to the Present Day*. Petersburg, Virginia: Dicken and Pesod, 1804-5.

Campbell, Mildred. *The English Yeoman Under Elizabeth and the Early Stuarts*. New Haven: Yale University Press, 1942.

Catterall, Helen, ed. *Judicial Cases Concerning American Slavery and the Negro, volumes I and IV*. Washington, D.C.: Carnegie Institute of Washington, 1926 and 1936.

Cliffe, J.T. *The Puritan Gentry, the Great Puritan Families of Early Stuart England*. London: Routledge & Kegan Paul, 1984.

Collinson, Patrick. *The Birthpangs of Protestant England, Religious and Cultural Change in the Sixteenth and Seventeenth Centuries*. New York: St. Martin's Press, 1988.

_____. *De Republica Anglorum, or History with the Politics Put Back*. Cambridge: Cambridge University Press, 1990.

Coward, Barry. *The Stuart Age: a history of England, 1603-1714*. London and New York: Longman, 1980.

Craven, Wesley Frank. *The Colonies in Transition, 1660-1713*. New York: Harper & Row, 1967/8.
_____. *The Dissolution of the Virginia Company*. 1932; reprint, Mass: Peter Smith, 1964.
_____. *The Southern Colonies in the Seventeenth Century, 1607-1689*. Baton Rouge: Louisiana State Univ. Press, 1949.
_____. *White, Red, & Black*. Charlottesville: Univ. Press of Virginia, 1971.
Curtin, Philip D. *The Atlantic Slave Trade, a Census*. Madison, Wisconsin: Univ. of Wisconsin Press, 1969.
_____. *The Rise and Fall of the Plantation Complex, Essays in Atlantic History*. Cambridge: Cambridge University Press, 1990.
Davidson, Basil. *The African Slave Trade, a revised and expanded edition*. Boston: Little, Brown & Co., 1980.
Davies, Margaret Gay. *The Enforcement of English Apprenticeship, a Study in Applied Mercantilism, 1563-1642*. Cambridge, Massachusetts: Harvard University Press, 1956.
Davis, David Brion. *The Problem of Slavery in Western Culture*. Ithaca, N.Y.: Cornell Univ., 1966.
Davis, Richard Beale. *Literature and Society in Early Virginia, 1608-1840*. Baton Rouge: Louisiana State University Press, 1973.
De Tocqueville, Alexis. *Democracy in America*. 1835; New York: Random House, Vintage Books, 1945.
Du Bois, W. E. B. *The Suppression of the African Slave Trade to the United States, 1638-1870*. 1896; reprint, New York: Shocken Books, 1969.
Eckenrode, H.J. *Separation of Church and State in Virginia*. Richmond: Virginia State Library, 1910.
Farrington, David P., Lloyd E. Ohlin, and James Q. Wilson. *Understanding and Controlling Crime: Toward a New Research Stategy*. New York: Springer-Verlag, 1986.
Federal Bureau of Investigation. *Uniform Crime Reports*. Washington, D.C., Federal Bureau of Investigation, 1991.
Fields, Barbara Jeanne. *Slavery and Freedom on the Middle Ground: Maryland during the Nineteenth Century*. New Haven: Yale University Press, 1985.
Fischer, David Hackett. *Albion's Seed: Four British Folkways in America*. New York: Oxford Univ. Press, 1989.
Fiske, John. *Old Virginia and Her Neighbors, 2 volumes*. Boston: Houghton, Mifflin and Company, 1897.
Fissel, Mark Charles. *The Bishops' Wars, Charles I's campaigns against Scotland, 1638-1640*. Cambridge: Cambridge University Press, 1994.
Foss, Michael. *Undreamed Shores, England's Wasted Empire in America*. London: Harrap, 1974.
Funk & Wagnall's New "Standard" Dictionary of the English Language, 1962 ed.
Gardiner, Samuel R. *History of England from the Accession of James I to the Outbreak of the Civil War, 1603-1642, Volumes 8-10*. New York: AMS Press, 1965.
_____. *History of the Great Civil War, 1642-1649, 4 volumes*. New York: AMS Press, Inc., 1965.
_____. *History of the Commonwealth and Protectorate, 1649-1656, 4 volumes*. New York: AMS Press, Inc., 1965.
Gilder, George. *Men and Marriage*. Gretna, Louisiana: Pelican, 1986.
Greene, Evarts B. and Virginia D. Harrington. *American Population Before the Federal Census of 1790*. New York: Columbia University Press, 1932.
Greene, Jack P. *Pursuits of Happiness, the Social Development of Early Modern British Colonies and the Formation of American Culture*. Chapel Hill: University of North Carolina Press, 1988.

Bibliography

Greene, Lorenzo. *The Negro in Colonial New England, 1620-1776*. New York: Columbia University Press, 1942; reprint, Port Washington, New York: Kennikat Press, 1966.

Greven, Philip J. Jr., "Family Structure in Seventeenth Century Andover, Massachusetts." *William and Mary Quarterly*, 3rd Ser., 23 (April, 1966): 234-256.

Hale, Nathaniel C. *Virginia Venturer, a Historical Biography of William Claiborne, 1600-1677*. Richmond, Virginia: Dietz Press, 1951.

Hambrick-Stowe, Charles E. *The Practice of Piety, Puritan Devotional Disciplines in Seventeenth-Century New England*. Chapel Hill: University of North Carolina Press, 1982.

Hamilton, Alexander, James Madison, and John Jay. *The Federalist Papers*. New York: Mentor, 1961.

Handlin, Oscar. *Race and Nationality in American Life*. Boston: Little, Brown & Co., 1957.

Harbage, Alfred. *Cavalier Drama, an Historical and Critical Supplement to the Study of the Elizabethan and Restoration Stage*. London: Oxford University Press, 1936.

Hatch, Charles E. *The 1st Seventeen Years, Virginia 1607-1624*. Williamsburg, Virginia: Virginia's 350th Anniversary Celebration Corporation, 1957.

Hayden, Horace. *Virginia Genealogies*. Baltimore: Genealogical Publishing Co., 1966.

Higginbotham, A. Leon. *In the Matter of Color: Race and the American Legal Process, the Colonial Period*. New York: Oxford University Press, 1978.

Hill, Christopher. *Change and Continuity in Seventeenth-Century England*, revised edition. New Haven: Yale University Press, 1991.

_____. *A Nation of Change and Novelty, Radical politics, religion and literature in seventeenth-century England*. London: Routledge, 1990.

Hume, Ivor Noel. *Discoveries in Martin's Hundred, Colonial Williamsburg Archaeological Series, no. 10*. Williamsburg, Virginia: Colonial Williamsburg Foundation, 1983.

Inwood, Stephen. *A History of London*. New York: Carroll & Graf Publishers, 1998.

Jester, Annie Lash, ed. *Adventurers of Purse and Person, Virginia, 1607-1625, 2nd Edition*. Virginia: Order of First Families of Virginia, 1964.

Jones, Howard Munford. *The Literature of Virginia in the Seventeenth Century*. Charlottesville: Univ. Press of Virginia, 1968.

Jones, Rufus. *The Quakers in the American Colonies*. London: MacMillan & Co., 1911.

Jordan, Winthrop D. *White Man's Burden, Historical Origins of Racism in the United States*. London: Oxford Univ. Press, 1974.

_____. *White Over Black: American attitudes toward the Negro, 1550-1812*. Chapel Hill: University of North Carolina, 1968.

Karraker, Cyrus Harreld. *The Seventeenth-Century Sheriff: a Comparative Study of the Sheriff in England and the Chesapeake Colonies, 1607-1689*. Chapel Hill: Univ. of North Carolina Press, 1930.

Kukla, Jon. *Political Institutions in Virginia, 1619-1660*. New York: Garland Publishing, 1989.

Kulikoff, Allan. *Tobacco and Slaves: the development of Southern cultures in the Chesapeake*. Chapel Hill: University of North Carolina Press, 1986.

Kupperman, Karen Ordahl. *Providence Island, 1630-1641, the Other Puritan Colony*. Cambridge: Cambridge University Press, 1993.

Lake, Peter. *Anglicans and Puritans? Presbyterianism and English Conformist Thought from Whitgift to Hooker*. London: Unwin Hyman, 1988.

_____. *Moderate Puritans and the Elizabethan Church*. Cambridge: Cambridge University Press, 1982.

Lefler, Hugh T., and William S. Powell. *Colonial North Carolina, A History*. New York: Charles Scribner's Sons, 1973.

Lockridge, Kenneth A. *Settlement and Unsettlement in Early America*. Cambridge: Cambridge University Press, 1981.

Lockyer, Roger. *The Early Stuarts, a Political History of England, 1603-1642*. London: Longman, 1989.
Lovejoy, Paul E. *Transformations in Slavery, a History of Slavery in Africa*. Cambridge: Cambridge University Press, 1983.
Malone, Dumas, ed. *Dictionary of American Biography*. New York: Scribners, 1934.
McManus, Edgar. *Black Bondage in the North*. Syracuse, N.Y.: Syracuse Univ. Press, 1973.
_____. *A History of Negro Slavery in New York*. Syracuse, New York: Syracuse Univ. Press, 1966.
Moore, George. *Notes on the History of Slavery in Massachusetts*. New York: D. Appleton & Co., 1866.
Morgan, Edmund. *American Slavery, American Freedom, the Ordeal of Colonial Virginia*. New York: Norton 1975.
_____. *The Puritan Family; Religion & Domestic Relations in Seventeenth Century New England*. New York: Harper & Row, 1966.
Morton, Richard L. *Colonial Virginia, Vol. I*. Chapel Hill: Univ. of North Carolina Press, 1960.
Murray, Charles. *Losing Ground, American Social Policy, 1950-1980*. New York: Basic Books, 1984.
Moss, Richard Shannon. *Slavery on Long Island, a study in local institutional and early African-American communal life*. New York: Garland Publishing, 1993.
Moynihan, Daniel Patrick. *Family and Nation*. New York: Harcourt, Brace, Jovanovich, 1986.
Nash, Gary B., and Jean R. Soderland. *Freedom by Degrees, Emancipation in Pennsylvania and its Aftermath*. New York: Oxford University Press, 1991.
Newman, Peter. *Atlas of the English Civil War*. New York: MacMillian Publishing Company, 1985.
Nicoll, Allardyce. *Stuart Masques and the Renaissance Stage*. New York: Benjamin Blom, 1938.
Northrup, David, ed. *The African Slave Trade*. Massachusetts: D.C. Heath and Company, 1994.
Orgel, Stephen and Roy Strong. *Inigo Jones: the Theatre of the Stuart Court, 2 volumes*. London: Sotheby Parke Bernet, 1973.
Quitt, Martin H. *Virginia House of Burgesses, 1660-1706: the Social, Educational and Economic Bases of Political Power*. New York: Garland Publishing, 1989.
Padover, Saul K., ed. *Thomas Jefferson on Democracy*. New York: New American Library, 1939.
Parks, George Bruner. *Richard Hakluyt and the English Voyages*. New York: Frederick Ungar Publishing, 1961.
Parry, Graham. *The Seventeenth Century, the Intellectual and Cultural Context of English Literature, 1603-1700*. London: Longman, 1989.
Peck, Linda Levy. *Court Patronage and Corruption in Early Stuart England*. Boston: Unwin Hyman, 1990.
Perry, James R. *The Formation of a Society on Virginia's Eastern Shore, 1615-1655*. Chapel Hill: Univ. of North Carolina, 1990.
Phillips, Ulrich B. *American Negro Slavery*. New York & London: D. Appleton & Co., 1918.
Rainwater, Lee & William Yancey. *The Moynihan Report and the Politics of Controversy*. Cambridge, Massachusetts: M.I.T. Press, 1967.
Robinson, W. Stitt. *Mother Earth, Land Grants in Virginia, 1607-1699*. Williamsburg, Virginia: Virginia's 350th Anniversary Celebration Corporation, 1957.
_____. *The Southern Colonial Frontier, 1607-1763*. Albuquerque: Univ. of New Mexico Press, 1979.
Rountree, Helen C., ed. *Powhatan, Foreign Relations, 1500-1722*. Charlottesville: Univ. Press of Virginia, 1993.
Russell, John. *Free Negro in Virginia, 1619-1865*. Baltimore: John Hopkins Press, 1913.

Bibliography

Rutman, Darrett Bruce. *A Militant New World, 1607-1640*. New York: Arno Press, 1979.
Rutman, Anita H. and Darrett B. *A Place in Time, Middlesex County, Virginia, 1650-1750*. New York: Norton & Co., 1984.
_____. *A Place in Time, Explicatus*. New York: Norton & Co., 1984.
Scott, Jonathan French. *Historical Essays on Apprenticeship and Vocational Education*. Ann Arbor: Ann Arbor Press, 1914.
Shaw, George Bernard. *Bernard Shaw: Selected Plays*. New York: Dodd, Mead and Company, 1981.
Sibley, John. *Biographical Sketches of Graduates of Harvard University, Vol 1, 1642-58*. Cambridge: Charles William Sever, 1873.
Simon, Joan. *Education and Society in Tudor England*. Cambridge: Cambridge University Press, 1966.
Smith, Abbot Emerson. *Colonists in Bondage: White Servitude and Convict Labor in America, 1607-1776*. Chapel Hill: Univ. of North Carolina, 1947.
Smith, Irwin. *Shakespeare's Blackfriar's Playhouse, its History and its Design*. New York: New York University Press, 1964.
Smuts, Malcolm R. *Court Culture and the Origins of a Royalist Tradition in Early Stuart England*. Philadelphia: University of Pennsylvania Press, 1987.
Squier, Charles L. *Sir John Suckling*. Boston: Twayne Publishers, 1978.
Stampp, Kenneth M. *The Peculiar Institution: Slavery in the Antebellum South*. 1956; reprint, New York: Vintage Books, 1989.
Stephen, Leslie and Sidney Lee, eds. *Dictionary of National Biography*. Oxford: Oxford University Press, 1917.
Stone, Lawrence. *Uncertain Unions and Broken Lives, Marriage and Divorce in England, 1660-1857*. Oxford: Oxford University Press, 1995.
Swem, E. G., John W. Jennings, and James A. Servies. *A Selected Bibliography of Virginia, 1607-1699*. Williamsburg, Virginia: Virginia's 350th Anniversary Celebration Corporation, 1957.
Thorton, John. *Africa and Africans in the making of the Atlantic world, 1400-1680*. Cambridge: Cambridge University Press, 1992.
Tonry, Michael, Lloyd E. Ohlin, and David Farrington. *Human Development and Criminal Behavior – New Ways of Advancing Knowledge*. New York: Springer-Verlag, 1991.
Twombly, Robert C. and Robert H. Moore. "Black Puritan: The Negro in Seventeenth-Century Massachusetts." *William and Mary Quarterly*, 3rd Ser., 23 (April 1967): 224-242.
Tyler, Lyon Gardiner, ed. *Encyclopedia of Virginia Biography, volumes 1, 4, 5*. New York: Lewis Historical Publishing Company, 1915.
_____. ed. *William and Mary Quarterly, Series 1, Volumes 1-27* (1892-1919)
Ulrich, Laurel Thatcher. *Good Wives, Image and Reality in the Lives of Women in Northern New England, 1650-1750*. New York: Vintage Books, 1991.
U.S. Bureau of the Census. *Historical Statistics of the United States, Colonial Times to 1970, Part 2*. Washington, D.C.: Government Printing Office, 1975.
Ver Steeg, Clarence. *The Formative Years, 1607-1763*. New York: Hill and Wang, 1964.
Wagandt, Charles Lewis. *The Mighty Revolution: Negro Emancipation in Maryland, 1862-1864*. Baltimore: John Hopkins Press, 1964.
Washburn, Wilcomb E. *The Governor and the Rebel*. Chapel Hill: Univ. of North Carolina, 1957.
_____. *Virginia Under Charles I and Cromwell, 1625-1660*. Williamsburg, Virginia: Virginia's 350th Anniversary Celebration Corporation, 1957.
Webb, Stephen Saunders. *1676, End of American Independence*. New York: Knopt, 1984.
Wertenbaker, Thomas. *Planters of Colonial Virginia; or, The Origin and Development of the Social Classes of the Old Dominion*. New York: Russell & Russell, 1959.

_____. *Torchbearer of the Revolution, the Story of Bacon's Rebellion and its Leader.* Princeton: Princeton Univ., 1940.

_____. *Virginia under the Stuarts, 1607-1688.* New York: Russell & Russell, 1959.

Whitelaw, Ralph T. *Virginia's Eastern Shore, A History of Northampton and Accomack Counties,* 2 vols. Richmond: Virginia Historical Society, 1951; reprint, Gloucester, Massachusetts: Peter Smith, 1968.

Wilson, James Q. *Thinking About Crime,* rev. ed. New York: Basic Books, 1983.

Wilson, James Q., and Richard J. Herrnstein. *Crime and Human Nature.* New York: Simon and Schuster, 1985.

Wood, Peter H. *Black Majority, Negroes in Colonial South Carolina from 1670 through the Stono Rebellion.* New York: W. W. Norton & Co., 1974.

Wright, James. *The Free Negro in Maryland, 1634-1860.* New York: Longmans, Green & Co, 1921.

Wright, Louis. *The First Gentlemen of Virginia; Intellectual Qualities of the Early Colonial Ruling Class.* San Marino, California: Huntington Library, 1940.

Ziner, Feenie. *The Pilgrims and Plymouth Colony.* New York: American Heritage Publishing, 1961.

Secondary Sources-Articles, Monographs

Ames, Susie M. "The Reunion of Two Virginia Counties." *Journal of Southern History,* VIII (1942): 536-548.

_____. "Colonel Edmund Scarborough," *Alumnae Bulletin of Randolph Macon Woman's College* 26, No. 1 (November 1932): 16-23.

Anderson, Virginia DeJohn. "Religion, the Common Thread of Motivation." In *Major Problems in American Colonial History,* ed. Karen Ordahl Kupperman, 145-157. Lexington, Massachusetts: D. C. Heath and Company, 1993.

"Anthony Johnson, Free Negro, 1622." *Journal of Negro History* 56 (1971): 71-73.

Bailyn, Bernard. "Politics and Social Structure in Virginia." In *Seventeenth Century America, Essays in Colonial History,* ed. James M. Smith, 90-115. Chapel Hill: Univ. of North Carolina, 1959.

_____. "The *Apologia* of Robert Keayne." *William and Mary Quarterly,* 3rd Ser., 7 (October 1950): 568-587.

Bald, R.C. "A Note on Suckling's *A Sessions of the Poets*." In *Modern Language Notes,* 10 (November 1943): 550-551.

Barry, Jonathan. "Introduction." In *The Middling Sort of People, Culture, Society and Politics in England, 1550-1800,* ed. Jonathan Barry and Christopher Brooks, 1-27. New York: St. Martin's Press, 1994.

_____. "Popular Culture in Seventeenth-Century Bristol." In *Popular Culture in Seventeenth-Century England,* ed. Barry Reay, 59-90. New York: St. Martin's Press, 1985.

Bassett, John Spencer. "Slavery in the State of North Carolina." Baltimore: John Hopkin Univ. Press, 1899; reprinted in *Early Studies of Slavery by States.* Northbrook, Illinois: Metro Books, 1972.

Beeman, Richard R. "Deference, Republicanism, and the Emergence of Popular Politics in Eighteenth-Century America." *William and Mary Quarterly,* 3rd Ser., 49 (July 1992): 401-430.

Billings, Warren M. "The Cases of Fernando and Elizabeth Key: a Note on the Status of Blacks in Seventeenth Century Virginia." *William and Mary Quarterly,* 3rd Ser., 30 (July 1973): 467-474.

_____. "The Growth of Political Institutions in Virginia, 1634 to 1676." *William and Mary Quarterly,* 3rd Ser., 31 (April 1974): 225-242.

_____. "Justices, Books, Laws and Courts in Seventeenth-Century Virginia." *Law Library Journal* 85, No. 2 (Spring 1993): 277-296.

Bibliography

_____. "The Law of Servants and Slaves in Seventeenth Century Virginia." *Virginia Magazine of History and Biography* 99 (January 1991): 45-62.

_____. "Sir William Berkeley – Portrait by Fischer: A Critique." *William and Mary Quarterly*, 3rd Ser., 48 (October 1991): 598-607.

Boogaart, Ernst van den. "The Dutch Participation in the Atlantic Slave Trade, 1596-1650." In *The Uncommon Market, Essays in the Economic History of the Atlantic Slave Trade*, ed. Henry A. Gemery and Jan S. Hogendorn, 353-375. New York: Academic Press, 1979.

Breen, T. H. "A Changing Labor Force and Race Relations in Virginia, 1660-1710." In *Shaping Southern Society, the Colonial Experience*, ed. T.H. Breen, 116-134. New York: Oxford University Press, 1976.

_____. "The Culture of Agriculture: the Symbolic World of the Tidewater Planter, 1760-1790." In *Saints & Revolutionaries: Essays in Early American History*, ed. David D. Hall, John M. Murrin, and Thad. W. Tate, 247-284. New York: Norton, 1983.

Brewer, James H. "Negro Property Owners in Seventeenth Century Virginia." *William and Mary Quarterly*, 3rd Ser., 12 (October 1955): 575-580.

Brewer, John, and John Styles. "Introduction," In *An Ungovernable People, the English and their law in the seventeenth and eighteenth centuries*, ed. John Brewer and John Styles, 11-21. New Brunswick, New Jersey: Rutgers University Press, 1980.

Brooks, Christopher. "Apprenticeship, Social Mobility and the Middling Sort, 1550-1800." In *The Middling Sort of People, Culture, Society and Politics in England, 1550-1800*, ed. Jonathan Barry and Christopher Brooks, 52-83. New York: St. Martin's Press, 1994.

Burke, Peter. "Popular Culture in Seventeenth-Century London." In *Popular Culture in Seventeenth-Century England*, ed. Barry Reay, 31-58. New York: St. Martin's Press, 1985.

Butler, Jon. "Thomas Teackle's 333 Books: A Great Library on Virginia's Eastern Shore, 1697." *William and Mary Quarterly*, 3rd Ser., 49 (July 1992): 449-491.

Campbell, Mildred. "Social Origins of Some Early Americans." In *Seventeenth Century America, Essays in Colonial History*, ed. James M. Smith, 63-89. Chapel Hill: Univ. of North Carolina, 1959.

Carr, Lois Green. "Diversification in the Colonial Chesapeake: Somerset County, Maryland, in Comparative Perspective." In *Colonial Chesapeake Society*, ed. Lois Carr, Philip D. Morgan, and Jean B. Russo, 342-388. Chapel Hill: University of North Carolina Press, 1988.

Carr, Lois Green and Lorena S. Walsh. "The Planter's Wife: the Experience of White Women in Seventeenth-century Maryland." *William and Mary Quarterly*, 3rd Ser., 34 (October 1977): 542-571.

_____. "The Standard of Living in the Colonial Chesapeake." *William and Mary Quarterly*, 3rd Ser., 45 (January 1988): 135-159.

Carroll, Kenneth L. "Quakerism on the Eastern Shore of Virginia." in *Virginia Magazine of History and Biography* 74 (April 1966): 170-189.

Cooley, Henry. "A Study of Slavery in New Jersey." Baltimore: John Hopkins Univ. Press, 1896; reprinted in *Early Studies of Slavery by States*. Northbrook, Illinois: Metro Books, 1972.

Curtis, George B. "The Colonial County Court, Social Forum and Legislative Precedent, Accomack County, Virginia, 1633-1639." *Virginia Magazine of History and Biography* 85 (July 1977): 274-288.

Cust, Richard. "Politics and the Electorate in the 1620s." In *Conflict in Early Stuart England, Studies in Religion and Politics, 1603-1642*, ed. Richard Cust and Ann Hughes, 134-167. London: Longman Group, 1989.

Cust, Richard, and Ann Hughes. "Introduction, after Revisionism." In *Conflict in Early Stuart England, Studies in Religion and Politics, 1603-1642*, ed. Richard Cust and Ann Hughes, 1-46. London: Longman Group, 1989.

Deal, Douglas. "A Constricted World: Free Blacks on Virginia's Eastern Shore, 1680-1750." In *Colonial Chesapeake Society*, ed. Lois Carr, Philip D. Morgan, and Jean B. Russo, 275-305. Chapel Hill: University of North Carolina Press, 1988.

Degler, Carl. "Slavery and the Genesis of American Race Prejudice." *Comparative Studies in Society and History*, vol. 2 (1959); reprinted in *Essays in American Colonial History*, ed. Paul Goodman, 223-249. New York: Holt, Rinehard & Winston, 1967.

Diamond, Sigmund. "Virginia in the Seventeenth Century." American Journal of Sociology, vol. 63 (1958); reprinted in *Essays in American Colonial History*, ed. Paul Goodman, 108-136. New York: Holt, Rinehart & Winston, 1967.

Dunn, Richard S. "Seventeenth Century English Historians of America." In *Seventeenth Century America, Essays in Colonial History*, ed. James M. Smith, 195-225. Chapel Hill: Univ. of North Carolina, 1959.

Earle, Carville V. "Environment, Disease, and Mortality in Early Virginia." In *The Chesapeake in the 17th Century: Essays on Anglo-American Society*, ed. Thad W. Tate, and David L. Ammerman, 96-125. Chapel Hill: Institute of Early American History and Culture, Univ. of North Carolina, 1979.

Fields, Barbara Jeanne. "Slavery, Race, and Ideology in the United States of America." *New Left Review*, 181 (1990): 95-118.

Fischer, David Hackett. "A Rejoinder." *William and Mary Quarterly*, 3rd Ser., 48 (October 1991): 608-611.

Fletcher, Anthony. "The Protestant Idea of Marriage in Early Modern England." In *Religion, Culture, and Society in Early Modern Britain*, eds. Anthony Fletcher and Peter Roberts, 161-181. Cambridge: Cambridge University Press, 1994.

Galenson, David W. "Economic Aspects of the Growth of Slavery in the Seventeenth-century Chesapeake." In *Slavery and the Rise of the Atlantic System*, ed. Barbara L. Solow, 265-292. Cambridge: Cambridge University Press, 1991.

Gemery, Henry A. "Emigration from the British Isles to the New World, 1630-1700: Inferences from Colonial Populations." In *Research in Economic History, a Research Annual, Volume 5*, ed. Paul Uselding, 179-231. Greenwich, Connecticut: Jai Press, 1980.

Gilsdorf, Joy B. and Robert R. "Elites and Electorates: Some Plain Truths for Historians of Colonial America." In *Saints & Revolutionaries: Essays on Early American History*, eds. David D. Hall, John M. Murrin, and Thad W. Tate, 207-244. New York: W. W. Norton & Co., 1984.

Graham, Michael. "Meetinghouse and Chapel: Religion and Community in Seventeenth-Century Maryland." In *Colonial Chesapeake Society*, ed. Lois Carr, Philip D. Morgan, and Jean B. Russo, 242-274. Chapel Hill: University of North Carolina Press, 1988.

Haffenden, Philip S. "The Anglican Church in Restoration Colonial Policy." In *Seventeenth Century America, Essays in Colonial History*, ed. James M. Smith, 166-191. Chapel Hill: Univ. of North Carolina, 1959.

Handler, Oscar and Mary. "Origins of the Southern Labor System." *William and Mary Quarterly*, 3rd Ser., 7 (April 1950): 199-222.

Henige, David. "When Did Smallpox Reach the New World (and Why Does It Matter?)." In *Africans in Bondage: Studies in Slavery and the Slave Trade*, ed. Paul E. Lovejoy, 11-26. Madison, Wisconsin: African Studies Program, University of Wisconsin-Madison, 1986.

Heston, Alfred Miller. "Slavery and Servitude in New Jersey." Baltimore: John Hopkins Univ. Press, 1896; reprinted in *Early Studies of Slavery by States*. Northbrook, Illinois: Metro Books, 1972.

Bibliography

Hirschi, Travis. "Crime and the Family." In *Crime and Public Policy*, ed. James Q. Wilson, 53-68. San Francisco: ICS Press, 1983.

Horn, James. "Adapting to a New World: A Comparative Study of Local Society in England and Maryland, 1650-1700." In *Colonial Chesapeake Society*, ed. Lois Carr, Philip D. Morgan, and Jean B. Russo, 133-175. Chapel Hill: University of North Carolina Press, 1988.

_____. "Servant Emigration to the Chesapeake in the Seventeenth Century." In *The Chesapeake in the 17th Century: Essays on Anglo-American Society*, ed. Thad W. Tate, and David L. Ammerman, 51-95. Chapel Hill: Institute of Early American History and Culture, Univ. of North Carolina, 1979.

Ingram, Martin. "The Reform of Popular Culture? Sex and Marriage in Early Modern England." In *Popular Culture in Seventeenth-Century England*, ed. Barry Reay, 129-165. New York: St. Martin's Press, 1985.

Jordan, David. "Elections and Voting in Early Colonial Maryland," *Maryland Historical Magazine*, 77 (1982, No. 3): 238-265.

Jordan, Winthrop D. "Modern Tensions and the Origins of American Slavery." *Journal of Southern History*, vol. 28 (1962); reprinted in *Essays in American Colonial History*, ed. Paul Goodman, 250-260. New York: Holt, Rinehart & Winston, 1967.

Kelly, Kevin P. "'In dispers'd Country Plantations': Settlement Patterns in Seventeenth-Century Surry County, Virginia." In *The Chesapeake in the 17th Century: Essays on Anglo-American Society*, ed. Thad W. Tate and David L. Ammerman, 183-205. Chapel Hill: Institute of Early American History and Culture, Univ. of North Carolina, 1979.

Kimmel, Ross M. "Free Blacks in Seventeenth-Century Maryland." *Maryland Historical Magazine* 71, No. 1 (Spring 1976): 19-25.

Kuran, Timur. "Islam and Economic Performance: Historical and Contemporary Links." *Journal of Economic Literature*, vol. 56 (2018), in press.

Lake, Peter. "Anti-popery: the Structure of a Prejudice." In *Conflict in Early Stuart England, Studies in Religion and Politics, 1603-1642*, ed. Richard Cust and Ann Hughes, 72-106. London: Longman Group, 1989.

Laslett, Peter. "The Gentry of Kent in 1640." In *Shaping Southern Society, the Colonial Experience*, ed. T.H. Breen, 32-46. New York: Oxford University Press, 1976.

Leonard, Sister Joan de Lourdes. "Operation Checkmate: The Birth and Death of a Virginia Blueprint for Progress, 1666-1676." *William and Mary Quarterly*, 3rd Ser., 24 (January 1967): 44-74.

Menard, Russell R. "The Africanization of the Lowcountry Labor Force, 1670-173." In *Race and Family in the Colonial South*, ed. Winthrop D. Jordan and Sheila L. Skemp, 81-103. Jackson, Miss.: Univ. Press of Miss., 1987.

_____. "British Migration to the Chesapeake Colonies in the Seventeenth Century." In *Colonial Chesapeake Society*, ed. Lois Carr, Philip D. Morgan, and Jean B. Russo, 99-132. Chapel Hill: University of North Carolina Press, 1988.

_____. "From Servant to Freeholder: Status Mobility and Property Accumulation in Seventeenth Century Maryland." *William and Mary Quarterly*, 3rd Ser., 30 (January 1973): 37-64.

_____. "The Maryland Slave Populations, 1658-1730: a Demographic Profile of Blacks in Four Counties." *William and Mary Quarterly*, 3rd Ser., 32 (January 1975): 29-54.

_____. "A Note on Chesapeake Tobacco Prices, 1618-1660." In *Virginia Magazine of History and Biography*, 84 (October 1976): 401-410.

_____. "The Tobacco Industry in the Chesapeake Colonies, 1617-1730: an Interpretation." In *Research in Economic History, a Research Annual, Volume 5*, ed. Paul Uselding, 109-177. Greenwich, Connecticut: Jai Press, 1980.

Miller, Henry M. "An Archaeological Perspective on the Evolution of Diet in the Colonial Chesapeake, 1620-1745." In *Colonial Chesapeake Society*, ed. Lois Carr, Philip D. Morgan, and Jean B. Russo, 176-199. Chapel Hill: University of North Carolina Press, 1988.

Miller, Perry. "Religious Impulse in the Founding of Virginia: Religion and Society in the Early Literature." *William and Mary Quarterly*, 3rd Ser., 5 (October 1948): 492-522.

Miller, Perry. "Religion and Society in the Early Literature: the Religious Impulse in the Founding of Virginia." *William and Mary Quarterly*, 3rd Ser., 6 (January 1949): 24-41.

Moller, Herbert. "Sex Composition and Correlated Culture Patterns of Colonial America." In *William and Mary Quarterly*, 3rd Ser., 2 (April 1945): 113-153.

Montague, Ludwell Lee. "Richard Lee, the Emmigrant, 1613(?)-1664." In *Virginia Magazine of History and Biography*, 62 (January 1954): 3-49.

Morgan, Edmund. "Headrights and Head Counts." *Virginia Magazine of History and Biography* 80 (July 1972): 361-367.

Morgan, Philip D. "Slave Life in Piedmont Virginia, 1720-1800." In *Colonial Chesapeake Society*, ed. Lois Carr, Philip D. Morgan, and Jean B. Russo, 433-484. Chapel Hill: University of North Carolina Press, 1988.

Mosse, George L. "Puritanism and Reason of State in Old and New England." *William and Mary Quarterly*, 3rd Ser., 9 (January 1952): 67-80.

Pagan, John R. "Dutch Maritime and Commerical Activity in Mid-Seventeenth-Century Virginia." In *Virginia Magazine of History and Biography*, 90 (October 1982): 485-501.

Philips, William D. Jr. "The Old World Background of Slavery in the Old World." In *Slavery and the Rise of the Atlantic System*, ed. Barbara L. Solow, 43-62. Cambridge: Cambridge University Press, 1991.

Porter, Harry Culverwell. "Alexander Whitaker: Cambridge Apostle to Virginia." *William and Mary Quarterly*, 3rd Ser., 14 (1957): 309-343.

Rainbolt, John C. "The Alteration in the Relationship between Leadership and Constituents in Virginia, 1660-1720." *William and Mary Quarterly*, 3rd Ser., 27 (July 1970): 411-434.

Reay, Barry. "Popular Religion." In *Popular Culture in Seventeenth-Century England*, ed. Barry Reay, 129-165. New York: St. Martin's Press, 1985.

Rowse, A. L. "Tudor Expansion: The Transition from Medieval to Modern History." *William and Mary Quarterly*, 3rd Ser., 14 (1957); reprinted in *Essays in American Colonial History*, ed. Paul Goodman, 250-260. New York: Holt, Rinehard & Winston, 1967.

Russo, Jean B. "Self-sufficiency and Local Exchange: Free Craftsmen in the Rural Chesapeake Economy." In *Colonial Chesapeake Society*, ed. Lois Carr, Philip D. Morgan, and Jean B. Russo, 389-432. Chapel Hill: University of North Carolina Press, 1988.

Rutman, Darrett B. and Anita H. Rutman. "'Now-Wives and Sons-in-Law': Parental Death in a Seventeenth-Century Virginia County." In *The Chesapeake in the 17th Century: Essays on Anglo-American Society*, ed. Thad W. Tate, and David L. Ammerman, 153-182. Chapel Hill: Institute of Early American History and Culture, Univ. of North Carolina, 1979.

Seiler, William H. "The Anglican Parish in Virginia." In *Seventeenth-Century America, Essays in Colonial History*, ed. James M. Smith, 119-142. Chapel Hill: Univ. of North Carolina, 1959.

Shirley, John W. "George Percy at Jamestown, 1607-1612." *Virginia Magazine of History and Biography* 57 (July 1949): 226-243.

Bibliography

Smith, Daniel Blake. "In Search of the Family." In *Race and Family in the Colonial South*, ed. Winthrop D. Jordan and Sheila L. Skemp, 21-36. Jackson, Miss.: Univ. Press of Miss., 1987.

Solow, Barbara L. "Slavery and Colonization." In *Slavery and the Rise of the Atlantic System*, ed. Barbara L. Solow, 21-42. Cambridge: Cambridge University Press, 1991.

Sommerville, Johann. "Ideology, Property and the Constitution." In *Conflict in Early Stuart England, Studies in Religion and Politics, 1603-1642*, ed. Richard Cust and Ann Hughes, 47-71. London: Longman Group, 1989.

Spencer, Nicholas. "Nicholas Spencer of Nominy in Westmoreland Co. in Virginia, Will 25 April 1688, proved 15 January 1699." *New England Historical and Genealogical Register* 45 (1891): 67-68.

Steiner, Bernard. "History of Slavery in Connecticut." Baltimore: John Hopkins Press, 1893; reprinted in *Early Studies of Slavery by States*. Northbrook, Illinois: Metro Books, 1972.

Tate, Thad W. "Defining the Colonial South." In *Race and Family in the Colonial South*, ed. Winthrop D. Jordan and Sheila L. Skemp, 3-19. Jackson, Miss.: Univ. Press of Miss., 1987.

_____. "The Seventeenth-Century Chesapeake and its Modern Historians." In *The Chesapeake in the 17th Century: Essays on Anglo-American Society*, ed. Thad W. Tate, and David L. Ammerman, 3-50. Chapel Hill: Institute of Early American History and Culture, Univ. of North Carolina, 1979.

Thorton, J. Mills III. "The Thrusting Out of Governor Harvey, a 17th Century Rebellion." *Virginia Magazine of History and Biography* 76 (January 1968): 11-26.

Tyler, Lyon G. "Maj. Edmund Chisman, Jr." In *William and Mary Quarterly*, 3rd Ser., 1 (1892-3): 89-94.

Vaughan, Alden T. "Blacks in Virginia: a Note on the First Decade." *William and Mary Quarterly*, 3rd Ser., 29 (July 1972): 469-478.

Voorhis, Manning C. "Crown Versus Council in the Virginia Land Policy." In *William and Mary Quarterly*, 3rd Ser., 3 (October 1946): 499-514.

Walsh, Lorena S. "Community Networks in Early Chesapeake." In *Colonial Chesapeake Society*, ed. Lois Carr, Philip D. Morgan, and Jean B. Russo, 200-241. Chapel Hill: University of North Carolina Press, 1988.

_____. "'Till Death Us Do Part': Marriage and Family in Seventeenth-Century Maryland." In *The Chesapeake in the 17th Century: Essays on Anglo-American Society*, ed. Thad W. Tate, and David L. Ammerman, 126-152. Chapel Hill: Institute of Early American History and Culture, Univ. of North Carolina, 1979.

Walter, John. "Grain Riots and Popular Attitudes to the Law: Maldon and the crisis of 1629." In *An Ungovernable People, the English and their law in the seventeenth and eighteenth centuries*, ed. John Brewer and John Styles, 47-85. New Brunswick, New Jersey: Rutgers University Press, 1980.

Wiecek, William M. "The Statutory Law of Slavery and Race in the Thirteen Mainland Colonies." *William and Mary Quarterly*, 3rd Ser., 34 (April 1977): 258-280.

Wrightson, Keith. "Two Concepts of Order: Justices, Constables, and Jurymen in Seventeenth-Century England." In *An Ungovernable People, the English and their law in the seventeenth and eighteenth centuries*, ed. John Brewer and John Styles, 21-46. New Brunswick, New Jersey: Rutgers University Press, 1980.

_____. "'Sorts of People' in Tudor and Stuart England." In *The Middling Sort of People, Culture, Society and Politics in England, 1550-1800*, ed. Jonathan Barry and Christopher Brooks, 28-51. New York: St. Martin's Press, 1994.

Secondary Sources-Unpublished materials

Billings, Warren M. "Sir William Berkeley and His Papers – Introduction, 1995(?)" TMs (photocopy).
_____. "Virginia's Deplored Condition, 1660-1676: the Coming of Bacon's Rebellion." Ph.D. diss., Northern Illinois University, 1968.
Brown, Kathleen M. "Gender and the Genesis of Race and Class System in Virginia, 1630-1750." Ph.D. diss., University of Wisconsin-Madison, 1990.
Carson, Jane Dennison. "Sir William Berkeley, Governor of Virginia: A Study of Colonial Policy." Ph.D. diss., University of Virginia, 1951.
Deal, Joseph Douglas, III. "Race and Class in Colonial Virginia: Indians, Englishmen, and Africans on the Eastern Shore During the Seventeenth Century." Ph.D. diss., University of Rochester, New York, 1981.
Essah, Patience. "Slavery and Freedom in the First State: the History of Blacks in Delaware from the Colonial Period to 1865." Ph.D. diss., University of California at Los Angeles, 1985.
Haight, Elizabeth Stanton. "Heirs of Tradition/Creators of Change: Virginia's Eastern Shore, 1633-1663." Ph.D. diss., University of Virginia, 1987.
Hallman, Clive Raymond, Jr. "Vestry as a Unit of Local Government in Colonial Virginia." Ph.D. diss., University of Georgia, 1987.
Hecht, Irene W.D. "Virginia Colony 1607-1640: A Study in Frontier Growth." Ph.D. diss., University of Washington, 1969.
Hudgins, Carter. "Patrician Culture, Public Ritual and Political Authority in Virginia, 1680-1740." Ph.D. diss., College of William and Mary, 1984.
Lawrence-McIntyre, Charshee Charlotte. "Free Blacks: a Troublesome and Dangerous Population in Antebellum America." Ph.D. diss., State University of New York at Stony Brook, 1984.
McCormick, Jo Anne. "The Quakers of Colonial South Carolina, 1670-1807." Ph.D. diss., University of South Carolina, 1984.
Meaders, Daniel. "Fugitive Slaves and Indentured Servants before 1800." Ph.D. diss., Yale University, 1990.
Parent, Anthony S. Jr. "'Either a Fool or a Fury': The Emergence of Paternalism in Colonial Virginia Slave Society." Ph.D. diss., University of California, Los Angeles, 1982.
Richter, Caroline Julia. "A Community and its Neighborhoods: Charles Parish, York County, Virginia, 1630-1740." Ph.D. diss., College of William and Mary, 1992.
Small, Clara Louise. "Three Generations of the Ennis Family: A Demographic Study on the Lower Eastern Shore." Ph.D. diss., University of Delaware, 1990.
Soderlund, Jean Ruth. "Conscience, Interest, and Power, the Development of Quaker Opposition to Slavery in the Delaware Valley, 1688-1780." Ph.D. diss., Temple University, 1981.
Sprinkle, John Harold, Jr. "Loyalists and Baconians: the Participants in Bacon's Rebellion in Virginia, 1676-1677." Ph.D. diss., College of William and Mary, 1992.
Toliver, Susan Diane. "The Black Family in Slavery, the Foundation of Afro American Culture: Its Importance to Members of the Slave Community." Ph.D. diss., University of California, Berkeley, 1982.
Voorhis, Manning C. "Land Grant Policy of Colonial Virginia, 1607-1774." Ph.D. diss., University of Virginia, 1940.
Wares, Lydia Jean. "Dress of African American Woman in Slavery and Freedom, 1500 to 1935." Ph.D. diss., Purdue University, 1981.
Wawrzyczek, Irmina Violetta. "Unfree Labor in Early Modern English Culture: England and Colonial Virginia." Ph.D. diss., University of Maria Curia-Sklodowska, Pl. M., 1988.

Bibliography

Endnotes

[1] Alexis de Tocqueville, *Democracy in America* (1835; New York: Random House, Vintage Books, 1945), 1:371.

[2] Lawrence B. Evans, ed, *Writings of George Washington* (New York: Putnam's, 1908), 550.

[3] Virginia De John Anderson, "Religion, the Common Thread of Motivation," in *Major Problems in American Colonial History*, ed Karen Ordahl Kupperman (Lexington, Massachusetts: D. C. Heath and Company, 1993), 156.

[4] Lyon G. Tyler, ed. *Narratives of Early Virginia, 1606-1625* (New York: Scribner's, 1907), 9-10.

[5] Richard Rich, "A Ballad of Virginia" in *A Library of American Literature*, vol 1, (New York: Bartleby.com, 2013); Wesley Frank Craven. *The Dissolution of the Virginia Company* 1932; reprint, Mass: Peter Smith, 1964), 1.

[6] Tyler, *Narratives of Early Virginia*, 125-126.

[7] John W. Shirley, "George Percy at Jamestown, 1607-1612," *Virginia Magazine of History and Biography* 57 (July 1949): 226-243; Dumas Malone, ed, *Dictionary of American Biography* (New York: Scribners, 1934), 836-37; Leslie Stephen and Sidney Lee, eds. *Dictionary of National Biography* (Oxford: Oxford University Press, 1917), 462.

[8] Samuel Daniel, *Poems and A Defense of Ryme*, Arthur Colby Sprague, ed, (Chicago: University of Chicago Press, 1965), 96.

[9] M. H. Abrams and others, eds., *The Norton Anthology of English Literature, revised, volume 1* (New York: W.W. Norton & Co., 1968), 1208.

[10] Warren M. Billings, *The Old Dominion in the Seventeenth Century, A Documentary History of Virginia, 1606-1689* (Chapel Hill: University of North Carolina Press, 1975), 14.

[11] Billings, *Old Dominion*, 14

[12] Billings, *Old Dominion*, 15

[13] Perry Miller, "Religion and Society in the Early Literature: the Religious Impulse in the Founding of Virginia," *William and Mary Quarterly*, 3rd ser., 6 (January 1949): 24-41.

[14] Anthony Fletcher, "The Protestant Idea of Marriage in Early Modern England," in *Religion, Culture, and Society in Early Modern Britain*, eds. Anthony Fletcher and Peter Roberts (Cambridge: Cambridge University Press, 1994), 161-181.

[15] Tyler, *Narratives of Early Virginia*, 136

[16] Tyler, *Narratives of Early Virginia*, 157

[17] Tyler, *Narratives of Early Virginia*, 137

[18] Tyler, *Narratives of Early Virginia*, 37

[19] Craven, *Dissolution*, 1.

[20] Tyler, *Narratives of Early Virginia*, 21

[21] Evarts B. Greene and Virginia D. Harrington, *American Population Before the Federal Census of 1790* (New York: Columbia University Press, 1932) 135; U.S. Bureau of the Census, *Historical Statistics of the United States, Colonial Times to 1970, Part 2* (Washington, D.C.: Government Printing Office, 1975) 1168.

Endnotes

22 Peter Force, ed. *Tracts and Other Papers, relating principally to the Origin, Settlement, and Progress of the Colonies in North America, from the Discovery of the Country to the Year 1776* (New York: Peter Smith, 1947), vol. 3, 2:9.

23 Tyler, *Narratives of Early Virginia*, 312

24 Tyler, *Narratives of Early Virginia*, 330

25 Greene/Harrington, 135; *Historical Statistics*, 1168, 1171.

26 Craven, *Dissolution*, 81-89; Sigmund Diamond, "Virginia in the Seventeenth Century," American Journal of Sociology, vol. 63 (1958); reprinted in *Essays in American Colonial History*, ed. Paul Goodman, 108-136 (New York: Holt, Rinehart & Winston, 1967), 108-9.

27 Greene/Harrington, 135; *Historical Statistics*, 1168;

28 Force, vol. 3, 6:21

29 This headright system was not something the English government had originally planned for Virginia. The original proposal by the officers in England for the Virginia Company implied that the settlers would get the land, not the providers of the transportation. When the men in Virginia modified the policy, the men in England conveniently avoided explaining this to the king, and the king conveniently did not ask. See Manning C. Voorhis, "Crown Versus Council in the Virginia Land Policy," in *William and Mary Quarterly*, 3rd ser., 3 (October 1946): 499-502.; Manning C. Voorhis, "Land Grant Policy of Colonial Virginia, 1607-1774" (Ph.D. diss., University of Virginia, 1940), 30-87.

30 Tyler, *Narratives of Early Virginia*, 17

31 Joan Simon, *Education and Society in Tudor England* (Cambridge: Cambridge University Press, 1966), 333-337; Margaret Gay Davies, *The Enforcement of English Apprenticeship, a Study in Applied Mercantilism, 1563-1642* (Cambridge, Massachusetts: Harvard University Press, 1956), 1-17; Jonathan French Scott, *Historical Essays on Apprenticeship and Vocational Education* (Ann Arbor: Ann Arbor Press, 1914), 7-44, 50-59.

32 Mildred Campbell, *The English Yeoman Under Elizabeth and the Early Stuarts*, (New Haven: Yale University Press, 1942), 211-8, 275-81; Abbot Emerson Smith, *Colonists in Bondage: White Servitude and Convict Labor in America, 1607-1776*, (Chapel Hill: Univ. of North Carolina, 1947), 227-41, 279-80.

33 See Irene W. D. Hecht's unpublished Ph.D. dissertation, "Virginia Colony 1607-1640: A Study in Frontier Growth" (University of Washington, 1969) for a detailed discussion of the historical research behind a single crop colony and its need to use forced labor to make that crop profitable.

34 Warren M. Billings, John E. Selby, and Thad. W. Tate, eds. *Colonial Virginia: a History* (White Plains, N.Y.: KTO Press, 1986), 40n.

35 John Rolfe, *A True Relation of the State of Virginia*, (New Haven: Yale, 1951), 37; Tyler, *Narratives of Early Virginia*, 346.

36 *Historical Statistics*, 1196; Edmund Morgan, *American Slavery, American Freedom, the Ordeal of Colonial Virginia* (New York: Norton 1975), 108-10.

37 Force, vol. 3, 6:1-26; Craven, *Dissolution*, 47-80.

38 Greene/Harrington, 135, 143-4; *Historical Statistics*, 1168;

39 Charles E. Hatch, *The 1st Seventeen Years, Virginia 1607-1624* (Williamsburg, Virginia: Virginia's 350th Anniversary Celebration Corporation, 1957), 86-9.

[40]Lyon Gardiner Tyler, ed., *Encyclopedia of Virginia Biography, vol. 1-5* (New York: Lewis Historical Publishing Company, 1915), 1:47.

[41]*Historical Statistics*, 1168.

[42]Helen Catterall, ed, *Judicial Cases Concerning American Slavery and the Negro, volume 1* (Washington, D.C.: Carnegie Institute of Washington, 1926), 1, 9.

[43]T.H. Breen, and Stephen Innes, *Myne Owne Ground* (New York: Oxford Univ. Press, 1980), 8-11; "Anthony Johnson, Free Negro, 1622," *Journal of Negro History* 56 (1971): 71-73; James H. Brewer, "Negro Property Owners in Seventeenth Century Virginia," *William and Mary Quarterly*, 3rd ser., 12 (October 1955): 575-580; Ross M. Kimmel, "Free Blacks in Seventeenth-Century Maryland," *Maryland Historical Magazine* 71, No. 1 (Spring 1976): 19-25.

[44]The ambiguities of British law concerning slavery are discussed at length by several scholars. See in particular Helen Catterall, ed., *Judicial Cases Concerning American Slavery and the Negro, volumes I and IV* (Washington, D.C.: Carnegie Institute of Washington, 1926 and 1936); Leon A. Higginbotham, *In the Matter of Color: Race and the American Legal Process, the Colonial Period* (New York: Oxford University Press, 1978); and David Brion Davis, *The Problem of Slavery in Western Culture* (Ithaca, N.Y.: Cornell Univ., 1966)

[45]Morgan, *American Slavery*, 111n, 407; Herbert Moller, "Sex Composition and Correlated Culture Patterns of Colonial America," in *William and Mary Quarterly*, 3rd ser., 2 (April 1945): 117. Moller's statistics for the years 1934-35 are especially revealing, with 1636 men coming versus only 271 women, a six to one ratio.

[46]Morgan, *American Slavery*, 101; Craven, *Dissolution*, 148-175.

[47]Wesley Frank Craven, *White, Red, & Black* (Charlottesville: Univ. Press of Virginia, 1971), 14.

[48]Samuel Mathews, "The Mutiny in Virginia, 1635: Letter from Capt. Sam'l Mathews concerning the eviction of Harvey, Governor of Va." *Virginia Magazine of History and Biography* 1 (April 1894): 416-424.

[49]"Report on Petition of Capt. Samuel Mathews," *Virginia Magazine of History and Biography* 10 (April 1903): 428.

[50]Jon Kukla, *Political Institutions in Virginia, 1619-1660* (New York: Garland Publishing, 1989), 81-99; J. Mills Thorton III, "The Thrusting Out of Governor Harvey, a 17th Century Rebellion," *Virginia Magazine of History and Biography* 76 (January 1968): 11-26.

[51]Nugent, 1:20-53.

[52]Morgan, *American Slavery*, 129; See also Abbot Emerson Smith, *Colonists in Bondage: White Servitude and Convict Labor in America, 1607-1776* (Chapel Hill: Univ. of North Carolina, 1947), 8-9, 147-151.

Endnotes

[53] A survey of the laws passed by the General Assembly in William Waller Hening, ed., *The Statutes at Large: Being a collection of all the Laws of Virginia from the 1st Session of the Legislature in the Year 1619*, vols. 1-2 (Richmond 1809-1823) shows numerous examples of how local justices were given enormous freedom for running the court as they wished. See also Kukla, *Political Institutions*, 42-48; Warren Billings, "The Growth of Political Institutions in Virginia, 1634 to 1676," *William and Mary Quarterly*, 3rd ser., 31 (April 1974): 225-242; Warren Billings, "Justices, Books, Laws and Courts in Seventeenth-Century Virginia," *Law Library Journal* 85, No. 2 (Spring 1993): 277-296; Warren Billings, "The Law of Servants and Slaves in Seventeenth Century Virginia," *Virginia Magazine of History and Biography* 99 (January 1991): 45-62; George B. Curtis, "The Colonial County Court, Social Forum and Legislative Precedent, Accomack County, Virginia, 1633-1639," *Virginia Magazine of History and Biography* 85 (July 1977): 274-288.

[54] Force, vol. 3, 15:3-4.

[55] Phillip Bruce, *Economic History of Virginia in the 17th Century*, 2 vols (Gloucester, Mass.: Peter Smith, 1935), 525.

[56] Kevin P. Kelly, "'In dispers'd Country Plantations': Settlement Patterns in Seventeenth-Century Surry County, Virginia," in *The Chesapeake in the 17th Century: Essays on Anglo-American Society*, ed. Thad W. Tate, and David L. Ammerman (Chapel Hill: Institute of Early American History and Culture, Univ. of North Carolina, 1979), 183-205.

[57] Edmund Morgan, *American Slavery, American Freedom, the Ordeal of Colonial Virginia*, (New York: Norton 1975), 407; Herbert Moller, "Sex Composition and Correlated Culture Patterns of Colonial America," in *William and Mary Quarterly*, 3rd ser., 2 (April 1945): 117-118.

[58] Morgan, *American Slavery*, 128.

[59] Morgan, *American Slavery*, 158-64, 406-7, 412; Carville V. Earle, "Environment, Disease, and Mortality in Early Virginia," in *The Chesapeake in the 17th Century: Essays on Anglo-American Society*, ed. Thad W. Tate, and David L. Ammerman, 96-125 (Chapel Hill: Institute of Early American History and Culture, Univ. of North Carolina, 1979), 120-121.

[60] Darrett B. Rutman and Anita H. Rutman, "'Now-Wives and Sons-in-Law': Parental Death in a Seventeenth-Century Virginia County," in *The Chesapeake in the 17th Century: Essays on Anglo-American Society*, ed. Thad W. Tate, and David L. Ammerman (Chapel Hill: Institute of Early American History and Culture, Univ. of North Carolina, 1979), 162.

[61] Gerald Eades Bentley, *The Jacobean and Caroline Stage, Dramatic Companies and Players, Vol I-V* (Oxford: Clarendon Press, 1941), 49-56, 652-666, 669-670; Irwin Smith, *Shakespeare's Blackfriar's Playhouse, its History and its Design*, (New York: New York University Press, 1964), 280-281;

[62] Bentley, 2:664.

[63] Sir William Berkeley, *A Lost Lady* (1638; reprint, London: Malone Society, 1987), 2.

[64] Jane Dennison Carson, "Sir William Berkeley, Governor of Virginia: A Study of Colonial Policy," (Ph.D. diss. University of Virginia, 1951), 1-4; *VMHB* 18 (1910) 441-443.

[65] Warren M. Billings, "Sir William Berkeley and His Papers--Introduction, 1995(?)" (TMs (photocopy)), 2-3; David Hackett Fischer, *Albion's Seed: Four British Folkways in America* (New York: Oxford Univ. Press, 1989), 208-9.

[66] G.E. Aylmer, *The King's Servants: The Civil Service of Charles I, 1625-1642* (New York: Columbia University Press, 1961), 28, 127, 153, 168-171, 207, 220-221.

[67] Roger Lockyer, *The Early Stuarts, a Political History of England, 1603-1642* (London: Longman, 1989) 354.

[68] G. Huehns, ed., *Clarendon, Selections from "The History of the Rebellion and Civil Wars" and "The Life by Himself."* (London: Oxford University Press, 1955), 67-83.

[69] Lockyer, 217-239, 325-353; Johann Sommerville, "Ideology, Property and the Constitution," in *Conflict in Early Stuart England, Studies in Religion and Politics, 1603-1642*, ed. Richard Cust and Ann Hughes (London: Longman Group, 1989), 47-71.

[70] Malcolm R. Smuts, *Court Culture and the Origins of a Royalist Tradition in Early Stuart England* (Philadelphia: University of Pennsylvania Press, 1987), 118-133, 183-209; Graham Parry, *The Seventeenth Century, the Intellectual and Cultural Context of English Literature, 1603-1700* (London: Longman, 1989) 9-71.

[71] Billings, "Introduction," 3, 37.

[72] As an example, in 1644 he even proposed giving away the Channel islands in exchange for French aid, a plan vigorously opposed by more moderate Royalists. "Jermyn's freedom from personal scruples and political principles made him a useful instrument of the king's foreign policy." From the *Dictionary of National Biography*, Lesilie Stephen and Sidney Lee, eds., (Oxford: Oxford University Press, 1917), 10:779-781.

[73] Leslie Stephen and Sidney Lee, eds., *Dictionary of National Biography* (Oxford: Oxford University Press, 1917), 5:551-558.

[74] Charles L. Squier, *Sir John Suckling* (Boston: G.K. Hall, 1978), 13-32.

[75] Mark Charles Fissel, *The Bishops' Wars, Charles I's campaigns against Scotland, 1638-1640* (Cambridge: Cambridge University Press, 1994), 78.

[76] Thomas Clayton, *The Works of John Suckling*, (Oxford: Clarendon Press, 1971), 73.

[77] R.C. Bald, "A Note on Suckling's *A Sessions of the Poets*," in *Modern Language Notes*, 10 (November 1943): 550-551.

[78] Billings, "Introduction," 6, 39.

[79] DNB, 10:372.

[80] Billings, "Introduction," 6, 39.

[81] DNB, 19:1072-4.

[82] Samuel R. Gardiner, *History of England from the Accession of James I to the Outbreak of the Civil War, 1603-1642, Volume 10*, (New York: AMS Press, 1965), 10:28; Carson, 1.

[83] DNB, 2:361-363; Carson, 12-14.

[84] Ronald B. Bond, *Certain Sermons or Homilies (1547) and A Homily against Disobedience and Wilful Rebellion (1570, a critcal edition* (Toronto: University of Toronto Press, 1987), 161.

[85] A superb discussion of how the royalist concept of a Great Chain was transformed into bigotry in the eighteenth century can be found in Winthrop Jordan's classic work, *White Over Black: American attitudes toward the Negro, 1550-1812* (Chapel Hill: University of North Carolina, 1968) 219-234.

[86] J.W. Allen, *English Political Thought, 1603-1660. volume 1*, (London: Methuen & Co., 1938), 3-25; Sommerville, 50-7; Christopher Hill, *Change and Continuity in Seventeenth-Century England*, revised edition (New Haven: Yale University Press, 1991), 181-204.

[87] Gardiner, *1642-49*, 4:322.

Endnotes

[88] Sommerville, 55-7.

[89] Alexander Brown, *The Genesis of the United States*, 2 vols (Cambridge: Riverside Press, 1890), 2:827-828; *VMHB* 18 (1910): 443.

[90] Carson, 3.

[91] "Virginia Gleanings in England," *Virginia Magazine of History and Biography*, October, 1910, 18:441-443.

[92] Dumas Malone, ed., *Dictionary of American Biography* (New York: Scribners, 1934), 2:339-343; Beckett, 9-13; Peter Laslett, "The Gentry of Kent in 1640," in *Shaping Southern Society, the Colonial Experience*, ed. T.H. Breen (New York: Oxford University Press, 1976), 32-46.

[93] Greene, 104-12; Beckett, 91-131.

[94] Bond, 161-2.

[95] *Clarendon*, 320.

[96] *Clarendon*, 302-3.

[97] Gardiner, *1642-49*, 3:136.

[98] *Funk & Wagnall's New "Standard" Dictionary of the English Language*, 1962 ed., s.v. "class."

[99] *Funk & Wagnall's*, 1962 ed., s.v. "caste."

[100] Steven Orgel and Roy Strong, *Inigo Jones: the Theatre of the Stuart Court, 2 volumes*, (London: Sotheby Parke Bernet, 1973), 80-81.

[101] Orgel, 25-27.

[102] Berkeley, *A Lost Lady*, 15.

[103] Berkeley, *A Lost Lady*, 60.

[104] Berkeley, *A Lost Lady*, 69-70.

[105] William Haller, ed., *Tracts on Liberty in the Puritan Revolution, 1638-1647*, 3 vols. (New York: Columbia University Press, 1934), 1:10-14.

[106] Haller, 2:7.

[107] Haller, 2:17.

[108] Haller, 2:25.

[109] Mildred Campbell, *The English Yeoman Under Elizabeth and the Early Stuarts*, (New Haven: Yale University Press, 1942) 156-220, 372-374; Jack P. Greene, *Pursuits of Happiness, the Social Development of Early Modern British Colonies and the Formation of American Culture*, (Chapel Hill: University of North Carolina Press, 1988) 107-12; J.V. Beckett, *The Aristocracy in England, 1660-1914*, (Oxford: Basil Blackwell, 1986), 16-42; Mildred Campbell, "Social Origins of Some Early Americans," in *Seventeenth Century America, Essays in Colonial History*, ed. James M. Smith (Chapel Hill: Univ. of North Carolina, 1959), 64-66; Jonathon Barry, "Introduction," in *The Middling Sort of People, Culture, Society and Politics in England, 1550-1800*, ed. Jonathan Barry and Christopher Brooks, (New York: St. Martin's Press, 1994), 12-14.

[110] The unstable and complicated nature of this stratified culture is well illustrated in Keith Wrightson's essay, "'Sorts of People' in Tudor and Stuart England," in *The Middling Sort of People, Culture, Society and Politics in England, 1550-1800*, ed. Jonathan Barry and Christopher Brooks, (New York: St. Martin's Press, 1994), 28-51.

[111] Sir Thomas Smith, *De Pepublica Anglorum, a Discourse on the Commonwealth of England*, (1583; reprint, Cambridge: Cambridge University Press, 1906), 32-33.

[112] Christopher Brooks, "Apprenticeship, Social Mobility and the Middling Sort, 1550-1800," in *The Middling Sort of People, Culture, Society and Politics in England, 1550-1800*, ed. Jonathan Barry and Christopher Brooks (New York: St. Martin's Press, 1994), 52-83.

[113] Beckett, 17.

[114] Gardiner, *History of the Great Civil War, 1642-1649*, 4 volumes (New York: AMS Press, Inc., 1965), 3:136.

[115] Margaret Guy Davies, *The Enforcement of English Apprenticeship, a Study in Applied Mercantilism, 1563-1642* (Cambridge, Massachusetts: Harvard University Press, 1956), 1-16.

[116] Campbell, *The English Yeoman*, 68-104.

[117] See in particular three essays in *Popular Culture in Seventeenth-Century England*, ed. Barry Reay (New York: St. Martin's Press, 1985): Barry Reay, "Popular Culture in Early Modern England," 4; Peter Burke, "Popular Culture in Seventeenth-Century London," 49; and Jonathan Barry, "Popular Culture in Seventeenth-Century Bristol," 62-67. See also Joan Simon's study of the development of education under Elizabeth, *Education and Society in Tudor England* (Cambridge: Cambridge University Press, 1966), 369-403.

[118] John Brewer and John Styles, "Introduction," in *An Ungovernable People, the English and their law in the seventeenth and eighteenth centuries*, ed. John Brewer and John Styles, (New Brunswick, New Jersey: Rutgers University Press, 1980), 14. This book has several essays illustrating the defiant and proud attitude of ordinary British citizens to the law. See especially Keith Wrightson, "Two Concepts of Order: Justices, Constables, and Jurymen in Seventeenth-Century England," 21-46, and John Walter, "Grain Riots and Popular Attitudes to the Law: Maldon and the crisis of 1629," 47-85.

[119] Laurel Thatcher Ulrich, *Good Wives, Image and Reality in the Lives of Women in Northern New England, 1650-1750* (New York: Vintage Books, 1991), 7.

[120] Barry, "Introduction," 14-17.

[121] Anthony Fletcher, "The Protestant Idea of Marriage in Early Modern England," in *Religion, Culture, and Society in Early Modern Britain*, eds. Anthony Fletcher and Peter Roberts, (Cambridge: Cambridge University Press, 1994) 161-181.

[122] Lawrence Stone, *Uncertain Unions and Broken Lives, Marriage and Divorce in England, 1660-1857*, (Oxford: Oxford University Press, 1995), 36. Stone states that informal break-ups probably increased this number to 5 percent of all marriages, but he admits that this is still tiny "as compared with the 33 percent that in 1994 are projected to end in divorce in England."

[123] Martin Ingram, "The Reform of Popular Culture? Sex and Marriage in Early Modern England," in *Popular Culture in Seventeenth-Century England*, ed. Barry Reay (New York: St. Martin's Press, 1985), 150-161.

[124] Campbell, *The English Yeoman*, 314-360.

125 Barry Reay, "Popular Religion," in *Popular Culture in Seventeenth-Century England*, ed. Barry Reay (New York: St. Martin's Press, 1985), 91-94; Patrick Collinson, *The Birthpangs of Protestant England, Religious and Cultural Change in the Sixteenth and Seventeenth Centuries* (New York: St. Martin's Press, 1988), 55-56, 60-68.

126 Malcolm R. Smuts describes the contradictory nature of Charles' religious philosophy in *Court Culture and the Origins of a Royalist Tradition in Early Stuart England* (Philadelphia: University of Pennsylvania Press, 1987), 217-238. While advocating approaches strongly reminiscent of catholicism, the Stuart court was remarkably apathetic to religion. This apathy might help explain the disinterest the royalists felt to truly establishing a strong church in Virginia.

127 A thorough examination of the Puritan concept of this British society is found in J.T. Cliffe *The Puritan Gentry, the Great Puritan Families of Early Stuart England* (London: Routledge & Kegan Paul, 1984). Their attempt to transplant this social order is well described by T. H. Breen, *Puritans and Adventurers, Change and Persistence in Early America* (New York: Oxford University Press, 1980), 46-80.

128 David Hackett Fischer, *Albion's Seed: Four British Folkways in America* (New York: Oxford Univ. Press, 1989), 18-24; Charles E. Hambrick-Stowe, *The Practice of Piety, Puritan Devotional Disciplines in Seventeenth-Century New England* (Chapel Hill: University of North Carolina Press, 1982), 50-52; Virginia DeJohn Anderson, "Religion, the Common Thread of Motivation," in *Major Problems in American Colonial History*, ed. Karen Ordahl Kupperman (Lexington, Massachusetts: D. C. Heath and Company, 1993), 145-157.

129 Edmund Morgan, ed., *Puritan Political Ideas, 1558-1794* (Indianapolis: Bobbs-Merrill, 1965), 92-93.

130 Rufus Jones, *The Quakers in the American Colonies* (London: MacMillan & Co., 1911), 16-21, 71, 76-89.

131 Edmund Morgan, *The Puritan Family; Religion & Domestic Relations in Seventeenth Century New England* (New York: Harper & Row, 1966), 133-160; Philip J. Greven, Jr., "Family Structure in Seventeenth Century Andover, Massachusetts," *William and Mary Quarterly*, 3rd Ser., 23 (April, 1966): 234-256.

132 Abbot Emerson Smith, *Colonists in Bondage: White Servitude and Convict Labor in America, 1607-1776* (Chapel Hill: Univ. of North Carolina, 1947), 28-29, 147-151, 228-9, 336-337.

133 Morgan, *The Puritan Family*, 109-132.

134 See in particular the work of Winthrop D. Jordan and David Brion Davis.

135 See especially David Brion Davis' *The Problem of Slavery in Western Culture* (Ithaca, N.Y.: Cornell Univ., 1966).

136 Basil Davidson, *The African Slave Trade, a revised and expanded edition* (Boston: Little, Brown & Co., 1980), 59-63.

137 Davidson, 61.

138 *De Pepublica Anglorum*, 106.

139 Burke, 41-54.

140 Graham Parry, *The Seventeenth Century, the Intellectual and Cultural Context of English Literature, 1603-1700* (London: Longman, 1989), 216-218; Christopher Hill, *A Nation of Change and Novelty, Radical politics, religion and literature in seventeenth-century England* (London: Routledge, 1990), 114-132.

[141] Parry, 217. See also J.W. Allen, *English Political Thought, 1603-1660. volume 1* (London: Methuen & Co., 1938), 26-47; Richard Cust and Ann Hughes, "Introduction, after Revisionism," in *Conflict in Early Stuart England, Studies in Religion and Politics, 1603-1642* (London: Longman Group, 1989), 1-47; Sommerville, 47-71; Cust, 134-167.

[142] See Karen Ordahl Kupperman's "Author's Note" in *Providence Island, 1630-1641, the Other Puritan Colony* (Cambridge: Cambridge University Press, 1993), xiii.

[143] Gardiner, *1642-1649*, 4:39.

[144] Patrick Collinson, *The Birthpangs of Protestant England, Religious and Cultural Change in the Sixteenth and Seventeenth Centuries* (New York: St. Martin's Press, 1988), 150.

[145] Hill, *A Nation of Change and Novelty*, 152-193.

[146] Parry, 151.

[147] Allen, 269-276.

[148] Clayton Coleman Hall, ed., *Narratives of Early Maryland, 1633-1684*, (New York: Scribners, 1910), 16.

[149] Peter Lake, "Anti-popery: the Structure of a Prejudice," in *Conflict in Early Stuart England, Studies in Religion and Politics, 1603-1642*, ed. Richard Cust and Ann Hughes (London: Longman Group, 1989), 72-106.

[150] Rufus Jones, *The Quakers in the American Colonies* (London: MacMillan & Co., 1911) xv-xvii; David Hackeet Fischer, *Albion's Seed: Four British Folkways in America* (New York: Oxford Univ. Press, 1989) 420-9; Mary Maples Dunn and Richard S. Dunn, ed., *The Papers of William Penn, Volume 1, 1644-1679*, (Pennsylvania: University of Pennsylvania Press, 1981), 83-4, 355-8.

[151] The Puritan impulse to apply moral and spiritual questions to government is well documented. See for example George L. Mosse, "Puritanism and Reason of State in Old and New England," *William and Mary Quarterly*, 3rd Ser., 9 (January 1952): 67-80.

[152] See especially William Haller's introduction to *Tracts on Liberty in the Puritan Revolution, 1638-1647, 3 vols.*, 1-9; also, Hill's *A Nation of Change and Novelty*, 24-56.

[153] Gardiner, *1642-49*, 1:292.

[154] Morgan, *Puritan Political Ideas*, 20.

[155] Morgan, *Puritan Political Ideas*, 155.

[156] Gardiner, *1642-49*, 4:302.

[157] Hill, *A Nation of Change and Novelty*, 118.

[158] Gardiner, *1642-49*, 3:120.

[159] G. Huehns, ed, *Clarendon, Selections from* The History of the Rebellion and Civil Wars *and* The Life by Himself (London: Oxford University Press, 1955), 230-244.

[160] Warren M. Billings, "Sir William Berkeley and His Papers--Introduction, 1995(?)" (TMs (photocopy)), 3-8.

[161] Force, Peter, ed., *Tracts and Other Papers, relating principally to the Origin, Settlement, and Progress of the Colonies in North America, from the Discovery of the Country to the Year 1776, 5 vols*, (New York: Peter Smith, 1947), vol 2, 7:4-5.

[162] Jane Dennison Carson, "Sir William Berkeley, Governor of Virginia: A Study of Colonial Policy," (Ph.D. diss., University of Virginia, 1951), 28-29.

Endnotes

[163] Edmund Morgan, *American Slavery, American Freedom, the Ordeal of Colonial Virginia*, (New York: Norton 1975), 407; Herbert Moller, "Sex Composition and Correlated Culture Patterns of Colonial America," in *William and Mary Quarterly*, 3rd Ser., 2 (April 1945): 117-118.

[164] "James Revel Describes the Servant's Plight, ca. 1680," in *The Old Dominion in the Seventeenth Century, A Documentary History of Virginia, 1606-1689*, Warren M. Billings, ed. (Chapel Hill: University of North Carolina Press, 1975), 139. Though this poem describes the life of a transportee (sent to Virginia as punishment for a crime), it describes quite vividly the life of the farm servant throughout this century.

[165] Force, vol. 2 8:3, 8.

[166] Morgan, *American Slavery*, 404.

[167] David Hackett Fischer, *Albion's Seed: Four British Folkways in America* (New York: Oxford Univ. Press, 1989), 130-134, 344-349. See especially his notes p. 130-1 and 345-6.

[168] See Patience Essah, "Slavery and Freedom in the First State: the History of Blacks in Delaware from the Colonial Period to 1865" (Ph.D. diss., University of California at Los Angeles, 1985), 13-16; John Pagan, "Dutch Maritime and Commercial Activity in Mid-Seventeenth-Century Virginia," in *Virginia Magazine of History and Biography*, 90 (October 1982): 485-501.

[169] Susie M. Ames, *Studies of the Virginia Eastern Shore in the 17th Century* (Richmond, Va.: Dietz Press, 1940), 104-106.

[170] Joseph Douglas Deal, III, "Race and Class in Colonial Virginia: Indians, Englishmen, and Africans on the Eastern Shore During the Seventeenth Century" (Ph.D. diss., University of Rochester, New York, 1981), 312. The court records only refer to "Anthony." Considering the respect Johnson garnered among whites on the Eastern Shore, as well as his successful ability to use the courts to benefit himself, I feel it reasonable to consider this Anthony to be Anthony Johnson.

[171] Phillip Bruce, *Social Life of Virginia in the 17th Century* (Richmond: Whittet & Shepperson, 1907), 177-181, 188-217; Morgan, *American Slavery*, 113, 151-152.

[172] Billings, *Old Dominion*, 51-58.

[173] William Waller Hening, ed., *The Statutes at Large; Being a collection of all the Laws of Virginia from the 1st Session of the Legislature in the Year 1619, 13 vols*, (Richmond 1809-1823) 1:230-236; Jon Kulka, *Political Institutions in Virginia, 1619-1660* (New York: Garland Publishing, 1989) 106-7.

[174] Kukla, *Political Institutions*, 108-122.

[175] Lyon Gardiner Tyler, ed., *Encyclopedia of Virginia Biography, 5 volumes*, (New York: Lewis Historical Publishing Company, 1915), 106-7.

[176] *VMHB*, 36 (1928), 98-101.

[177] Tyler, *Encyclopedia*, 1:104.

[178] Ralph T. Whitelaw, *Virginia's Eastern Shore, A History of Northampton and Accomack Counties, 2 vols*, (Richmond: Virginia Historical Society, 1951; reprint, Gloucester, Massachusetts: Peter Smith, 1968), 624-5.

[179] Nell Marion Nugent, *Cavaliers and Pioneers, Abstracts of Virginia Land Patents and Grants, volumes 1 and 2* (Volume 1: Baltimore: Genealogical Publishing Co., 1979; Volume 2: Richmond: Virginia State Library, 1977), 5, 101, 119, 134.

[180]Tyler, *Encyclopedia*, 1:47, 106; *VMHB* 8:302-306, 11:170; Anita H. and Darrett B. Rutman, *A Place in Time, Middlesex County, Virginia, 1650-1750* (New York: Norton & Co., 1984), 48-49.

[181]Tyler, *Encyclopedia*, 123-4; *WMQ* (1) 14:117-119; Warren M. Billings, "Virginia's Deplored Condition, 1660-1676: the Coming of Bacon's Rebellion," (Ph.D. diss., Northern Illinois University, 1968) 271.

[182]Tyler, *Encyclopedia*, 1:199; Nugent, 1:97, 185, 215, 219-220; Martin H. Quitt, *Virginia House of Burgesses, 1660-1706: the Social, Educational and Economic Bases of Political Power* (New York: Garland Publishing, 1989) 27-28; Rutman, *A Place in Time*, 48.

[183]*WMQ* (1) 10:35-38.

[184]Tyler, *Encyclopedia*, 1: 121; Whitelaw, 177, 629, 639; Annie Lash Jester, ed., *Adventurers of Purse and Person, Virginia, 1607-1625, 2nd Edition* (Virginia: Order of First Families of Virginia, 1964) 256-7; Elizabeth Stanton Haight, "Heirs of Tradition/ Creators of Change: Virginia's Eastern Shore, 1633-1663" (Ph.D. diss., University of Virginia, 1987) 440.

[185]Kukla, *Political Institutions*, 132-137; Carson, 81.

[186]Kukla, *Political Institutions*, 84, 110, 118.

[187]Kulka, *Political Institutions*, 145.

[188]William Hand Browne, ed., *Archives of Maryland, Proceedings and Acts of the General Assembly of Maryland, January 1637/8-September 1664* (Baltimore: Maryland Historical Society, 1883) 4:281, 390, 458-459; Nathaniel C. Hale, *Virginia Venturer, a Historical Biography of William Claiborne, 1600-1677* (Richmond, Virginia: Dietz Press, 1951); Tyler, *Encyclopedia*, 1:96-97;

[189]Tyler, *Encyclopedia*, 1:114-5, 279; "Speech of Sir William Berkeley and Declaration of the Assembly, March 1651," *Virginia Magazine of Biography and History*, 1 (1893): 75-81.

[190]Nugent, 1:120; Tyler, *Encyclopedia*, 1:106; Jester, 334-335; *VMHB* 9:173-174.

[191]Tyler, *Encyclopedia*, 1:122; Jester, 362-363, 402.

[192]Bernard Bailyn, "Politics and Social Structure in Virginia," in *Seventeenth Century America, Essays in Colonial History*, ed. James M. Smith (Chapel Hill: Univ. of North Carolina, 1959), 95.

[193]J. Mills Thorton, III, "The Thrusting Out of Governor Harvey, a 17th Century Rebellion," *Virginia Magazine of History and Biography* 76 (January 1968): 11-26; Kukla, *Political Institutions*, 81-99; Bailyn, "Politics," 95-98.

[194]Thomas Ludwell, "A Description of the Government of Virginia," *Virginia Magazine of History and Biography* 21 (January 1913): 37; Kukla, *Political Institutions*, 112a.

[195]Billings, *Old Dominion*, 51-8; Hening, 1:240-82.

[196]Billings, *Old Dominion*, 51.

[197]Hening, 1:268-9, 277.

[198]Hening, 1:242.

[199]Hening, 1:252-5, 257, 259, 274-5.

[200]John C. Rainbolt, "The Alteration in the Relationship between Leadership and Constituents in Virginia, 1660-1720," *William and Mary Quarterly*, 3rd Ser., 27 (July 1970): 416.

Endnotes

[201] Warren M. Billings, "Sir William Berkeley and His Papers--Introduction, 1995(?)" TMs (photocopy), 11-12.

[202] "Acts, Orders, and Resolutions of the General Assembly of Virginia, at sessions of March 1643-1646," *Virginia Magazine of History and Biography* 23 (July 1915) 237-238.

[203] Browne, *Archives*, 4:281, 390, 458-459; Billings, "Introduction," 15.

[204] Carson, 125-7.

[205] Edward Johnson, "A Chapter from a Puritan Writer." *William and Mary Quarterly*, 1st Ser., 10 (January 1901): 35-8; Clayton Coleman Hall, ed., *Narratives of Early Maryland, 1633-1684* (New York: Scribners, 1910), 235, 254-5, 302-3; Edward James, ed., "The Church in Lower Norfolk County," chapters in *Lower Norfolk County Virginia Antiquary, volumes 1-5* (New York: Peter Smith, 1951), 2:11-12.

[206] Hening, 1:289-91, 298.

[207] Hening, 1:290-1.

[208] Hening, 1:290.

[209] Rutman, *A Place in Time*, 48; Montague, Ludwell Lee, "Richard Lee, the Emigrant," in *Virginia Magazine of History and Biography* 62 (January 1954) 41-42.

[210] Montague, 13. Montague takes the position that Lee might have had a few parliamentarian sympathies, despite his numerous close royalist ties.

[211] Carson, 36.

[212] Hening, 1:241-2.

[213] Said in England by a parson when a group of layman attempted to read the Bible together. In Patrick Collinson, *The Birthpangs of Protestant England, Religious and Cultural Change in the Sixteenth and Seventeenth Centuries* (New York: St. Martin's Press, 1988), 37.

[214] Billings, *Old Dominion*, 52-3.

[215] Hening, 1:279.

[216] Manning C. Voorhis, "Land Grant Policy of Colonial Virginia, 1607-1774" (Ph.D. diss., University of Virginia, 1940), 30-87.

[217] Nugent, 1:61, 87, 104-5, 120, 181-2.

[218] Evarts B. Greene and Virginia D. Harrington, *American Population Before the Federal Census of 1790*, (New York: Columbia University Press, 1932), 136.

[219] A list of the remaining Assembly members who patented land on blacks in the years before 1650 is as follows:

Bartholomew Hoskins (1648)-6
Thomas Mathews (1639)-5
Francis Eps, Sr. (1635)-5
Adam Thorowgood, Sr. (1637)-3
Rice Hoe (1643)-4
Richard Townsend (1639)-3
John Robbins, Sr. (1638)-3
John Upton (1635)-2
Thomas Harris (1635)-2
John Chesman (1636)-2
Richard Cocke, Sr. (1636 & 1639)-2
Leonard Yeo (1637)-2
John Chew (1637)-2
Richard Bennett (1635)-1
Argoll Yardly (1639)-2
Stephen Hamelyn (1638)-1
Arthur Smith (1637)-1
Hugh Gwin (1642)-1
William Davis (1639)-1
Benjamin Harrison (1643)-1
Moore Fantleroy (1643)-1
George Reade (1648)-1
Phillip Bennett (1648)-1.

A total of 129 blacks were imported into the colony by Assembly members, out of a total population of less than 300 before 1650. More than half of these black slaves, however, were imported by just five men: Browne, Burnham, Kemp, Menifie, and Wormeley.

[220] Hening, 1:242.

[221] Hening, 1:293.

[222] Hening, 1:361.

[223] Nugent, 1:97, 144, 185, 215, 219-20; W.G. Stanard, "Lancaster County Book: Estates, Deeds, etc., 1654-1702," *William and Mary Quarterly*, 1st Ser., 2 (1893): 270.

[224] T.H. Breen and Stephen Innes, *Myne Owne Ground*, (New York: Oxford Univ. Press, 1980), 78; Deal, "Race and Class in Colonial Virginia: Indians, Englishmen, and Africans on the Eastern Shore During the Seventeenth Century," 293-297, 330. Deal notes in detail how Charleton seemed an almost "reluctant" slave-holder, exhibiting strongly mixed feelings to the custom.

[225] Johann Sommerville, "Ideology, Property and the Constitution," in *Conflict in Early Stuart England, Studies in Religion and Politics, 1603-1642*, ed. Richard Cust and Ann Hughes (London: Longman Group, 1989), 47-71.

[226] See David Brion Davis, *The Problem of Slavery in Western Culture* (Ithaca, N.Y.: Cornell Univ., 1966)

[227] The history of the English Civil War is well known. I have used as my main source the work of Samuel Rawson Gardiner, including his *History of England from the Accession of James I to the Outbreak of the Civil War, 1603-1642, Volumes 8-10* (New York: AMS Press, 1965), *History of the Great Civil War, 1642-1649, 4 volumes* (New York: AMS Press, Inc., 1965), and *History of the Commonwealth and Protectorate, 1649-1656, 4 volumes* (New York: AMS Press, Inc., 1965)

Endnotes

[228] Peter Newman, *Atlas of the English Civil War* (New York: MacMillian Publishing Company, 1985) 82-85.

[229] Gardiner, *1642-1649*, 3:267n.

[230] Gardiner, *1642-1649*, 3:392.

[231] Gardiner, *1942-1649*, 3:392-394.

[232] Gardiner, *1642-1649*, 4:308.

[233] Clayton Coleman Hall, ed., *Narratives of Early Maryland, 1633-1684* (New York: Scribners, 1910) 235, 254-5; Lyon Gardiner Tyler, ed., *Encyclopedia of Virginia Biography*, (New York: Lewis Historical Publishing Company, 1915), 1:47, 184, 279, 289; Edward James, ed., "The Church in Lower Norfolk County," chapters in *Lower Norfolk County Virginia Antiquary, volumes 1-5* (New York: Peter Smith, 1951) 2:14-5, 61-2, 83-8, 120-8, 3:29-34.

[234] Jon Kukla, *Political Institutions in Virginia, 1619-1660* (New York: Garland Publishing, 1989), 132-135,145.

[235] Mildred Campbell, "Social Origins of Some Early Americans," in *Seventeenth Century America, Essays in Colonial History*, ed. James M. Smith (Chapel Hill: Univ. of North Carolina, 1959), 63-89; Martin H. Quitt, *Virginia House of Burgesses, 1660-1706: the Social, Educational and Economic Bases of Political Power* (New York: Garland Publishing, 1989), 10-94, 303-312; Warren M. Billings, "Virginia's Deplored Condition, 1660-1676: the Coming of Bacon's Rebellion," (Ph.D. diss., Northern Illinois University, 1968), 133.

[236] Tyler, *Encyclopedia*, 1:49.

[237] Tyler, *Encyclopedia*, 50-51.

[238] Tyler, i, 1:126.

[239] Peter Force, ed., *Tracts and Other Papers, relating principally to the Origin, Settlement, and Progress of the Colonies in North America, from the Discovery of the Country to the Year 1776, 5 vols.* (New York: Peter Smith, 1947); Jane Dennison Carson, "Sir William Berkeley, Governor of Virginia: A Study of Colonial Policy," (Ph.D. diss., University of Virginia, 1951.) 38-39; Quitt, 35-36.

[240] Phillip Bruce, *Social Life of Virginia in the 17th Century* (Richmond: Whittet & Shepperson, 1907) 30; Richard L. Morton, *Colonial Virginia, Vol. I* (Chapel Hill: Univ. of North Carolina Press, 1960) 166-167; Tyler, *Encyclopedia*, 1:54.

[241] Bernard Bailyn, "Politics and Social Structure in Virginia." In *Seventeenth Century America, Essays in Colonial History*, ed. James M. Smith (Chapel Hill: Univ. of North Carolina, 1959), 98. Though Bailyn notes that "Most of Virginia's great eighteenth-century names, such as Bland, Burwell, Byrd, Carter, Digges, Ludwell, and Mason, appear for the first time within ten years either side of 1655," he ignores the obvious connection these names have with this Royalist migration.

[242] Tyler, *Encyclopedia*, 1:54; "Note," *William and Mary Quarterly*, 1st Ser., (January, 1915), 177; Billings, "Deplored," 270.

[243] Tyler, *Encyclopedia*, 1:122; Bruce, *Social Life*, 73:

[244] Tyler, *Encyclopedia*, 1:130; Morton, 166-167; *VMHB*, 5:65;

[245] Tyler, *Encyclopedia*, 1:136; John Harold Sprinkle, Jr., "Loyalists and Baconians: the Participants in Bacon's Rebellion in Virginia, 1676-1677" (Ph.D. diss., College of William and Mary, 1992), 159; Billings, "Deplored," 286; Nell Marion Nugent, *Cavaliers and Pioneers, Abstracts of Virginia Land Patents and Grants, volumes 1 and 2 (Volume 1:* Baltimore: Genealogical Publishing Co., 1979; Volume 2: Richmond: Virginia State Library, 1977) 1:252, 279, 340; H.R. McIlwaine, ed., *Minutes of the Council and General Court of Colonial Virginia* (Richmond, 1926) 454-455.

[246] Leslie Stephen and Sidney Lee, eds., *Dictionary of National Biography* (Oxford: Oxford University Press, 1917) 5:973-975.

[247] Tyler, *Encyclopedia*, 1:47-48; Billings, "Deplored," 67, 271; *VMHB* 19:350, 357-358; William Waller Hening, ed., *The Statutes at Large; Being a collection of all the Laws of Virginia from the 1st Session of the Legislature in the Year 1619, 13 vols.* (Richmond 1809-1823) 1:426; Nathaniel C. Hale, *Virginia Venturer, a Historical Biography of William Claiborne, 1600-1677* (Richmond, Virginia: Dietz Press, 1951), 300.

[248] Tyler, *Encyclopedia*, 1:132; McIlwaine, *Minutes*, 221; Billings, "Deplored," 271.

[249] Tyler, *Encyclopedia*, 1:294; James, "Church," 3:141, 4:79, 105; Billings, "Deplored," 278.

[250] Tyler, *Encyclopedia*, 1:220; Billings, "Deplored," 276.

[251] Tyler, *Encyclopedia*, 1:174; Billings, "Deplored," 280.

[252] Billings, "Deplored," 280-282.

[253] Bailyn, "Politics," 98.

[254] "Speech of Sir William Berkeley and Declaration of the Assembly, March 1651," *Virginia Magazine of Biography and History 1* (1893): 78.

[255] "Speech of Sir William Berkeley:" 77.

[256] Evarts B. Greene and Virginia D. Harrington, *American Population Before the Federal Census of 1790* (New York: Columbia University Press, 1932) 136; Edmund Morgan, *American Slavery, American Freedom, the Ordeal of Colonial Virginia*, (New York: Norton 1975), 180; Billings, "Deplored", 38.

[257] Russell R. Menard, "The Tobacco Industry in the Chesapeake Colonies, 1617-1730: an Interpretation," in *Research in Economic History, a Research Annual, Volume 5*, ed. Paul Uselding (Greenwich, Connecticut: Jai Press, 1980), 132-3.

[258] William H. Seiler, "The Anglican Parish in Virginia," in *Seventeenth-Century America, Essays in Colonial History*, ed. James M. Smith, 119-142 (Chapel Hill: Univ. of North Carolina, 1959), 133-4; Clive Raymond Hallman, "Vestry as a Unit of Local Government in Colonial Virginia," (Ph.D. diss., University of Georgia, 1987), 32-9, 61-4, 84-5.

[259] Force, vol. 2 8:3, 8.

[260] Hening, 1:252.

[261] Morgan, *American Slavery*, 407.

[262] Morgan, *American Slavery*, 180-4; Force, vol. 2, 7:5.

[263] Edmund Morgan gives an excellent description of the patent process in "Headrights and Head Counts," *Virginia Magazine of History and Biography* 80 (July 1972): 362.

[264] Clayton Coleman Hall, ed., *Narratives of Early Maryland, 1633-1684* (New York: Scribners, 1910), 206-8.

Endnotes

[265] Leo Francis Stock, *Proceedings and Debates of the British Parliaments respecting North America, Volume I, 1542-1688* (Washington, D.C.: Carnegie Institution, 1924), 226; Nathaniel C. Hale, *Virginia Venturer, a Historical Biography of William Claiborne, 1600-1677* (Richmond, Virginia: Dietz Press, 1951), 274-275.

[266] Warren M. Billings, "Sir William Berkeley and His Papers--Introduction, 1995(?)" TMs (photocopy): 17-8.

[267] Ludwell Lee Montague, "Richard Lee, the Emmigrant, 1613(?)-1664," in *Virginia Magazine of History and Biography*, 62 (January 1954), 24-25.

[268] William Waller Hening, ed., *The Statutes at Large; Being a collection of all the Laws of Virginia from the 1st Session of the Legislature in the Year 1619*, 13 vols (Richmond 1809-1823), 1:363-8.

[269] Lyon Gardiner Tyler, ed., *Encyclopedia of Virginia Biography, volumes I, 4, 5* (New York: Lewis Historical Publishing Company, 1915), 1: 47.

[270] William Hand Browne, ed. *Archives of Maryland, Proceedings and Acts of the General Assembly of Maryland, January 1637/8-September 1664* (Baltimore: Maryland Historical Society, 1883), 340-2.

[271] Lyon G. Tyler, "Maj. Edmund Chisman, Jr," in *William and Mary Quarterly*, 1st Ser., 1 (1892-3): 93-94; *William and Mary Quarterly*, 1st Ser., 11:30-33; Caroline Julia Richter, "A Community and its Neighborhoods: Charles Parish, York County, Virginia, 1630-1740," (Ph.D. diss., College of William and Mary, 1992), 212.

[272] Tyler, *Encyclopedia*, 1:124-125; Hening 1:421, 427, 512, 552.

[273] Nell Marion Nugent, *Cavaliers and Pioneers, Abstracts of Virginia Land Patents and Grants, volumes 1 and 2* (Volume 1: Baltimore: Genealogical Publishing Co., 1979; Volume 2: Richmond: Virginia State Library, 1977), 1:222; Tyler, *Encyclopedia*, 1:347.

[274] Nugent, 1:185; Tyler, *Encyclopedia*, 1:116-7; Martin H. Quitt, *Virginia House of Burgesses, 1660-1706: the Social, Educational and Economic Bases of Political Power* (New York: Garland Publishing, 1989), 36-9.

[275] Tyler, *Encyclopedia*, 1:362-3; Annie Lash Jester, ed., *Adventurers of Purse and Person, Virginia, 1607-1625, 2nd Edition* (Virginia: Order of First Families of Virginia, 1964), 365-7; Hening, 2:12.

[276] Jon Kukla, *Political Institutions in Virginia, 1619-1660* (New York: Garland Publishing, 1989), 163.

[277] Ralph T. Whitelaw, *Virginia's Eastern Shore, A History of Northampton and Accomack Counties, 2 vols* (Richmond: Virginia Historical Society, 1951; reprint, Gloucester, Massachusetts: Peter Smith, 1968), 28-31, 528-9.

[278] Tyler, *Encyclopedia*, 1:115; Tyler, "Maj. Edmund Chisman, Jr.," 97-99.

[279] Kukla, *Political Institutions*, 147-8; Tyler, *Encyclopedia*, 1:118-9.

[280] Warren M. Billings, ed., "Some Acts not in Hening's Statutes, the Acts of Assembly, April 1652, November 1652, July 1653," *Virginia Magazine of History and Biography* 83 (January 1975): 22-76.

[281] Billings, "Acts" 31-2.

[282] Billings, "Acts," 37-41, 47, 49-51, 59-60, 66-8; Hening, 1:286.

283For a concise explanation of deference and its influence on colonial politics, see Joy B. and Robert R. Gilsdorf, "Elites and Electorates: Some Plain Truths for Historians of Colonial America," in *Saints & Revolutionaries: Essays on Early American History*, eds. David D. Hall, John M. Murrin, and Thad W. Tate (New York: W. W. Norton & Co., 1984), 207-244; Richard R. Beeman, "Deference, Republicanism, and the Emergence of Popular Politics in Eighteenth-Century America," *William and Mary Quarterly*, 3rd ser., 49 (July 1992): 403-412.

284Whitelaw, 31.

285Billings, "Acts," 33.

286Hening, 1:380.

287Rufus Jones, *The Quakers in the American Colonies* (London: MacMillan & Co., 1911), 265-9.

288Hening, 1:412.

289Hening, 1:403.

290Warren M. Billings, *The Old Dominion in the Seventeenth Century, A Documentary History of Virginia, 1606-1689* (Chapel Hill: University of North Carolina Press, 1975), 165-9; Warren M. Billings, "The Cases of Fernando and Elizabeth Key: a Note on the Status of Blacks in Seventeenth Century Virginia," *William and Mary Quarterly*, 3rd ser., 30 (July 1973): 467-474.

291Joseph Douglas Deal, III, "Race and Class in Colonial Virginia: Indians, Englishmen, and Africans on the Eastern Shore During the Seventeenth Century" (Ph.D. diss., University of Rochester, New York, 1981), 310-315.

292T.H. Breen, and Stephen Innes, *Myne Owne Ground* (New York: Oxford Univ. Press, 1980), 80, 85-86; Deal, "Race and Class in Colonial Virginia," 420-427. Deal gives a different interpretation, saying that Longo's treatment was more harsh than that usually afforded whites.

293Deal, "Race and Class in Colonial Virginia," 326-336.

294Billings, "The Cases of Fernando and Elizabeth Key," 468-469; Kathleen M. Brown, "Gender and the Genesis of Race and Class System in Virginia, 1630-1750," (Ph.D. diss., University of Wisconsin-Madison, 1990), 219-220.

295Deal, "Race and Class in Colonial Virginia," 332-334.

296Breen and Innes, 83-84; Deal, "Race and Class in Colonial Virginia," 315-316.

297Breen and Innes, 80, 107; Deal, "Race and Class in Colonial Virginia," 386. There are minor conflicts between what Breen/Innes and Deal describe of this event. I have generally accepted Deal in my interpretation.

298Breen and Innes, 96.

299The failed Puritan colony of New Providence Island off the coast of Mexico illustrates this fact quite clearly. Its collapse in 1641 was due to the reluctance of any Puritans to immigrate there, partly because the colony depended so heavily on slave labor. See Karen Ordahl Kupperman's *Providence Island, 1630-1641, the Other Puritan Colony* (Cambridge: Cambridge University Press, 1993), 138-141, 168-169, 243-244.

300T.H. Breen and Stephen Innes, *Myne Owne Ground* (New York: Oxford Univ. Press, 1980), 12-15.

301Breen and Innes, 13-14.

302Susie M. Ames, *Studies in the Virginia in the Eastern Shore in the 17th Century* (Richmond, Va.: Dietz Press, 1940), 102-4; Breen and Innes, 1-15; Ross M. Kimmel, "Free Blacks in Seventeenth-Century Maryland," *Maryland Historical Magazine* 71, No. 1 (Spring 1976): 23-4; "Anthony Johnson, Free Negro, 1622," *Journal of Negro History* 56 (1971): 72.

303Lyon Gardiner Tyler, ed., *Encyclopedia of Virginia Biography, volumes I, 4, 5* (New York: Lewis Historical Publishing Company, 1915) 1:47-8; Nathaniel C. Hale, *Virginia Venturer, a Historical Biography of William Claiborne, 1600-1677* (Richmond, Virginia: Dietz Press, 1951) 300; Lyon Gardiner Tyler, "Pedigree of a Representative Virginia Planter, Edward Digges, Esq." *William and Mary Quarter*, 1st Ser., 1 (1892): 208-13; Warren M. Billing, "Virginia's Deplored Condition, 1660-1676: the Coming of Bacon's Rebellion," (Ph.D. diss., Northern Illinois University, 1968), 67.

304Lyon Gardiner Tyler, "Virginia Under the Commonwealth," *William and Mary Quarterly*, 1st Ser., 1 (April 1893): 193.

305Tyler, *Encyclopedia*, 1:125-126; Harry Culverwell Porter, "Alexander Whitaker: Cambridge Apostle to Virginia," *William and Mary Quarterly*, 3rd Ser., 14 (1957): 309-343.

306The Assembly listed no reason for Davis's dismissal, though the timing seems prophetic. Hening, 2:15. See also Annie Lash Jester, ed., *Adventurers of Purse and Person, Virginia, 1607-1625, 2nd Edition* (Virginia: Order of First Families of Virginia, 1964).

307William Hand Browne, ed., *Archives of Maryland, Proceedings and Acts of the General Assembly of Maryland, January 1637/8-September 1664* (Baltimore: Maryland Historical Society, 1883), 119.

308Tyler, *Encyclopedia*, 1:229; Warren M. Billings, "Virginia's Deplored Condition, 1660-1676: the Coming of Bacon's Rebellion" (Ph.D. diss., Northern Illinois University, 1968) 275; *VMHB* 72 (1964): 42-49.

309Tyler, *Encyclopedia*, 1:190; William Waller Hening, ed., *The Statutes at Large; Being a collection of all the Laws of Virginia from the 1st Session of the Legislature in the Year 1619, 13 vols.* (Richmond 1809-1823) 2:39.

310Hening, 1:374.

311Hening, 1:376-377.

312Hening, 1:388.

313Jon Kukla, *Political Institutions in Virginia, 1619-1660* (New York: Garland Publishing, 1989) 170-176.

314Hening, 1:377.

315Hening, 1:380, 384.

316Susie M. Ames, "Colonel Edmund Scarborough," *Alumnae Bulletin of Randolph Macon Woman's College* 26, No. 1 (November 1932): 17; Ralph T. Whitelaw, *Virginia's Eastern Shore, A History of Northampton and Accomack Counties, 2 vols.* (Richmond: Virginia Historical Society, 1951; reprint, Gloucester, Massachusetts: Peter Smith, 1968) 628-31.

317Ames, *Eastern Shore*, 95; Nell Marion Nugent, *Cavaliers and Pioneers, Abstracts of Virginia Land Patents and Grants, volumes 1 and 2* (Volume 1: Baltimore: Genealogical Publishing Co., 1979; Volume 2: Richmond: Virginia State Library, 1977) 1:328.

[318] Patience, Essah, "Slavery and Freedom in the First Stage: the History of Blacks in Delaware from the Colonial Period to 1865" (Ph.D. diss., University of California at Los Angeles, 1985), 13-16.

[319] Billings, *Old Dominion*, 149; Philip D. Curtin, *The Atlantic Slave Trade, a Census* (Madison, Wisconsin: Univ. of Wisconsin Press, 1969), 52-60, 84-85.

[320] Nugent, 1:35, 119, 101, 134, 183, 225.

[321] Nugent, 1:286.

[322] Manning C. Voorhis, "Land Grant Policy of Colonial Virginia, 1607-1774" (Ph.D. diss., University of Virginia, 1940), 30-87.

[323] Nugent, 1:195, 214, 215, 219-220, 240, 286, 298.

[324] Ames, "Scarborough," 17-18.

[325] Whitelaw, 631.

[326] Nugent, 1:328.

[327] Nugent, 1:547-548.

[328] Nugent, 1:559.

[329] Nugent, 1:524-525; Tyler, *Encyclopedia*, 1:319.

[330] Tyler, *Encyclopedia*, 1:307; Nugent, 1:526.

[331] Nugent, 1:358, 444, 465, 558.

[332] Hening, 1:399-400.

[333] Hening, 1:418.

[334] Hening, 1:418.

[335] Rufus Jones, *The Quakers in the American Colonies* (London: MacMillan & Co., 1911), 265-9.

[336] An excellent example of this was the parish in Lower Norfolk County. After tossing the Puritans out in 1649 (including their minister Thomas Harrison (see p. 87), the citizens of Lower Norfolk County still lacked a clergyman five years later. Rufus Jones, *The Quakers in the American Colonies* (London: MacMillan & Co., 1911), 265-269; Edward James, ed., "The Church in Lower Norfolk County," chapters in *Lower Norfolk County Virginia Antiquary, volumes 1-5* (New York: Peter Smith, 1951), 3:29-34.

[337] "Two John Smiths," *William and Mary Quarterly*, 1st Ser., 23 (April 1915): 292-3.

[338] Hening, 1:421.

[339] Hening, 1:432-94.

[340] Hening, 1:495.

[341] H. R. McIlwaine, ed., *Journals of the House of Burgesses of Virginia 1619-1776. 13 vols* (Richmond, 1925-1930) 107.

[342] Hening, 1:502.

[343] Hening, 1:499-505.

[344] Tyler, *Encyclopedia*, 1:128.

Endnotes

345Tyler, *Encyclopedia*, 1:238-239; Billings, "Deplored," 284.

346William Hand Browne, ed., *Archives of Maryland, Proceedings and Acts of the General Assembly of Maryland, January 1637/8-September 1664* (Baltimore: Maryland Historical Society, 1883), 238-239, 266; William Hand Browne, ed., *Archives of Maryland, Judicial and Testamentary Business of the Provincial Court, 1637-1650* (Baltimore: Maryland Historical Society, 1887), 314-316, 321-2; Nathaniel C. Hale, *Virginia Venturer, a Historical Biography of William Claiborne, 1600-1677* (Richmond, Virginia: Dietz Press, 1951), 261-266; Jon Kukla, *Political Institutions in Virginia, 1619-1660* (New York: Garland Publishing, 1989), 145-146.

347William Waller Hening, ed., *The Statutes at Large; Being a collection of all the Laws of Virginia from the 1st Session of the Legislature in the Year 1619, 13 vols* (Richmond 1809-1823), 1:321.

348Hening, 1:387.

349Hening, 1:422-423.

350Nell Marion Nugent, *Cavaliers and Pioneers, Abstracts of Virginia Land Patents and Grants, 1623-1666, volumes 1 and 2* (Volume 1: Baltimore: Genealogical Publishing Co., 1979; Volume 2: Richmond: Virginia State Library, 1977), 1:93, 324, 382, 405, 457, 460.

351Browne, *Provincial Court*, 340-341, 389, 408, 444, 512-513.

352Lyon Gardiner Tyler, ed., *Encyclopedia of Virginia Biography, volumes 1, 4, 5* (New York: Lewis Historical Publishing Company, 1915), 1:119.

353Edmund Morgan, *American Slavery, American Freedom, the Ordeal of Colonial Virginia* (New York: Norton 1975), 404.

354Hening, 2:515; Morgan, *American Slavery*, 404; Evarts B. Greene and Virginia D. Harrington, *American Population Before the Federal Census of 1790* (New York: Columbia University Press, 1932), 136. The higher number is an actual estimate of servants from 1665. The lower estimate of indentured servants is based on the percentage of servants to population expressed by William Berkeley in 1670, adjusted to recognize that Berkeley was probably trying to exaggerate the number of free men in Virginia and using the total population numbers estimated by Morgan.

355Warren M. Billings, *The Old Dominion in the Seventeenth Century, A Documentary History of Virginia, 1606-1689* (Chapel Hill: University of North Carolina Press, 1975), 72.

356Russell R. Menard, "The Tobacco Industry in the Chesapeake Colonies, 1617-1730: an Interpretation," in *Research in Economic History, a Research Annual, Volume 5*, ed. Paul Uselding (Greenwich, Connecticut: Jai Press, 1980), 111.

357Warren M. Billings, John E. Selby, and Thad. W. Tate, eds., *Colonial Virginia: a History* (White Plains, N.Y.: KTO Press, 1986), 59-63; David Hackett Fischer, *Albion's Seed: Four British Folkways in America* (New York: Oxford Univ. Press, 1989), 264-274.

358Edmund Morgan, *American Slavery*, 405-410; Abbot Emerson Smith, *Colonists in Bondage: White Servitude and Convict Labor in America, 1607-1776* (Chapel Hill: Univ. of North Carolina, 1947), 323-331, 336.

359Morgan, *American Slavery*, 406-407.

360Peter Force, ed., *Tracts and Other Papers, relating principally to the Origin, Settlement, and Progress of the Colonies in North America, from the Discovery of the Country to the Year 1776, 5 vols* (New York: Peter Smith, 1947), vol. 2, 7:5.

361Anitia H. Rutman and Darrett B., *A Place in Time, Explicatus*, (New York: Norton & Co., 1984), 79-81.

[362] Menard, *Research*, 111, 133-4.

[363] Menard, *Research*, 153.

[364] Hening, 1:412-414, 420, 491-492; Warren M. Billings, ed., "Some Acts not in Hening's Statures, the Acts of Assembly, April 1652, November 1652, July 1653," *Virginia Magazine of History and Biography* 83 (January 1975): 70.

[365] Hening, 1:505-525.

[366] T. H. Breen, *Puritans and Adventurers, Change and Persistence in Early America* (New York: Oxford University Press, 1980), 106-126; Carville V. Earle, "Environment, Disease, and Mortality in Early Virginia," in *The Chesapeake in the 17th Century: Essays on Anglo-American Society*, ed. Thad W. Tate, and David L. Ammerman, (Chapel Hill: Institute of Early American History and Culture, Univ. of North Carolina, 1979), 96-125; Darrett B. Rutman and Anita H. Rutman, "'Now-Wives and Sons-in-Law': Parental Death in a Seventeenth-Century Virginia County," in *The Chesapeake in the 17th Century: Essays on Anglo-American Society*, ed. Thad W. Tate, and David L. Ammerman (Chapel Hill: Institute of Early American History and Culture, Univ. of North Carolina, 1979), 153-182; Daniel Blake Smith, "In Search of the Family," in *Race and Family in the Colonial South*, ed. Winthrop D. Jordan and Sheila L. Skemp, 21-36. Jackson, Miss.: Univ. Press of Miss., 1987, 21-36; Fischer, *Albion*, 275-332; Jack P. Greene, *Pursuits of Happiness, the Social Development of Early Modern British Colonies and the Formation of American Culture* (Chapel Hill: University of North Carolina Press, 1988), 12-13, 15-17, 27.

[367] Lorena S. Walsh, "'Till Death Us Do Part': Marriage and Family in Seventeenth-Century Maryland," in *The Chesapeake in the 17th Century: Essays on Anglo-American Society*, ed. Thad W. Tate, and David L. Ammerman (Chapel Hill: Institute of Early American History and Culture, Univ. of North Carolina, 1979), 127.

[368] Herbert Moller, "Sex Composition and Correlated Culture Patterns of Colonial America," in *William and Mary Quarterly*, 3rd Ser., 2 (April 1945): 116-118.

[369] *Smith, Race and Family*, 24.

[370] *Smith, Race and Family*, 26.

[371] Clive Raymond Hallman, Jr., "Vestry as a Unit of Local Government in Colonial Virginia" (Ph.D. diss., University of Georgia, 1987), 145-6.

[372] Force, vol. 3, 15:6.

[373] Samuel Eliot Morison, "Virginians and Marylanders at Harvard College in the Seventeenth Century," *William and Mary Quarterly*, 2nd ser., 13 (January 1933): 1-9.

[374] Tyler, *Encyclopedia*, 1:134.

[375] Fischer, *Albion's*, 130-134, 344-349; Peter Burke, "Popular Culture in Seventeenth-Century London," in *Popular Culture in Seventeenth-Century England*, ed. Barry Reay (New York: St. Martin's Press, 1985), 49-54; Jonathan Barry, "Popular Culture in Seventeenth-Century Bristol," in *Popular Culture in Seventeenth-Century England*, ed. Barry Reay, (New York: St. Martin's Press, 1985), 61-70.

[376] Force, vol. 3, 15:4.

[377] William H. Seiler, "The Anglican Parish in Virginia," in *Seventeenth-Century America, Essays in Colonial History*, ed. James M. Smith (Chapel Hill: Univ. of North Carolina, 1959), 124-135; Hallman, 32-39, 46-47, 63.

[378] James Q. Wilson and Richard J. Herrnstein, *Crime and Human Nature* (New York: Simon and Schuster, 1985), 242-4.

[379] Travis Hirschi, "Crime and the Family," in *Crime and Public Policy*, James Q. Wilson, ed. (San Francisco: ICS Press, 1983), 53-68; Michael Tonry, Lloyd E. Ohlin, and David Farrington, *Human Development and Criminal Behavior–New Ways of Advancing Knowledge* (New York: Springer-Verlag, 1991), 100-3, 122-3, 137-8, 157-8; George Gilder, *Men and Marriage* (Gretna, Louisiana: Pelican, 1986), 49-98; see also David P. Farrington, Lloyd E. Ohlin, and James Q. Wilson, *Understanding and Controlling Crime: Toward a New Research Stategy* (New York: Springer-Verlag, 1986), 2-3, 39-61, and Wilson/Herrnstein, 213-264, 289-311, for a complete review of this literature.

[380] Smith, *Race and Family*, 24.

[381] Daniel Patrick Moynihan, *Family and Nation* (New York: Harcourt, Brace, Jovanovich, 1986), 9.

[382] Breen, Puritans, 106-126.

[383] Warren M. Billings, "Virginia's Deplored Condition, 1660-1676: the Coming of Bacon's Rebellion," (Ph.D. diss., Northern Illinois University, 1968), 273.

[384] Hening, 2:364.

[385] Billings, "Deplored," 74-76.

[386] John Harold Sprinkle, Jr., "Loyalists and Baconians: the Participants in Bacon's Rebellion in Virginia, 1676-1677," (Ph.D. diss., College of William and Mary, 1992), 129-130; "Defense of Edward Hill," *Virginia Magazine of History and Biography* 3 (1896): 240.

[387] Ralph T. Whitelaw, *Virginia's Eastern Shore, A History of Northampton and Accomack Counties, 2 vols* (Richmond: Virginia Historical Society, 1951; reprint, Gloucester, Massachusetts: Peter Smith, 1968), 177, 639; Annie Lash Jester, ed., *Adventurers of Purse and Person, Virginia, 1607-1625, 2nd Edition* (Virginia: Order of First Families of Virginia, 1964), 256-7.

[388] Nugent, 1:152, 224-225, 401, 407.

[389] H. R. McIlwaine, ed., *Minutes of the Council and General Court of Colonial Virginia* (Richmond, 1926), 344.

[390] Nugent, 1:554.

[391] Nugent, 1:524-525.

[392] Tyler, *Encyclopedia*, 1:178, 5:1017; Martin H. Quitt, *Virginia House of Burgesses, 1660-1706: the Social, Educational and Economic Bases of Political Power* (New York: Garland Publishing, 1989) 29-30; Billings, "Deplored," 275-276.

[393] Nugent, 176, 353, 355.

[394] Nugent 2:186, 216, 283, 322, 342, 353, 354, 371, 373, 401.

[395] Tyler, *Encyclopedia*, 1:260.

[396] Jester, 211-213.

[397] Nugent, 1:37, 86-87, 110, 119, 148.

[398] Hening, 1:262.

[399] Billings, *Old Dominion*, 170.

[400] Billings, *Old Dominion*, 170.

[401] Nugent, 1:185, 298.

[402] Tyler, *Encyclopedia*, 2:308.

[403] Tyler, *Encyclopedia*, 1:293.

[404] Nugent, 1:67, 173, 383, 385.

[405] Jester, 250-253.

[406] Nugent, 1:57, 181, 287; Tyler, *Encyclopedia*, 1:362-363; Jester, 365-366.

[407] Edward W. James, ed., "Henry Woodhouse," *William and Mary Quarterly*, 1st Ser., 1 (1892-3): 229.

[408] "Virginia Gleanings in England," *Virginia Magazine of History and Biography*, 13 (October 1905): 202-3.

[409] Kukla, *Political Institutions*, 205-6.

[410] Hening, 1:512.

[411] *VMHB*, 19:189, 359-360; Tyler, *Encyclopedia*, 1:131, 265.

[412] Kukla, *Political Institutions*, 213.

[413] Kukla, *Political Institutions*, 209-216.

[414] William Berkeley, *A Lost Lady* (1638; reprint, London: Malone Society, 1987), 67.

[415] William Waller Hening, ed., *The Statutes at Large; Being a collection of all the Laws of Virginia from the 1st Session of the Legislature in the Year 1619, 13 vols* (Richmond 1809-1823), 1:517, 537.

[416] Jon Kukla, "Some Acts Not in Hening's Statues, Acts of Assembly October 1660," *Virginia Magazine of History and Biography* 83 (January 1975): 84.

[417] Hening, 1:523-524, 531-533; Kukla, "Acts," 85.

[418] Kukla, "Acts," 77-97.

[419] Hening, 1:546.

[420] Richard Morton, *Colonial Virginia, Vol. I* (Chapel Hill: Univ. of North Carolina Press, 1960), 188.

[421] Lyon G. Tyler, "Virginia under the Commonwealth," *William and Mary Quarterly*, 1st Ser., 1 (1892): 195-6.

[422] Kukla, "Acts," 86-87.

[423] Kukla, "Acts," 90-91.

[424] Kukla, "Acts," 89, 91, 94; Hening, 2:11.

[425] Hening, 2:9-12.

[426] Kukla, "Acts," 93-94.

[427] Hening, 2:17-18; Wesley Frank Craven, *The Colonies in Transition, 1660-1713* (New York: Harper & Row, 1967/8), 10-13, 32-44.

[428] Hening, 2:17-32.

[429] Hening, 2:25.

[430] Hening, 2:31-32.

Endnotes

[431] Hening, 2:20-21; Kukla, "Acts," 96.

[432] Hening, 1:486-487.

[433] Hening, 2:25.

[434] Edward James, ed., "The Church in Lower Norfolk County." Chapters in *Lower Norfolk County Virginia Antiquary, volumes 1-5* (New York: Peter Smith, 1951), 1:81-5, 139-46 2:11-17; William H. Seiler, "The Anglican Parish in Virginia," in *Seventeenth-Century America, Essays in Colonial History*, ed. James M. Smith (Chapel Hill: Univ. of North Carolina, 1959), 136-9.

[435] Hening, 2:39.

[436] Hening, 2:19, 21, 24.

[437] Historian Warren Billings has theorized that during this period power actually became de-centralized, moving from Berkeley downward to the assembly and to the lower courts. While this is true, power was also being focused upward, centered among that exclusive and tightknit core of royalist elites. See Billings' "Virginia's Deplored Condition, 1660-1676: the Coming of Bacon's Rebellion" (Ph.D. diss., Northern Illinois University, 1968).

[438] Hening, 2:17, 23, 26-27, 35.

[439] Hening, 2:41.

[440] Hening, 2:34.

[441] Hening, 2:42-52.

[442] Hening, 2:48.

[443] Warren M. Billings, ed. "A Quaker in Seventeenth-Century Virginia: Four Remonstrances by George Wilson," *William and Mary Quarterly*, 3rd Ser., 33 (January 1976): 127-140.

[444] Hening, 2:48; H. R. McIlwaine, ed., *Minutes of the Council and General Court of Colonial Virginia* (Richmond, 1926), 507; Phillip Bruce, *Institutional History of Virginia in the 17th Century*, 2 vols (Gloucester, Mass.: Peter Smith, 1964), 234-7; Rufus Jones, *The Quakers in the American Colonies* (London: MacMillan & Co., 1911) 276; Kenneth L. Carroll, "Quakerism on the Eastern Shore of Virginia," in *VMHB* 74:2 4/66, 175.

[445] Hening, 2:41-148.

[446] Craven, *Transition*, 32-44.

[447] William Berkeley, *A Discourse & View of Virginia* (1663; reprint, Norwalk, Conn.: W. H. Smith, 1914), 6-7.

[448] "Berkeley's Instructions." *Virginia Magazine of History and Biography* 3 (July 1895): 15-20; Sister Joan de Lourdes Leonard, "Operation Checkmate: The Birth and Death of a Virginia Blueprint for Progress, 1666-1676," *William and Mary Quarterly*, 3rd Ser., 24 (January 1967): 44-74.

[449] Despite this, the assembly still worked to benefit its friends. Edmund Scarburgh was involved in two separate decisions during this assembly, once as the plaintiff and once as the defendant. In both cases the assembly ruled in his favor. The assembly also ruled for Francis Moryson and John Stringer in two other cases. Hening, 2:36, 149-150, 158-161.

[450] Russell R. Menard, "The Tobacco Industry in the Chesapeake Colonies, 1617-1730: an Interpretation," in *Research in Economic History, a Research Annual, Volume 5*, ed. Paul Uselding, (Greenwich, Connecticut: Jai Press, 1980), 134-6.

[451] William Waller Hening, ed., *The Statutes at Large; Being a collection of all the Laws of Virginia from the 1st Session of the Legislature in the Year 1619, 13 vols* (Richmond 1809-1823), 1:403.

[452] "Speech of Sir William Berkeley and Declaration of the Assembly, March 1651," *Virginia Magazine of Biography and History* 1 (1893): 75-81.

[453] *Virginia Magazine of History and Biography* 2 (1895): 381-385.

[454] "Proclamations of Nathaniel Bacon," *Virginia Magazine of Biography and History* 1 (1893): 60; H. R. McIlwaine, *Minutes of the Council and General Court of Colonial Virginia* (Richmond, 1926), 454-5; "Colonel William Claiborne, Jr., Certificate of Loyalty," *William and Mary Quarterly* Ser. 1, 5 (1896): 109.

[455] McIlwaine, *Minutes*, 454-5.

[456] Cyrus Harreld Karraker, *The Seventeenth-Century Sheriff: a Comparative Study of the Sheriff in England and the Chesapeake Colonies, 1607-1689* (Chapel Hill: Univ. of North Carolina Press, 1930), 73-79.

[457] Lyon G. Tyler, "Virginia under the Commonwealth," *William and Mary Quarterly*, 1st Ser., 1 (1892): 195-6.

[458] "Berkeley's Instructions," *Virginia Magazine of History and Biography* 3 (July 1895): 15-20; Sister Joan de Lourdes Leonard, "Operation Checkmate: The Birth and Death of a Virginia Blueprint for Progress, 1666-1676," *William and Mary Quarterly*, 3rd Ser., 24 (January 1967): 44-74.

[459] Hening, 2:172-176.

[460] Joy B. and Robert R. Gilsdorf, "Elites and Electorates: Some Plain Truths for Historians of Colonial America," in *Saints & Revolutionaries: Essays on Early American History*, eds. David D. Hall, John M. Murrin, and Thad W. Tate (New York: W. W. Norton & Co., 1984), 207-244.

[461] David Jordan, "Elections and Voting in Early Colonial Maryland," *Maryland Historical Magazine*, 77 (1982, No. 3): 238-265.

[462] Edward James, ed., "The Church in Lower Norfolk County," chapters in *Lower Norfolk County Virginia Antiquary, volumes 1-5* (New York: Peter Smith, 1951), 3:141-6, 4:78-89, 183.

[463] Hening, 2:178.

[464] Hening, 2:164-165.

[465] Hening, 2:114-115.

[466] Hening, 2:167.

[467] Hening, 2:167-168.

[468] Hening, 2:170.

[469] Hening, 2:179.

[470] Hening, 2:165-166.

[471] Hening, 2: 170

[472] U.S. Bureau of the Census, *Historical Statistics of the United States, Colonial Times to 1970, Part 2* (Washington, D.C.: Government Printing Office, 1975), 1168; Edmund Morgan, *American Slavery, American Freedom, the Ordeal of Colonial Virginia* (New York: Norton 1975), 404.

Endnotes

[473] Warren M. Billings, "The Cases of Fernando and Elizabeth Key: a Note on the Status of Blacks in Seventeenth Century Virginia," *William and Mary Quarterly*, 3rd Ser., 30 (July 1973): 467-474.

[474] Indians were also enslaved, but much more rarely. Furthermore, the assembly in the Commonwealth era acted several times over the years to restrict and even outlaw this practice. See Hening 1:396, 410, 471, 481-2.

[475] *Virginia Magazine of History and Biography* 1 (July 1893): 326; James, "Church," 4:78-89.

[476] Hening, 2:198.

[477] Hening, 2:180-183.

[478] Ralph T. Whitelaw, *Virginia's Eastern Shore, A History of Northampton and Accomack Counties*, 2 vols (Richmond: Virginia Historical Society, 1951; reprint, Gloucester, Massachusetts: Peter Smith, 1968), 1414-18.

[479] Whitelaw, 1416.

[480] William H. Seiler, "The Anglican Parish in Virginia," in *Seventeenth-Century America, Essays in Colonial History*, ed. James M. Smith, 119-142 (Chapel Hill: Univ. of North Carolina, 1959), 129; Peter Force, ed., *Tracts and Other Papers, relating principally to the Origin, Settlement, and Progress of the Colonies in North America, from the Discovery of the Country to the Year 1776*, 5 vols. (New York: Peter Smith, 1947), vol. 3, 15:4.

[481] Kenneth L. Carroll, "Quakerism on the Eastern Shore of Virginia," in *Virginia Magazine of History and Biography* 74 (April 1966): 175.

[482] Martin H. Quitt, *Virginia House of Burgesses, 1660-1706: the Social, Educational and Economic Bases of Political Power* (New York: Garland Publishing, 1989), 194-5n.

[483] Karraker, 76-81.

[484] William Berkeley, A Lost Lady (1638; reprint, London: Malone Society, 1987), 8.

[485] "Anthony Johnson, Free Negro, 1622," *Journal of Negro History* 56 (1971): 72; Ross M. Kimmel, "Free Blacks in Seventeenth-Century Maryland," *Maryland Historical Magazine* 71, No. 1 (Spring 1976): 23; T.H. Breen and Stephen Innes, *Myne Owne Ground* (New York: Oxford Univ. Press, 1980), 15-16.

[486] William Waller Hening, ed., *The Statutes at Large; Being a collection of all the Laws of Virginia from the 1st Session of the Legislature in the Year 1619*, 13 vols (Richmond 1809-1823), 2:178.

[487] Hening, 2:185.

[488] Hening, 2:216.

[489] Hening, 2:190-191, 200-202, 224-226, 228-232, 251-252.

490Most recent historians have followed the position of Sister Joan de Lourdes Leonard from "Operation Checkmate: The Birth and Death of a Virginia Blueprint for Progress, 1666-1676," *William and Mary Quarterly*, 3rd Ser., 24 (January 1967): 44-74. In her paper, Ms. Leonard concluded that in the years after the Restoration Berkeley aggressively tried to use the law to re-shape the economy of Virginia. In my reading of the laws, however, I find much less economic and social tinkering than previous years, and in fact, significant action by the Long Assembly to repeal such economic interference. Since Ms. Leonard did not have access to Warren M. Billings, "Some Acts not in Hening's Statutes, the Acts of Assembly, April 1652, November 1652, July 1653," *Virginia Magazine of History and Biography* 83 (January 1975): 22-76 and Jon Kukla, "Some Acts Not in Hening's Statutes, Acts of Assembly October 1660," *Virginia Magazine of History and Biography* 83 (January 1975): 77-97, her perspective on the laws was unfortunately incomplete.

491Hening, 2:191, 241-242.

492Hening, 2:212.

493Hening, 2:267.

494Hening, 2:287.

495Hening, 2:238-239.

496Wesley Frank Craven, *The Colonies in Transition, 1660-1713* (New York: Harper & Row, 1967/8), 32-44.

497Richard L. Morton, *Colonial Virginia, Vol. I* (Chapel Hill: Univ. of North Carolina Press, 1960), 225.

498William Berkeley, *A Discourse & View of Virginia* (1663; reprint, Norwalk, Conn.: W. H. Smith, 1914), 3.

499Lyon Gardiner Tyler, ed., *Encyclopedia of Virginia Biography, volumes 1, 4, 5* (New York: Lewis Historical Publishing Company, 1915) 1:126-127; Hening, 2:39; Thomas Ludwell, "A Description of the Government of Virginia," *Virginia Magazine of History and Biography* 21 (January 1913): 36-42.

500Tyler, *Encyclopedia*, 1:352-353; Warren M. Billings, "Virginia's Deplored Condition, 1660-1676: the Coming of Bacon's Rebellion" (Ph.D. diss., Northern Illinois University, 1968) 285; Martin H. Quitt, *Virginia House of Burgesses, 1660-1706: the Social, Educational and Economic Bases of Political Power* (New York: Garland Publishing, 1989) 44-48; VMHB, 6:92 (?).

501Tyler, *Encyclopedia*, 1:139-140; Billings, "Deplored," 279.

502Tyler, *Encyclopedia*, 1:53; Quitt, 76-77n; WMQ (2) 134-136.

503Abbot Emerson Smith, *Colonists in Bondage: White Servitude and Convict Labor in America, 1607-1776* (Chapel Hill: Univ. of North Carolina, 1947), 67-86.

504Hening, 1:257.

505Warren M. Billings, ed., "Some Acts not in Hening's Statutes, the Acts of Assembly, April 1652, November 1652, July 1653," *Virginia Magazine of History and Biography* 83 (January 1975): 41.

506Hening, 1:441-442.

507Jon Kukla, "Some Acts Not in Hening's Statutes, Acts of Assembly October 1660," *Virginia Magazine of History and Biography* 83 (January 1975): 83.

508Hening, 2:113-114.

Endnotes

[509] Hening, 2:240.

[510] "Servant's Plot of 1663, Virginia Colonial Records," *Virginia Magazine of History and Biography* 15 (July 1907): 38-43; "Notes from the Council and General Court Records, 1641-1672, by the late Conway Robinson," *Virginia Magazine of History and Biography* 8 (January 1901): 240.

[511] Hening, 2:191, 204.

[512] Hening, 2:187-188, 195.

[513] Berkeley, *Discourse*, 3.

[514] Hening, 2:239.

[515] Hening, 2:266.

[516] Hening, 2:273-274, 277-279.

[517] Hening, 2:278.

[518] Hening, 2:280.

[519] Hening, 2:298.

[520] Berkeley, *Discourse*, 3.

[521] Berkeley, *Discourse*, 3-4.

[522] G. Huehns, G. ed., *Clarendon, Selections from* The History of the Rebellion and Civil Wars *and* The Life by Himself (London: Oxford University Press, 1955), 380-8; J. V. Beckett, *The Aristocracy in England, 1660-1914* (Oxford: Basil Blackwell, 1986), 22-6, 40-2; Christopher Hill, *A Nation of Change and Novelty, Radical politics, religion and literature in seventeenth-century England* (London: Routledge, 1990), 218-43.

[523] Hening, 2:260.

[524] Hening, 2:267.

[525] Hening, 2:270.

[526] Hening, 2:279-280.

[527] Hening, 2:280-281.

[528] Hening, 2:288.

[529] Hening, 2:299.

[530] Hening, 2:296.

[531] Nell Marion Nugent, *Cavaliers and Pioneers, Abstracts of Virginia Land Patents and Grants, volumes 1 and 2* (Volume 1: Baltimore: Genealogical Publishing Co., 1979; Volume 2: Richmond: Virginia State Library, 1977), 1:429, 2:92.

[532] Nugent, 2:32.

[533] Nugent, 1:547-548.

[534] Nugent, 1:536.

[535] Nugent, 1:559, 524-525, 526, 558, 465, 2:3.

[536] Nugent, 1:62, 79, 221, 241, 264, 301, 324, 358, 385-386, 389; Quitt, 45.

537Phillip Bruce, *Economic History of Virginia in the 17th Century*, 2 vols (Gloucester, Mass.: Peter Smith, 1935), 2:125; "Isle of Wight County Records," *William and Mary Quarterly*, 1st Ser., 7 (April 1899): 238; "Census of Tithables in Surry County in the Year 1668," *William and Mary Quarterly*, 1st Ser., 8 (January 1900): 161.

538Tyler, *Encyclopedia*, 1:126-127.

539Ralph T. Whitelaw, *Virginia's Eastern Shore, A History of Northampton and Accomack Counties, 2 vols* (Richmond: Virginia Historical Society, 1951; reprint, Gloucester, Massachusetts: Peter Smith, 1968), 217-9.

540Morton, 192-195; Hening, 2:264.

541Russell R. Menard, "The Tobacco Industry in the Chesapeake Colonies, 1617-1730: an Interpretation," in *Research in Economic History, a Research Annual, Volume 5*, ed. Paul Uselding (Greenwich, Connecticut: Jai Press, 1980), 135-6.

542Russell R. Menard, "British Migration to the Chesapeake Colonies in the Seventeenth Century," in *Colonial Chesapeake Society*, ed. Lois Carr, Philip D. Morgan, and Jean B. Russo (Chapel Hill: University of North Carolina Press, 1988), 101-132; Edmund Morgan, *American Slavery, American Freedom, the Ordeal of Colonial Virginia* (New York: Norton 1975), 404.

543Some historians have proposed these difficulties as an explanation for the growth of slavery. See for example Jean B. Russo, "Self-sufficiency and Local Exchange: Free Craftsmen in the Rural Chesapeake Economy," in *Colonial Chesapeake Society*, ed. Lois Carr, Philip D. Morgan, and Jean B. Russo (Chapel Hill: University of North Carolina Press, 1988), 390-1.

544 Hening, 2:517.

545Joseph Douglas Deal, III, "Race and Class in Colonial Virginia: Indians, Englishmen, and Africans on the Eastern Shore During the Seventeenth Century." (Ph.D. diss., University of Rochester, New York, 1981), 206.

546 William Sherwood, "Virginia's Deploured Condition," *Collections of the Massachusetts Historical Society*, 4th Ser., 9 (1871): 164.

547Warren M. Billings, "Virginia's Deplored Condition, 1660-1676: the Coming of Bacon's Rebellion" (Ph.D. diss., Northern Illinois University, 1968), 76-78; Cyrus Harrell Karraker, *The Seventeenth-Century Sheriff: a Comparative Study of the Sheriff in England and the Chesapeake Colonies, 1607-1689* (Chapel Hill: Univ. of North Carolina Press, 1930), 74-80.

548"Papers from the Records of Surry County," *William and Mary Quarterly* 1st Ser., 3 (October 1894): 121-126; Warren M. Billings, *The Old Dominion in the Seventeenth Century, A Documentary History of Virginia, 1606-1689* (Chapel Hill: University of North Carolina Press, 1975), 263-267.

549Billings, *Old Dominion*, 265.

550Billings, *Old Dominion*, 264.

551"Papers from the Records of Surry County," *William and Mary Quarterly* 1st Ser., 3 (October 1894): 123-125.

552Billings, *Old Dominion*, 266-267.

553Nell Marion Nugent, *Cavaliers and Pioneers, Abstracts of Virginia Land Patents and Grants, volumes 1 and 2* (Volume 1: Baltimore: Genealogical Publishing Co., 1979; Volume 2: Richmond: Virginia State Library, 1977), 1:53, 308-309, 372, 396, 442, 497, 2:99-100, 124.

Endnotes

554 Edmund Morgan, "Headrights and Head Counts," *Virginia Magazine of History and Biography* 80 (July 1972): 361-367.

555 Charles M. Andrews, ed., *Narratives of the Insurrections, 1675-1690* (New York: Scribners, 1915), 105.

556 Andrews, 105; Sherwood, 165.

557 Andrews, 16.

558 William Waller Hening, ed., *The Statutes at Large; Being a collection of all the Laws of Virginia from the 1st Session of the Legislature in the Year 1619, 13 vols* (Richmond 1809-1823), 2:150.

559 Andrews, 17.

560 Andrews, 108.

561 Wilcomb E. Washburn, *The Governor and the Rebel* (Chapel Hill: Univ. of North Carolina, 1957), 21-22.

562 Peter Force, ed., *Tracts and Other Papers, relating principally to the Origin, Settlement, and Progress of the Colonies in North America, from the Discovery of the Country to the Year 1776, 5 vols.* (New York: Peter Smith, 1947), vol. 1, 9:3-4; Andrews, 106; Washburn, *Governor*, 22-23.

563 George Bernard Shaw, *Bernard Shaw: Selected Plays* (New York: Dodd, Mead and Company, 1981), 148.

564 Andrews, 23.

565 "Causes of Discontent in Virginia, 1676," *Virginia Magazine of History and Biography* 3 (July 1895): 35-36.

566 A wonderfully clear description of this violent and oppressive world was written in 1678 by Reverend Paul Williams, transcribing the story of Thomas Hellier, an English man who had been sold into indentured servitude and had brutally murdered his master and his wife. See T. H. Breen, James H. Lewis, and Keith Schlesinger, eds. "Motive for Murder: A Servant's Life in Virginia, 1678." *William and Mary Quarterly*, 3rd Ser., 40 (January 1983): 106-120.

567 Washburn, *Governor*, 31.

568 Washburn, *Governor*, 18.

569 Edmund Morgan, *American Slavery, American Freedom, the Ordeal of Colonial Virginia* (New York: Norton 1975) 254; Washburn, *Governor*, 17-18; Thomas Wertenbaker, *Torchbearer of the Revolution, the Story of Bacon's Rebellion and its Leader* (Princeton: Princeton Univ., 1940), 39-50.

570 H. R. McIlwaine, ed., *Minutes of the Council and General Court of Colonial Virginia* (Richmond, 1926), 401.

571 Sherwood, 165-166; Morgan *American Slavery*, 254-255.

572 Washburn, *Governor*, 22.

573 Morgan, *American Slavery*, 255.

574 Hening, 2:326-339.

575 Hening, 2:330.

576 "Charles City County Grievances," *Virginia Magazine of History and Biography*, 3 (October, 1895): 137; Andrews, 109.

577 Andrews, 110-111.

578 Andrews, 111.

579 Washburn, *Governor*, 35-37.

580 Washburn, *Governor*, 40-41.

581 Morgan, *American Slavery*, 258.

582 Modern demographic research tells us that this declaration was actually an accurate description of the make-up of the now developing opposing camps, with Berkeley's Royalist camp generally members of the economic and social elite, while the rebels of Bacon's camp were men who had been shut out of the government during the reign of the Long Assembly and who came from a broader cross section of society. See John Harold Sprinkle, Jr., "Loyalists and Baconians: the Participants in Bacon's Rebellion in Virginia, 1676-1677," (Ph.D. diss., College of William and Mary, 1992), 51-60, 105, 132-150.

583 Andrews, 21, 113-114; Sherwood, 167-168; Washburn, *Governor*, 37-46.

584 Washburn, *Governor*, 47-48.

585 Billings, *Old Dominion*, 269-270; Washburn, *Governor*, 41.

586 Morgan, *American Slavery*, 260.

587 Washburn, *Governor*, 42.

588 Andrews, 113-114; Billings, *Old Dominion*, 273.

589 Sherwood, 170.

590 Andrews, 114.

591 Washburn, *Governor*, 51.

592 Andrews, 22-23, 54, 115.

593 Andrews, 55.

594 Andrews, 110; McIlwaine, *Minutes*, 455.

595 Andrews, 40-41, 95-97; McIlwaine, *Minutes*, 372.

596 Hening, 2:356-357.

597 Hening, 2:353-355.

598 Hening, 2:356.

599 Hening, 2:359.

600 Hening, 2:357.

601 Hening, 2:358.

602 Hening, 2:363-364.

603 Hening, 2:364-365.

604 Hening, 2:352-353.

605 Hening, 2:350-351.

[606] Hening, 2:361.

[607] Hening, 2:361.

[608] John Harold Sprinkle, Jr., "Loyalists and Baconians: the Participants in Bacon's Rebellion in Virginia, 1676-1677" (Ph.D. diss., College of William and Mary, 1992), 43-111, 132-150.

[609] The following events are vividly described in all the primary sources: Andrews, 28-34, 116-117; Force, vol. 1, 9:4-5; Sherwood, 170-173.

[610] Sherwood, 171.

[611] Andrews, 33, 117.

[612] Andrews, 34.

[613] Andrews, 34.

[614] Andrews, 56-57.

[615] Andrews, 34.

[616] "Proclamations of Nathaniel Bacon," *Virginia Magazine of Biography and History* 1 (1893): 55-63.

[617] Andrews, 35, 58-63, 121-122; Force, vol. 1, 9:5-7.

[618] John Fiske, *Old Virginia and Her Neighbors, 2 volumes* (Boston: Houghton, Mifflin and Company, 1897), 2:82-86.

[619] Andrews, 123-128.

[620] Andrews, 72; Washburn, *Governor*, 70, 84, 102.

[621] Andrews, 66-71, 130-136; Force, vol. 1, 9:8.

[622] Andrews, 38-39, 74-75, 139.

[623] Andrews, 92-95; Washburn, *Governor*, 87-89.

[624] McIlwaine, *Minutes*, 454-455.

[625] However, the Council minutes show twenty death sentences. See H. R. McIlwaine, *Minutes of the Council and General Court of Colonial Virginia* (Richmond, 1926), 454-461. Either Berkeley was under-reporting the numbers, or the minutes don't reflect actual executions. For Berkeley's numbers, see Force, vol. 1, 10:1-4.

[626] Morgan, *American Slavery*, 273.

[627] William Berkeley, *A Lost Lady* (1638; reprint, London: Malone Society, 1987), 67.

[628] Wilcomb E. Washburn, ed. "Sir William Berkeley's 'A History of Our Miseries,'" *William and Mary Quarterly*, 3rd Ser., 14 (July 1957): 404-406.

[629] John Burk, *The History of Virginia from its First Settlement to the Present Day* (Petersburg, Virginia: Dicken and Pesod, 1804-5), 255.

[630] Wertenbaker, *Torchearer*, 205.

[631] Hening, 3:86, 102, 140, 179.

[632] Saul K. Padover, ed, *Thomas Jefferson on Democracy* (New York: New American Library, 1939), 99.

[633] Warren M. Billings, "Sir William Berkeley and His Papers – Introduction, 1995(?)" TMs (photocopy): 34.

[634] Wilcomb E. Washburn, *The Governor and the Rebel* (Chapel Hill: Univ. of North Carolina, 1957), 139.

[635] Ralph T. Whitelaw, *Virginia's Eastern Shore, A History of Northampton and Accomack Counties, 2 vols* (Richmond: Virginia Historical Society, 1951; reprint, Gloucester, Massachusetts: Peter Smith, 1968), 1151.

[636] Joseph Douglas Deal III, "Race and Class in Colonial Virginia: Indians, Englishmen, and Africans on the Eastern Shore During the Seventeenth Century," (Ph.D. diss., University of Rochester, New York, 1981), 151, 185.

[637] Whitelaw, 808; David Hackett Fischer, *Albion's Seed: Four British Folkways in America* (New York: Oxford Univ. Press, 1989), 306-10.

[638] Whitelaw, 626-628; Susie M. Ames, "Colonel Edmund Scarborough," *Alumnae Bulletin of Randolph Macon Woman's College* 26, No. 1 (November 1932): 16-23.

[639] Carl Degler, "Slavery and the Genesis of American Race Prejudice," *Comparative Studies in Society and History, vol. 2* (1959); reprinted in *Essays in American Colonial History*, ed. Paul Goodman, (New York: Holt, Rinehart & Winston, 1967), 223-49; Oscar and Mary Handler, "Origins of the Southern Labor System." *William and Mary Quarterly*, 3rd Ser., 7 (April 1950): 199-222.

[640] See Hening, 1:486, 545, 2:16, 34, 289-290, 302, 308, 339, 400, 447, 464.

[641] See especially George Moore, *Notes on the History of Slavery in Massachusetts* (New York: D. Appleton & Co., 1866), 1-111 for a concise summary of the attitudes of New Englanders. While they would not at first make the practice illegal, and many actually considered it entirely moral, many others repeatedly railed against it, and acted to try and discourage the custom.

[642] Winthrop D. Jordan, *White Over Black: American attitudes toward the Negro, 1550-1812* (Chapel Hill: University of North Carolina, 1968), 187.

[643] Robert C. Twombly and Robert H. Moore, "Black Puritan: The Negro in Seventeenth-Century Massachusetts," *William and Mary Quarterly*, 3rd Ser., 23 (April 1967): 239.

[644] Lorenzo Greene, *The Negro in Colonial New England, 1620-1776* (New York: Columbia University Press, 1942; reprint, Port Washington, New York: Kennikat Press, 1966) 177-190; Robert C. Twombly and Robert H. Moore, "Black Puritan: The Negro in Seventeenth-Century Massachusetts," *William and Mary Quarterly*, 3rd Ser., 23 (April 1967): 242.

[645] William M. Wiecek, "The Statutory Law of Slavery and Race in the Thirteen Mainland Colonies," *William and Mary Quarterly*, 3rd Ser., 34 (April 1977): 260.

[646] Lorenzo Green, 109-110.

[647] Edgar McManus, *A History of Negro Slavery in New York* (Syracuse, New York: Syracuse Univ. Press, 1966), 29-30; Richard Shannon Moss, *Slavery on Long Island, a study in local institutional and early African-American communal life* (New York: Garland Publishing, 1993), 35-41. Though McManus cites an oversupply of slaves as an explanation for the inability of traders to sell their slaves, he also notes that when the supply of slaves dropped after 1770, traders could still not get rid of their cargos.

[648] Rufus Jones, *The Quakers in the American Colonies* (London: MacMillan & Co., 1911), 441; Gary B. Nash and Jean R. Soderland, *Freedom by Degrees, Emancipation in Pennsylvania and its Aftermath* (New York: Oxford University Press, 1991), 43.

[649] Jones, *Quakers*, 156-166, 510-521; Nash and Soderland, 41-73; Soderlund, 5-9.

Endnotes

650 Barbara Jeanne Fields, *Slavery and Freedom on the Middle Ground: Maryland during the Nineteenth Century* (New Haven: Yale University Press, 1985) 2; Patience Essah, "Slavery and Freedom in the First State: the History of Blacks in Delaware from the Colonial Period to 1865" (Ph.D. diss., University of California at Los Angeles, 1985)

651 Even a small sampling of the literature forces one to this conclusion. I have already referred to Edmund Morgan's classic work, *The Puritan Family; Religion & Domestic Relations in Seventeenth Century New England*. A few other examples include Laurel Thatcher Ulrich, *Good Wives, Image and Reality in the Lives of Women in Northern New England, 1650-1750* (New York: Vintage Books, 1991); Virginia DeJohn Anderson, "Religion, the Common Thread of Motivation," in *Major Problems in American Colonial History*, ed. Karen Ordahl Kupperman (Lexington, Massachusetts: D. C. Heath and Company, 1993), 145-157; Herbert Moller, "Sex Composition and Correlated Culture Patterns of Colonial America," in *William and Mary Quarterly*, 3rd Ser., 2 (April 1945): 113-115; T. H. Breen, *Puritans and Adventurers, Change and Persistence in Early America* (New York: Oxford University Press, 1980), 46-80.

652 John Winthrop, "Governor John Winthrop of Massachusetts Bay Gives a Model of Christian Charity, 1630," in *Major Problems in American Colonial History*, ed. Karen Ordahl Kupperman (Lexington, Massachusetts: D. C. Heath and Company, 1993), 125.

653 Bernard Bailyn, "The *Apologia* of Robert Keayne," *William and Mary Quarterly*, 3rd Ser., 7 (October 1950): 568-587; Fischer, *Albion's Seed*, 156-8.

654 Warren M. Billings, ed., "Some Acts not in Hening's Statues, the Acts of Assembly, April 1652, November 1652, July 1653," *Virginia Magazine of History and Biography* 83 (January 1975): 34; Hening, 1:397.

655 Note also that in New England, indentured servitude itself was a relatively unpopular custom. Of the 10,394 known immigrants coming from Bristol between 1654 to 1686, most of whom were indentured servants, only 162 came to New England. See Abbot Emerson Smith, *Colonists in Bondage: White Servitude and Convict Labor in America, 1607-1776* (Chapel Hill: Univ. of North Carolina, 1947), 4, 28-9, 308-10, 316-7.

656 Saul K. Padover, ed., *Thomas Jefferson on Democracy* (New York: New American Library, 1939), 89.

657 As already noted, the Puritan colony of Providence Island had failed because British Puritans had refused to immigrate there. According to Karen Ordahl Kupperman, the colony failed for the very reasons cited above, combined with the colony's over reliance on slave labor. See Kupperman's *Providence Island, 1630-1641, the Other Puritan Colony* (Cambridge: Cambridge University Press, 1993).

658 William Berkeley, *A Discourse & View of Virginia* (1663; reprint, Norwalk, Conn.: W. H. Smith, 1914), 3.

659 Alexander Hamilton, James Madison, and John Jay, *The Federalist Papers* (New York: Mentor, 1961), 325.

660 Berkeley, *Discourse*, 3.

661 Anthony S. Parent, Jr., "'Either a Fool or a Fury': The Emergence of Paternalism in Colonial Virginia Slave Society," (Ph.D. diss., University of California, Los Angeles, 1982), 22-24, 73-74.

662 See Peter H. Wood's superb study of South Carolina's slave history for a discussion of this charter, *Black Majority, Negroes in Colonial South Carolina from 1670 through the Stono Rebellion* (New York: W. W. Norton & Co., 1974) 14-20.

663 William L. Saunders, ed., *The Colonial Records of North Carolina, Volume 1, 1662 to 1712* (Raleigh: P. M. Hale, 1886) 48-75.

664 This constitution originally applied to both North and South Carolina. It was not until 1712 that these two colonies were officially separated. See William J. Saunders, ed., *The Colonial Records of North Carolina, Volume I, 1662 to 1712* (Raleigh: P. M. Hale, 1886) 204; Thomas Cooper, ed., *The Statues at Large of South Carolina, Vol. 1* (Columbia, South Carolina: A. S. Johnston, 1936) 55.

665 For a complete and detail history of the establishment of the slave society in South Carolina, see Peter H. Wood, *Black Majority, Negroes in Colonial South Carolina from 1670 through the Stono Rebellion* (New York: W. W. Norton & Co., 1974).

666 William L. Saunders, ed., *The Colonial Records of North Carolina, Volume 1, 1662 to 1712* (Raleigh: P. M. Hale, 1886) 53-57, 84, 86-90, 183-206; Alexander S. Salley, Jr., ed., *Narratives of Early Carolina, 1650-1708* (New York: Charles Scribner's Sons, 1911) 323-328.

667 Hugh T. Lefler and William S. Powell, *Colonial North Carolina, A History* (New York: Charles Scribner's Sons, 1973) 29-56.

668 Lois Green Carr, "Diversification in the Colonial Chesapeake: Somerset County, Maryland, in Comparative Perspective," in *Colonial Chesapeake Society*, ed. Lois Carr, Philip D. Morgan, and Jean B. Russo (Chapel Hill: University of North Carolina Press, 1988) 342-388; Lois Green Carr and Lorena S. Walsh, "The Planter's Wife: the Experience of White Women in Seventeenth-century Maryland," *William and Mary Quarterly*, 3rd Ser., 34 (October 1977): 542-571; Lois Green Carr and Lorena S. Walsh, "The Standard of Living in the Colonial Chesapeake." *William and Mary Quarterly*, 3rd Ser., 45 (January 1988): 135-159; Russell R. Menard, "British Migration to the Chesapeake Colonies in the Seventeenth Century," in *Colonial Chesapeake Society*, ed. Lois Carr, Philip D. Morgan, and Jean B. Russo, (Chapel Hill: University of North Carolina Press, 1988) 99-132.

669 Fields, 1-15; James Wright, *The Free Negro in Maryland, 1634-1860* (New York: Longmans, Green & Co, 1921) 36-93.

670 Charles Lewis Wagandt, *The Mighty Revolution: Negro Emancipation in Maryland, 1862-1864* (Baltimore: John Hopkins Press, 1964) 155-246.

671 W. E. B. Du Bois, *The Suppression of the African Slave Trade to the United States, 1638-1870* (1896; reprint, New York: Shocken Books, 1969), 199.

672 We do not know Harvey's exact reasons for refusing to issue patents, other than he was following the lead of his superiors in England. It is possible some were questioning the consequences of the headright system. It could also be that some had recognized that the system had been warped by the planters in Virginia, and wished to reform this. Either way, Harvey was making some effort to change this system, an effort that failed due to opposition of the planters who were benefiting from it.

673 "The Treaty on Principles Governing the Activities of States in the Exploration and Use of Outer Space, including the Moon and Other Celestial Bodies," normally referred to as the Outer Space Treaty, available as of May 15, 2018 at http://www.unoosa.org/oosa/en/ourwork/spacelaw/treaties/introouterspacetreaty.html.

674 Vanderklippe, Nathan, "Bright side of the moon: China launches pioneering mission to explore valuable source of solar energy." *The Globe and Mail*, May 24, 2016.

675 Huberty, Martine. "Doubts over space mining legal status" Delano News, April 14, 2017, available as of May 15, 2018 at http://delano.lu/d/detail/news/doubts-over-space-mining-legal-status/142567,

Endnotes

[676] Space Resources.Lu. "Luxembourg is the First European Nation to Offer a Legal Framework for Space Resources Utilization," available as of May 15, 2018 at http://www.spaceresources.public.lu/en/actualites/2017/Luxembourg-is-the-first-European-nation-to-offer-a-legal-framework-for-space-resources-utilization.html; Space Resources.Lu. Draft law, available [pdf] as of May 15, 2018 at http://www.spaceresources.public.lu/content/dam/spaceresources/news/Translation%20Of%20The%20Draft%20Law.pdf

[677] UK Space Agency, "New laws unlock exciting space era for UK," available as of May 15, 2018 at https://www.gov.uk/government/news/new-laws-unlock-exciting-space-era-for-uk; Nikkei Asian Review, "Japan to fuel space startups with nearly $1bn funding pool," available as of May 15, 2018 at https://asia.nikkei.com/Japan-Update/Japan-to-fuel-space-startups-with-nearly-1bn-funding-pool; Zimmerman, Robert, "India proposes new oppressive space law" available as of May 15, 2018 at http://behindtheblack.com/behind-the-black/points-of-information/india-proposes-new-oppressive-space-law/; Abbany, Zulfikar, *Deutsche Welle*, "Inter-agency meeting to 'improve' space law on mega-constellations of satellites," available as of May 15, 2018 at http://www.dw.com/en/inter-agency-meeting-to-improve-space-law-on-mega-constellations-of-satellites/a-38564904; Zimmerman, Robert, "Congress revises law governing commercial space," available as of May 15, 2018 at http://behindtheblack.com/behind-the-black/points-of-information/congress-revises-law-governing-commercial-space/

[678] Available as of February 3, 2021 at https://www.nasa.gov/sites/default/files/atoms/files/america_to_the_moon_2024_artemis_20190523.pdf

[679] Zimmerman, Robert. "U.S. and Japan formalize partnership on Gateway," January 13, 2021, available as of February 3, 2021 at https://behindtheblack.com/behind-the-black/points-of-information/u-s-and-japan-formalize-partnership-on-gateway/; Zimmerman, Robert. "Russia says it will oppose Artemis Accords," available as of February 3, 2021 at https://behindtheblack.com/behind-the-black/points-of-information/russia-says-it-will-oppose-artemis-accords/; Ji, Elliot, et. al. "What Does China Think About NASA's Artemis Accords?," *The Diplomat*, available as of February 3, 2021 at https://thediplomat.com/2020/09/what-does-china-think-about-nasas-artemis-accords/.

[680] Johnson-Freese, Joan, "Build on the outer space treaty," *Nature*, October 2, 2017, available as of May 15, 2018 at http://www.nature.com/news/build-on-the-outer-space-treaty-1.22789?error=cookies_not_supported&code=41813fd5-ec1d-4f5a-9738-dae198bf8375.

[681] Zimmerman, Robert, "Space, regulation, the Outer Space Treaty, and yesterday's Senate hearing," April 27, 2017, available as of May 25, 2018 at http://behindtheblack.com/behind-the-black/essays-and-commentaries/space-regulation-the-outer-space-treaty-and-yesterdays-senate-hearing/; Zimmerman, Robert, "Washington rallies around the Outer Space Treaty," May 24, 2017, available as of May 25, 2018 at http://behindtheblack.com/behind-the-black/essays-and-commentaries/washington-rallies-around-the-outer-space-treaty/.

[682] Robert Zimmerman, *Capitalism in Space: Private Enterprise and Competition Reshape the Global Aerospace Launch Industry* (Washington, D.C.: Center for a New American Security, 2017).

[683] Boorstin, Daniel. *The Discoverers, A history of man's search to know his world and himself* (New York: Random House, 1983), p. 249.

[684] Edmund Morgan, ed., *Puritan Political Ideas, 1558-1794* (Indianapolis: Bobbs-Merrill, 1965), 92-93.

[685] Kuran, Timur. "Islam and Economic Performance: Historical and Contemporary Links." *Journal of Economic Literature*, vol. 56 (2018), in press.

Index

A Lost Lady, 29, 30, 32, 37, 38, 133, 153
Abrahall, Robert, 83
Adams, John, 190
Allerton, Isaac, 169
Arnold, Anthony, 183
Ashton, Peter, 83
Bacon, Nathaniel, Jr., 171, 175, 179, 180
Bacon, Nathaniel, Sr., 81, 114, 155, 161, 171, 172, 183
Baker, Lawrence, 168
Ball, William, 125
Barnes, John, 167
Beckett, J.V., 40
Bennett, Phillip, 69, 72, 74
Bennett, Richard, 22, 55, 59, 65, 66, 68, 74, 80, 83, 87, 92, 93, 103, 110, 123, 125, 155, 157, 162, 177, 195, 198, 221
Berkeley, William, 29-31, 36, 38, 40, 51, 55, 85, 88, 93, 108, 112, 119, 131, 133, 138, 145, 148, 153, 155, 157, 165, 168, 171, 172, 175, 179, 183, 185, 200, 221
Best, Thomas, 26
Billings, Warren, 17
black population, *See* slave population
  Delaware, 192
  Massachusetts, 188
  New Jersey, 192
  Pennsylvania, 190
  Rhode Island, 188
Blackstone, William, 41
Bond, John, 107, 138
Bracewell, Robert, 94
Brent, Giles, 141, 169
Brewer, John, 41
Bridger, Joseph, 82, 111, 143, 172, 183

Brocas, William, 56, 88
Browne, Henry, 56, 72, 88, 110, 129
Bruce, Phillip, 25
Bullinger, Henry, 50
Burham, Rowland, 73, 110
Burke, Edmund, 13
Burnham, John, 172
Burnham, Roland, 72
Burnham, Rowland, 58
Byrd, William, 122
Carleton, Stephen, 73
Carter, John, Sr., 81, 111, 161
Casor, John, 103
Chain of Being, 33, 36, 40, 49, 148, 159, 202, *See* Royalist
Charles I, 24, 29, 33, 35, 38, 41, 43, 50, 51, 66, 69, 78, 80, 81, 82, 133, 155, 181, 183
Charles II, 56, 69, 81, 83, 84, 91, 116, 129, 131, 134, 141, 155, 181, 185, 199
Chickeley, Henry, 81, 105, 181, 183
Chiles, Walter, 64, 108
church. *See* religious institution
  Anglican, 66, 84, 92, 112, 123, 139, 151, 196
  Catholic, 17, 202
  government, 35, 49, 67, 68, 92, 139, 149, 196
  Puritan, 44, 194
  Virginia, 26, 54, 66, 84, 92, 112, 122, 151, 199
Claiborne, William, 61, 62, 68, 80, 83, 87, 88, 103, 108
Claiborne, William, Jr., 143, 172, 176
Coale, Josiah, 112
Collclough, George, 83, 98
Colleton, John, 200
Cooper, Anthony Ashley, 200

Corbin, Henry, 111, 116, 161
Coventry, Henry, 185
Cromwell, 49, 64, 78, 82, 83, 113, 129
Dade, Francis, 113
Dale, Edward, 83
Davenant, William, 31-33
Davis, Thomas, 106
death rate, 26, 53, 85, 122, *See* seasoning
Degler, Carl, 187
Delk, Roger, 168
Digges, Edward, 82, 92, 98, 103, 110, 112, 195
Driggus, Emmanual, 98
Duke of Albemarle, 200
Eastern Shore, 23, 55, 56, 60, 87, 97, 125, 202
Edlow, Joseph, 98
educational institution, 85, *See* schools
Ellyson, Robert, 106
Fantleroy, Moore, 116, 141
Fletcher, George, 108
Gardiner, Samuel Rawson, 51
Gardner, Thomas, 176
Gates, Thomas, 19
Gerrard, Thomas, 107
Gooch, William, 83
Goodrich, Thomas, 176
Governor Harvey, 56, 57, 65, 91, 92, 205, *See* Harvey, John
Grantham, Thomas, 181
Gray, Francis, 83
Great Charter, 20, 22, 25
Green Spring, 68, 88, 120, 131, 133, 158, 175, 181
Greene, John, 168
Greenstad, William, 98
Gunter, John, 158
Gussall, John, 98
Hammond, Manwaring, 81, 129, 135
Hancock, William, 167
Handlin, Mary, 187
Handlin, Oscar, 187
Harrison, Thomas, 68

Harvey, John, 55, 61, 110, *See* Governor Harvey
Hawkins, John, 44
headrights claimed on slaves, 58, 72-74, 110, 125, 162, 178
headrights, 20, 24, 25, 27, 91, 196, 205, 220
Hen, Robert, 169
Henrietta-Maria, 29, 30, 31
Higginbotham, Leon, 187
Hill, Edward, Sr., 74, 90, 116, 119, 128, 148, 198
Hill, Nicholas, 162
Hoe, Rice, 125
Holmwood, John, 83
Horsey, Stephen, 149, 153, 198
Horsmanden, Warham, 135
Hoskins, Bartholomew, 72
House of Burgesses, 20, 22, 56, 65, 69, 73, 88, 96, 115, 135, 138, 146, 151, 162, 176
Hunt, Robert, 17
Hyde, Edward, 32
indentured servants, 24, 53, 64, 120, 194, 220
indentured servitude, 54, 67, 156, 196, *See* indentured servants
Jamestown, 9, 10, 18-20, 23, 26, 53, 87, 88, 120, 131, 134, 177, 179, 181, 183
Jefferson, Thomas, 51, 185, 195
Jenings, Peter, 130
Jermyn, Henry, 31-33, 77
Johnson, Anthony, 23, 55, 75, 87, 97, 99, 101, 103, 110, 153, 165
Johnson, Ben, 31
Johnson, John, 99
Johnson, Mary, 23, 55
Jordan, Winthrop, 187
Joyce, Cornet, 78
Keayne, Robert, 194
Kemp, Richard, 56, 57, 68, 72, 110
Kendall, William, 98, 161

Key, Elizabeth, 97-100, 126, 147, 148
King James, 21, 81, 82, 91
Knight, Peter, 158
laws, 25
   1660 review of Commonwealth-era, 133
   black slaves, 160
   children of female black slaves, 147
   Commonwealth, 114
   economic, 67, 121
   economy, 135
   indentured servitude, 67, 156
   justice, 67
   Long Assembly, 153
   religion, 66, 140
   review, 138
   runaway servants, 158
   tithable, 72
Lear, John, 172
Leare, John, 156, 162
Lee, Richard, 69, 87, 91, 96, 123, 129, 184
Liliburne, John, 39, 48, 51
Little, William, 168
Lloyd, Edward, 69, 74, 80, 148
Locke, John, 200
Longo, Tony, 98
Lord Baltimore, 48, 56, 107, 119
Lord Culpepper, 156
Ludwell, Philip, 172, 183
Ludwell, Thomas, 66, 83, 155, 161, 162, 181
Lunsford, Thomas, 81
Madison, James, 197
Mason, George, 83, 141, 169
Mathews, Samuel, Sr., 61, 65, 80, 112, 126, 157
Mathews, Thomas, 169
Meares, Thomas, 69, 74, 80, 148
Minifie, George, 72
Molesworth, Guy, 130, 135
Mongon, Philip, 99
Montague, Peter, 126
Morgan, Edmund, 25, 187

Moryson, Francis, 81, 105, 135, 137, 138, 140, 148, 157, 197
Moseley, William, 83
Mottrum, John, 97
Moynihan, Daniel Patrick, 124
Page, John, 82, 105, 172
Parke, Daniel, Sr., 82, 181
Parker, Robert, 103
Parliamentarian, 59, 65, 66, 77, 78, 79, 92, 100, 105, 107, 109, 114, 126, 131, 133, 142, 145, 177
   philosophy, 75, 92, 95, 97, 107, 170
Payne, Francis, 98, 99
Percy, George, 15, 17-19, 26, 53, 120, 171
Peyton, Valintine, 83
Pilgrims, 9, 194, 205, 217, 219
   philosophy, 218
Plantagenet, Beauchamp, 53, 120
Pope, Alexander, 33
Porter, John, 144, 146, 147, 149
Pott, John, 24
Powell, John, 111, 161, 172
Presley, William, 98, 126, 143, 147, 148, 156, 161, 184
Price, Jenkin, 135
Puritans, 9, 21, 39, 44, 47, 49, 65, 68, 69, 80, 90-92, 100, 107, 112, 123, 155, 171, 187, 194-196, 202, 205, 211, 217, 219, 221
   philosophy, 14, 43, 48, 75, 92, 93, 95, 194, 218
Quakers, 9, 49, 83, 92, 107, 112, 123, 133, 140, 146, 147, 149, 171, 190, 195, 196, 198, 202, 221
   growth in popularity, 95
   philosophy, 48, 49, 218
Rainsborough, Thomas, 48
Ramsey, Edward, 144, 172
religious institution, 26, *See* church

Robins, Obedience, 59, 65, 74, 80, 91, 111, 125, 148, 198
Rolfe, John, 20
Sandys, Edwin, 20, 22, 24, 25, 82, 85, 120, 205, 206
Savage, John, 111, 125, 161
Scarburgh, Edmund, 56, 60, 65, 90, 94, 99, 105, 109, 124, 129, 137, 145, 149, 155, 161, 185, 202
schools, 54, 122, 137, 165, 195, 199, 202, *See* educational institution
seasoning, 120, *See* death rate
Seiler, William, 151
Shaw, George Bernard, 170
Sheppard, John, 167
Sherwood, William, 167, 179
slave population, 10, 186, *See* black population
Maryland, 202
New York, 190
North Carolina, 202
South Carolina, 200
Smith, Daniel Blake, 122
Smith, John, 18, 20, 105, *See* Dade, Francis
Smith, Lawrence, 181
Smith, Robert, 111, 161
Smith, Thomas, 40, 45
Soane, Henry, 91
Spencer, Nicolas, 156
Spencer, Robert, 168
Stegg, Thomas, 61, 62, 65, 80, 83, 87
Stith, John, 178
Styles, John, 41
Suckling, John, 29, 31, 32
Swan, Mathew, 167
Swan, Thomas, 168
Symms, Benjamin, 85
Taylor, Philip, 55, 98
Thurston, Thomas, 112
tobacco, 20, 21, 24, 53, 54, 55, 72, 84, 93, 98, 109, 112, 114, 120, 121, 122, 133, 135, 137, 140, 142, 147, 149, 154, 155, 163, 167, 186, 194, 202, 217
Todkill, Anas, 18
Townsend, Richard, 64
Tradescant, John, 33
Travers, Raleigh, 143, 162
Tyler, Margery, 99
Virginia Company, 16, 18-20, 22, 23, 56, 63, 205
Virginia population, 20, 22, 23, 26, 54, 73, 84, 93, 100, 111, 120, 127, 145, 147, 164, 175, 184, 186, 196
Walsh, Lorena, 122
Warner, Augustine, 90, 116, 161, 181
Warren, Thomas, 162
Washington, George, 12, 156, 218
Washington, John, 156, 161, 169, 184
Wertenbaker, Thomas, 183
West, Francis, 24
West, John, 172
White, Francis, 50, 183
Wilcox, John, 114
Williamson, Roger, 111, 161
Willis, Francis, 90, 105, 114, 116, 137, 158
Wilson, George, 140
Winstanley, Gerrard, 48
Winthrop, John, 43, 50, 194, 220
Wood, Abraham, 64, 106, 108, 114
Woodhouse, Henry, 91, 127, 135
Wormeley, Christopher, 56
Wormeley, Ralph, 72, 81, 110, 183
Wyatt, Francis, 25, 27, 51, 59
Yeardley, George, 22, 24, 110
Yeo, Leonard, 162

Made in the USA
Monee, IL
15 July 2024